Lecture Notes in Computer Science 5851

Commenced Publication in 1973
Founding and Former Series Editors:
Gerhard Goos, Juris Hartmanis, and Jan van Leeuwen

Thomas Stützle (Ed.)

Learning and Intelligent Optimization

Third International Conference, LION 3
Trento, Italy, January 14-18, 2009
Selected Papers

 Springer

Volume Editor

Thomas Stützle
IRIDIA
CoDE
Université Libre de Bruxelles
Avenue F. Roosevelt 50
CP 194/6
1050 Brussels, Belgium
E-mail: stuetzle@ulb.ac.be

Library of Congress Control Number: 2009940453

CR Subject Classification (1998): H.2.8, I.2.6, I.2.8, F.2.2, H.3.3, G.1.6

LNCS Sublibrary: SL 1 – Theoretical Computer Science and General Issues

ISSN 0302-9743

ISBN 978-3-642-11168-6 Springer Berlin Heidelberg New York

springer.com

© Springer-Verlag Berlin Heidelberg 2009

Typesetting: Camera-ready by author, data conversion by Scientific Publishing Services, Chennai, India
Printed on acid-free paper   SPIN: 12800858   06/3180   5 4 3 2 1 0

Preface

LION 3, the Third International Conference on Learning and Intelligent OptimizatioN, was held during January 14–18 in Trento, Italy. The LION series of conferences provides a platform for researchers who are interested in the intersection of efficient optimization techniques and learning. It is aimed at exploring the boundaries and uncharted territories between machine learning, artificial intelligence, mathematical programming and algorithms for hard optimization problems.

The considerable interest in the topics covered by LION was reflected by the overwhelming number of 86 submissions, which almost doubled the 48 submissions received for LION's second edition in December 2007. As in the first two editions, the submissions to LION 3 could be in three formats: (a) original novel and unpublished work for publication in the post-conference proceedings, (b) extended abstracts of work-in-progress or a position statement, and (c) recently submitted or published journal articles for oral presentations. The 86 submissions received include 72, ten, and four articles for categories (a), (b), and (c), respectively.

The articles for the post-conference proceedings were carefully selected after a rigorous refereeing process. Finally, 16 papers were accepted for publication in this proceedings volume (one of these being an extended version of an extended abstract), which gives an acceptance rate of about 20%. In addition, three of the papers presented at the Machine Learning and Intelligent Optimization in Bioinformatics (MALIOB) 2009 workshop are included in the post-conference proceedings; the MALIOB 2009 workshop was organized by Andrea Passerini of the University of Trento, Italy, as a LION 3 satellite workshop.

Apart from the oral presentations at LION 3, a number of promising contributions, which were published in electronic online proceedings, were presented during two poster sessions. The conference program was further enriched by four tutorials by leading researchers. The topics covered were constraint programming (Jean-Charles Regin, ILOG, France), algorithms for tackling the SAT problem (Youssef Hamadi, Microsoft Research Cambridge, UK), metaheuristics with a focus on simple metaheuristic strategies (Olivier Martin, Université Paris-Sud, France), and probabilistic reasoning techniques for combinatorial problem solving (Lukas Kroc, Ashish Sabharwal, and Bart Selman, Cornell University, USA). The poster sessions and the additional tutorials helped to make LION 3 a very lively meeting.

The Technical Program Committee Chair would like to acknowledge gratefully the contributions of multiple persons, in particular, the authors for submitting their work to LION 3 and the Program Committee and additional referees for their dedicated work to ensure a high-quality conference program. This conference was very successful also because of the numerous contributions of a number of people involved in its organization. Special thanks goes to Roberto Battiti, the Steering Committee and Local Organization Chair, who is also the main person behind the establishment of the LION conference series, to Bart Selman, the General Conference Chair, Andrea Passerini, for organizing the MALIOB workshop, Youssef Hamadi, the Tutorial Chair, for putting together an inspiring tutorial program, and Franco Mascia, the Web Chair, who was very helpful in all technical details arising before, during, and after the refereeing process. In addition, thanks also to Mauro Brunato, Elisa Cilia, Paolo Campigotto, Michela Dallachiesa, Cristina Di Risio, and Marco Cattani, members of the LION research group, for their help in the practical organization of the event.

Finally, we would like to thank the sponsors for their contribution to the conference: the Associazione Italiana per l'Intelligenza Artificiale, IEEE Computational Intelligence Society, and Microsoft Research for their technical co-sponsorship as well as the industrial sponsors, Eurotech Group S.P.A. and EnginSoft S.P.A.

September 2009   Thomas Stützle

Organization

Conference General Chair

Bart Selman   Cornell University, USA

Steering Committee and Local Organization Chair

Roberto Battiti   University of Trento, Italy

Technical Program Committee Chair

Thomas Stützle   Université Libre de Bruxelles (ULB), Belgium

Program Committee

Ethem Alpaydin	Bogazici University, Turkey
Roberto Battiti	University of Trento, Italy
Mauro Birattari	Université Libre de Bruxelles (ULB), Belgium
Christian Blum	Universitat Politècnica de Catalunya, Spain
Immanuel Bomze	University of Vienna, Austria
Andrea Bonarini	Politecnico di Milano, Italy
Juergen Branke	The University of Warwick, UK
Mauro Brunato	University of Trento, Italy
Carlos Cotta	University of Málaga, Spain
Karl Doerner	Universität Wien, Austria
Marco Dorigo	Université Libre de Bruxelles (ULB), Belgium
Michel Gendreau	Université de Montréal, Canada
Carla Gomes	Cornell University, USA
Marco Gori	University of Siena, Italy
Walter J. Gutjahr	Universität Wien, Austria
Youssef Hamadi	Microsoft Research, UK
Richard F. Hartl	Universität Wien, Austria
Geir Hasle	SINTEF Applied Mathematics, Norway
Pascal van Hentenryck	Brown University, USA
Franzisco Herrera	Universidad de Granada, Spain
Tomio Hirata	Nagoya University, Japan
Holger H. Hoos	University of British Columbia, Canada
Bernardo Huberman	Hewlett–Packard, USA
Márk Jelasity	University of Szeged, Hungary

Additional Referees

Mirela Andronescu   Chris Fawcett
Alena Shmygelska   Mohamed Saifullah Bin Hussin
Chris Thachuk

MALIOB Workshop Chair

Andrea Passerini   University of Trento, Italy

Tutorial Chair

Youssef Hamadi   Microsoft Research, UK

Web Chair

Franco Mascia   University of Trento, Italy

Steering Committee

Roberto Battiti   University of Trento, Italy
Mauro Brunato   University of Trento, Italy
Holger H. Hoos   University of British Columbia, Canada

Technical Co-sponsorship

Associazione Italiana per l'Intelligenza Artificiale
    http://www.aixia.it/
IEEE Computational Intelligence Society
    http://www.ieee-cis.org/
Microsoft Research
    http://research.microsoft.com/en-us/

Industrial Sponsorship

Eurotech Group S.P.A.
    http://www.eurotech.com/en/
EnginSoft S.P.A.
    http://www.enginsoft.com/

Table of Contents

MALIOB Workshop Papers

Evolutionary Dynamics of Extremal Optimization

Stefan Boettcher

Physics Department, Emory University, Atlanta, USA
stb@physics.emory.edu

Abstract. Dynamic features of the recently introduced extremal optimization heuristic are analyzed. Numerical studies of this evolutionary search heuristic show that it performs optimally at a transition between a jammed and a diffusive state. Using a simple, annealed model, some of the key features of extremal optimization are explained. In particular, it is verified that the dynamics of local search possesses a generic critical point under the variation of its sole parameter, separating phases of too greedy (non-ergodic, jammed) and too random (ergodic, diffusive) exploration. Analytic comparison with other local search methods, such as a fixed temperature Metropolis algorithm, within this model suggests that the existence of the critical point is the essential distinction leading to the optimal performance of the extremal optimization heuristic.

1 Introduction

We have introduced a new heuristic, Extremal Optimization (EO), in Refs. [1,2] and demonstrated its efficiency on a variety of combinatorial [3,4,5,6] and physical optimization problems [2,7,8]. Comparative studies with simulated annealing [1,3,9] and other Metropolis based heuristics [10,11,12,13,14] have established EO as a successful alternative for the study of NP-hard problems and its use has spread throughout the sciences. EO has found a large number of applications by other researchers, e. g. for polymer confirmation studies [15,16], pattern recognition [17,18,19], signal filtering [20,21], transport problems [22], molecular dynamics simulations [23], artificial intelligence [24,25,26], modeling of social networks [27,28,29], and $3d$−spin glasses [10,30]. Also, extensions [31,32,33,34] and rigorous performance guarantees [35,36] have been established. Ref. [37] provides a thorough description of EO and comparisons with other heuristics.

Here, we will apply EO to a spin glass model on a 3-regular random graph to elucidate some of its *dynamic* features as an evolutionary algorithm. These properties prove quite generic, leaving local search with EO virtually free of tunable parameters. We discuss the theoretical underpinning of its behavior, which is reminiscent of Kauffman's suggestion [38] that evolution progresses most rapidly near the "edge of chaos," in this case characterized by a critical transition between a diffusive and a jammed phase.

T. Stützle (Ed.): LION 3, LNCS 5851, pp. 1–14, 2009.

2   Spin Glass Ground States with Extremal Optimization

Disordered spin systems on sparse random graphs have been investigated as mean-field models of spin glasses or combinatorial optimization problems [39], since variables are long-range connected yet have a small number of neighbors. Particularly simple are α-regular random graphs, where each vertex possesses a fixed number α of bonds to randomly selected other vertices. One can assign a spin variable $x_i \in \{-1, +1\}$ to each vertex, and random couplings $J_{i,j}$, either Gaussian or ± 1, to existing bonds between neighboring vertices i and j, leading to competing constraints and "frustration" [40]. We want to minimize the energy of the system, which is the difference between violated bonds and satisfied bonds,

$$H = - \sum_{\{bonds\}} J_{i,j} x_i x_j. \tag{1}$$

EO performs a local search [6] on an existing configuration of n variables by changing preferentially those of poor *local* arrangement. For example, in case of the spin glass model in Eq. (1), $\lambda_i = x_i \sum_j J_{i,j} x_j$ assesses the local "fitness" of variable x_i, where $H = - \sum_i \lambda_i$ represents the overall energy (or cost) to be minimized. EO simply *ranks* variables,

$$\lambda_{\Pi(1)} \le \lambda_{\Pi(2)} \le \ldots \le \lambda_{\Pi(n)}, \tag{2}$$

where $\Pi(k) = i$ is the index for the kth-ranked variable x_i. Basic EO always selects the (extremal) lowest rank, $k = 1$, for an update. Instead, τ-EO selects the kth-ranked variable according to a scale-free probability distribution

$$P(k) \propto k^{-\tau}. \tag{3}$$

The selected variable is updated *unconditionally*, and its fitness and that of its neighboring variables are reevaluated. This update is repeated as long as desired, where the unconditional update ensures significant fluctuations with sufficient incentive to return to near-optimal solutions due to selection *against* variables with poor fitness, for the right choice of τ. Clearly, for finite τ, EO never "freezes" into a single configuration; it is able to return an extensive list of the best of the configurations visited (or simply their cost) instead [5].

For $\tau = 0$, this "τ-EO" algorithm is simply a random walk through configuration space. Conversely, for $\tau \to \infty$, the process approaches a deterministic local search, only updating the lowest-ranked variable, and is likely to reach a dead end. However, for finite values of τ the choice of a *scale-free* distribution for $P(k)$ in Eq. (3) ensures that no rank gets excluded from further evolution, while maintaining a clear bias against variables with bad fitness. As Sec. 3 will demonstrate, fixing $\tau - 1 \sim 1/\ln(n)$ provides a simple, parameter-free strategy, activating avalanches of adaptation [1,2].

3   EO Dynamics

Understanding the Dynamics of EO has proven a useful endeavor [41,14]. Such insights have lead to the implementation of τ-EO described in Sec. 2. Treating τ-EO as an evolutionary process allows us to elucidate its capabilities and to make further refinements. Using simulations, we have analyzed the dynamic pattern of the τ-EO heuristic. As described in Sec. 2, we have implemented τ-EO for the spin glass with Gaussian bonds on a set of instances of 3-regular graphs of sizes $n = 256$, 512, and 1024, and run each instance for $T_{\mathrm{run}} = 20n^3$ update steps. As a function of τ, we measured the ensemble average of the lowest-found energy density $\langle e \rangle = \langle H \rangle / n$, the first-return time distribution $R(\Delta t)$ of update activity to any specific spin, and auto-correlations $C(t)$ between two configurations separated by a time t in a single run. In Fig. 1, we show the plot of $\langle e \rangle$, which confirms the picture found numerically [2,4] and theoretically [41] for $\tau - \mathrm{EO}$. The transition at $\tau = 1$ we will investigate further below and theoretically in Sec. 5. The worsening behavior for large τ has been shown theoretically in Ref. [41] to originate with the fact that in any *finite*-time application, $T_{\mathrm{run}} < \infty$, $\tau - \mathrm{EO}$ becomes less likely to escape local minima for increasing τ and n. The combination of the purely diffusive search below $\tau = 1$ and the "jammed" state for large τ leads to the conclusion that the optimal value is approximated by $\tau - 1 \sim 1/\ln(n)$ for $n \to \infty$, consistent with Fig. 1 and experiments in Refs. [4,2].

In Fig. 2 we show the first-return probability for select values of τ. It shows that τ-EO is a fractal renewal process for all $\tau > 1$, and for $\tau < 1$ it is a Poisson process: when variables are drawn according to their "rank" k with probability $P(k)$ in Eq. (2), one gets for the first-return time distribution

$$R(\Delta t) \sim -\frac{P(k)^3}{P'(k)} \sim \Delta t^{\frac{1}{\tau} - 2}. \tag{4}$$

Neglecting correlations between variables, the number of updates of a variable of rank k is $\#(k) = T_{\mathrm{run}} P(k)$. Then, the typical life-time is $\Delta t(k) \sim T_{\mathrm{run}}/\#(k) = 1/P(k)$, which via $R(\Delta t)d\Delta t = P(k)dk$ immediately gives Eq. (4). The numerical

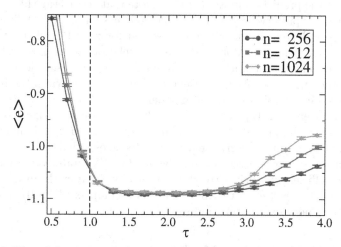

Fig. 1. Plot of the average lowest energy density found with $\tau - \mathrm{EO}$ over a fixed testbed of 3-regular graph instances of size n for varying τ. For $n \to \infty$, the results are near-optimal only in a narrowing range of τ just above $\tau = 1$. Below $\tau = 1$ results dramatically worsen, hinting at the phase transition in the search dynamics obtained in Sec. 5.

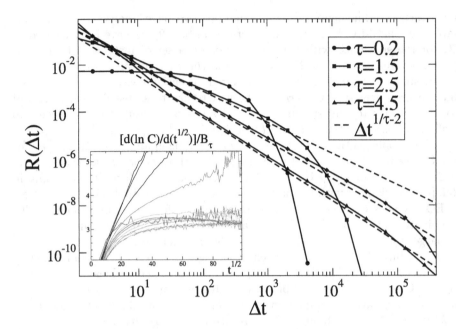

Fig. 2. Plot of the first-return time distribution $R(\Delta t)$ for τ-EO for various τ and $n = 256$. Poissonian behavior for $\tau < 1$ develops into a power-law regime limited by a cut-off for $\tau > 1$. The power-law scaling closely follows Eq. (4) (dashed lines). Inset: Data collapse (except for $\tau \leq 1$) of autocorrelations $C(t)$ according to the stretched-exponential fit given in the text. From top to bottom, $\tau = 0.5, 0.7, \ldots, 3.5$.

results in Fig. 2 fit the prediction in Eq. (4) well. Note that the average life-time, and hence the memory preserved by each variable, *diverges* for all $\tau(> 1)$, limited only by T_{run}, a size-dependent cut-off, and is widest for $\tau \to 1^+$, where τ-EO performs optimal. This finding affirms the subtle relation between searching configuration space widely while preserving the memory of good solutions.

Interestingly, the auto-correlations between configurations shown in the inset of Fig. 2 appear to decay with a *stretched*-exponential tail, $C(t) \sim \exp\{-B_\tau \sqrt{t}\}$ fitted with $B_\tau \approx 1.6 \exp\{-2.4\tau\}$, for all $\tau > 1$, characteristic of a super-cooled liquid [40] just *above* the glass transition temperature $T_g(> 0$ in this model). While we have not been able to derive that result, it suggests that τ-EO, driven far from equilibrium, never "freezes" into a glassy ($T < T_g$) state, yet accesses $T = 0$ properties efficiently. Such correlations typically decay with an agonizingly anemic power-law [40] for a thermal search of a complex ($T_g > 0$) energy landscape, entailing poor exploration and slow convergence.

4   Annealed Optimization Model

As described in Ref. [41], we can abstract certain combinatorial optimization problems into a simple, analytically tractable annealed optimization model

(AOM). To motivate AOM, we imagine a generic optimization problem, e. g. the spin glass in Sec. 2, as consisting of a number of variables $1 \leq i \leq n$, each of which contributes an amount $-\lambda_i$ to the overall cost per variable (or energy density),

$$\epsilon = -\frac{1}{n} \sum_{i=1}^{n} \lambda_i. \tag{5}$$

The "fitness" of each variable is $\lambda_i \leq 0$, where larger values are better and $\lambda_i = 0$ is optimal. The (optimal) ground state of the system is $\epsilon = 0$. In a realistic problem, variables are constrained such that not all of them can be simultaneously of optimal fitness. In AOM, those correlations are neglected.

We will consider that each variable i is in one of $\alpha+1$ different fitness states λ_i, where $\alpha_i = \alpha$ is fixed as a constant here. (For example, $\alpha = 2d$ on a d-dimensional hyper-cubic lattice or $\alpha = 3$ in Sec. 3.) We can specify occupation numbers n_a, $0 \leq a \leq \alpha$, for each state a, and define occupation densities $\rho_a = n_a/n$ ($a = 0, \ldots, \alpha$). Hence, any local search procedure with single-variable updates, say, can be cast simply as a set of evolution equations,

$$\dot{\rho}_b = \sum_{a=0}^{\alpha} T_{b,a} Q_a. \tag{6}$$

Here, Q_a is the probability that a variable in state a gets updated; any local search process (based on updating a finite number of variables) *defines* a unique set of $\mathbf{Q}$, as we will see below. The matrix $T_{b,a}$ specifies the net transition to state b *given* that a variable in state a is updated. This matrix allows us to *design* arbitrary, albeit annealed, optimization problems for AOM. Both, $\mathbf{T}$ and $\mathbf{Q}$, generally depend on the density vector $\rho(t)$ as well as on t explicitly.

We want to consider the different fitness states equally spaced, as in the spin glass example above, where variables in state a contribute $a\Delta E$ to the energy to the system. Here $\Delta E > 0$ is an arbitrary energy scale. The optimization problem is defined by minimizing the "energy" density

$$\epsilon = \sum_{a=0}^{\alpha} a\rho_a \geq 0. \tag{7}$$

Conservation of probability and of variables implies the constraints

$$\sum_{a=0}^{\alpha} \rho_a(t) = 1, \quad \sum_{a=0}^{\alpha} \dot{\rho}_a = 0, \quad \sum_{a=0}^{\alpha} Q_a = 1, \quad \sum_{a=0}^{\alpha} T_{a,b} = 0 \quad (0 \leq b \leq \alpha). \tag{8}$$

While AOM eliminates most of the relevant properties of a truly hard optimization problem, such as quenched randomness and frustration [40], two fundamental features of the evolution equations in Eq. (6) remain appealing: (1) The behavior for a large number of variables can be abstracted into a simple set of equations, describing their dynamics with merely a few unknowns, ρ, and (2) the separation of update preference, $\mathbf{Q}$, and update process, $\mathbf{T}$, lends itself to an analytical comparison between different heuristics, as we will show in Sec. 5.

5   Evolution Equations for Local Search Heuristics

The AOM developed above is quite generic for a class of combinatorial optimization problems. It was designed in particular to analyze EO [41], which we will
review next. Then we will present the update preferences $\mathbf{Q}$ through which each
local search heuristic enters into AOM. We also specify the update preferences
$\mathbf{Q}$ for Metropolis-based local searches, akin to simulated annealing.

5.1   Extremal Optimization Algorithm

EO is simply implemented in AOM: For a given configuration $\{x_i\}_{i=1}^n$, assign to
each variable x_i a "fitness" $\lambda_i = 0, -1, \ldots, -\alpha$ (e. g. $\lambda_i = -\{\#violated\ bonds\}$
in the spin glass), so that Eq. (5) is satisfied. Each variable falls into one of
only $\alpha + 1$ possible states. Say, currently there are n_α variables with the worst
fitness, $\lambda = -\alpha$, $n_{\alpha-1}$ with $\lambda = -(\alpha - 1)$, and so on up to n_0 variables with
the best fitness $\lambda = 0$ with $n = \sum_{b=0}^\alpha n_b$. Select an integer k, $1 \le k \le n$, from
some distribution, preferably with a bias towards lower values of k. Determine
$0 \le a \le \alpha$ such that $\sum_{b=a+1}^\alpha n_b < k \le \sum_{b=a}^\alpha n_b$. Note that lower values of
k would select a "pool" n_a with larger value of a, containing variables of lower
fitness. Finally, select one of the n_a variables in state a with equal chance and
update it *unconditionally*. As in Eq. (3), we prescribe a bias for selecting variables
of poor fitness on a slowly varying (power-law) scale over the *ranking* $1 \le k \le n$
of the variables by their fitnesses λ_i,

$$P_\tau(k) = \frac{\tau - 1}{1 - n^{1-\tau}} k^{-\tau} \quad (1 \le k \le n). \tag{9}$$

As an alternative, we can also study EO with threshold updating, which Ref. [35]
has shown rigorously to be optimal. Yet, the actual value of this threshold at
any point in time is typically not obvious (see also Ref. [36]). We will implement
a sharp threshold s $(1 \le s \le n)$ via

$$P_s(k) \propto \frac{1}{1 + e^{r(k-s)}} \quad (1 \le k \le n) \tag{10}$$

for $r \to \infty$. Since we can only consider fixed thresholds s, it is not apparent how
to shape the rigorous results into a successful algorithm.

5.2   Update Probabilities for Extremal Optimization

As described in Sec. 5.1, in each update of τ-EO a variable is selected based on
its rank according to the probability distribution in Eq. (9). When a rank $k(\le n)$
has been chosen, a variable is randomly picked from state α, if $k/n \le \rho_\alpha$, from
state $\alpha - 1$, if $\rho_\alpha < k/n \le \rho_\alpha + \rho_{\alpha-1}$, and so on. We introduce a new, continuous
variable $x = k/n$, for large n approximate sums by integrals, and rewrite $P(k)$
in Eq. (9) as

$$p_\tau(x) = \frac{\tau - 1}{n^{\tau-1} - 1} x^{-\tau} \quad \left(\frac{1}{n} \le x \le 1\right), \tag{11}$$

where the maintenance of the low-x cut-off at $1/n$ will turn out to be crucial. Now, the average likelihood in EO that a variable in a given state is updated is given by

$$Q_\alpha = \int_{1/n}^{\rho_\alpha} p(x)dx = \frac{1}{1-n^{\tau-1}} \left(\rho_\alpha^{1-\tau} - n^{\tau-1}\right),$$

$$Q_{\alpha-1} = \int_{\rho_\alpha}^{\rho_\alpha+\rho_{\alpha-1}} p(x)dx = \frac{1}{1-n^{\tau-1}} \left[(\rho_{\alpha-1}+\rho_\alpha)^{1-\tau} - \rho_\alpha^{1-\tau}\right],$$

$$\ldots$$

$$Q_0 = \int_{1-\rho_0}^{1} p(x)dx = \frac{1}{1-n^{\tau-1}} \left[1 - (1-\rho_0)^{1-\tau}\right], \tag{12}$$

where in the last line the norm $\sum_a \rho_a = 1$ was used. These values of the Q's completely describe the update preferences for τ-EO at arbitrary τ.

Similarly, we can proceed with the threshold distribution in Eq. (10) to obtain

$$p_s(x) \propto \frac{1}{1+e^{r(nx-s)}} \quad (\frac{1}{n} \le x \le 1), \tag{13}$$

with some proper normalization. While all the integrals to obtain $\mathbf{Q}$ in Eq. (12) are elementary, we do not display the rather lengthy results here.

Note that all the update probabilities in each variant of EO are *independent* of the matrix $\mathbf{T}$ in Eq. (6), i. e. of any particular model, which remains to be specified. This is special, as the following case of a Metropolis algorithm shows.

5.3   Update Probabilities for Metropolis Algorithms

It is more difficult to construct $\mathbf{Q}$ for a Metropolis-based algorithm, like simulated annealing [42,43]. Let's assume that we consider a variable in state a for an update. Certainly, Q_a would be proportional to ρ_a, since variables are randomly selected for an update. But as the actual update of the chosen variable may be accepted or rejected based on a Metropolis condition, further considerations are necessary. The requisite Boltzmann factor $e^{-\beta n \Delta \epsilon_a}$ for the potential update from time $t \to t+1$ of a variable in a, aside from the inverse temperature $\beta(t)$, only depends on the entries for $T_{a,b}$:

$$\Delta \epsilon_a = \left. \sum_{b=0}^{\alpha} b \left[\rho_b(t+1) - \rho_b(t)\right]\right|_a \sim \left. \sum_{b=0}^{\alpha} b\dot\rho_b \right|_a = \left. \sum_{b=0}^{\alpha} b \sum_{c=0}^{\alpha} T_{b,c} Q_c \right|_a = \sum_{b=0}^{\alpha} b T_{b,a},$$

where the subscript a expresses the fact that it is a *given* variable in state a considered for an update, i. e. $Q_c|_a = \delta_{a,c}$. Hence, from Metropolis we find for the average probability of an update of a variable in state a:

$$Q_a = \frac{1}{\mathcal{N}} \rho_a \min \left\{1, \exp\left[-\beta n \sum_{b=0}^{\alpha} b T_{b,a}\right]\right\}, \tag{14}$$

where the norm $\mathcal{N}$ is determined via $\sum_a Q_a = 1$. Unlike for EO, the update probabilities here are model-specific, i. e. they depend on the matrix $\mathbf{T}$.

5.4   Evolution Equations for a Simple Barrier Model

To demonstrate the use of these equations, we consider a simple model of an energetic barrier with only three states ($\alpha = 2$) and a constant flow matrix $T_{b,a} = [-\delta_{b,a} + \delta_{(2+b \bmod 3),a}]/n$, depicted in Fig. 3. Here, variables in ρ_1 can only reach their lowest-energy state in ρ_0 by first jumping *up* in energy to ρ_2. Eq. (6) gives

$$\dot{\rho}_0 = \frac{1}{n}\left(-Q_0 + Q_2\right), \quad \dot{\rho}_1 = \frac{1}{n}\left(Q_0 - Q_1\right), \quad \dot{\rho}_2 = \frac{1}{n}\left(Q_1 - Q_2\right), \qquad (15)$$

with some $\mathbf{Q}$ discussed in Sec. 5.2 for the variants of EO.

Given this $\mathbf{T}$, we can now also determine the specific update probabilities for Metropolis according to Eqs. (14). Note that for $a = 2$ we can evaluate the *min* as 1, since $\sum_b bT_{b,a=2} < 0$ always, while for $a = 0,1$ the *min* always evaluates to the exponential. Properly normalized, we obtain

$$Q_0 = \frac{\rho_0 e^{-\beta/2}}{(1 - e^{-\beta/2})\rho_2 + e^{-\beta/2}}, \quad Q_1 = \frac{\rho_1 e^{-\beta/2}}{(1 - e^{-\beta/2})\rho_2 + e^{-\beta/2}},$$

$$Q_2 = \frac{\rho_2}{(1 - e^{-\beta/2})\rho_2 + e^{-\beta/2}}. \qquad (16)$$

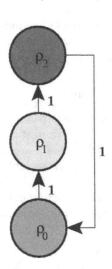

Flow up

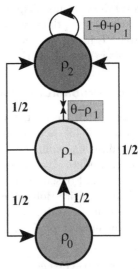

Flow jam

Fig. 3. Flow diagram with energetic barriers. Arrows indicate the net number of variables transferred, $nT_{b,a}$, into a state b, given that a variable in a gets updated. Diagonal elements $T_{a,a}$ correspondingly are negative, accounting for the outflow. Here, variables transferring from ρ_1 to ρ_0 must first jump up in energy to ρ_2.

Fig. 4. Same as Fig. 3, but with a model leading to a jam. Variables can only transfer from ρ_2 to ρ_0 through ρ_1, but only if $\rho_1 < \theta$. Once $\rho_1 = \theta$, flow down from ρ_2 ceases until ρ_1 reduces again.

It is now very simple to obtain the stationary solution: For $\dot{\rho} = 0$, Eqs. (15) yield $Q_0 = Q_1 = Q_2 = 1/3$, and we obtain from Eq. (12) for τ-EO:

$$\rho_0 = 1 - \left(\frac{1}{3}n^{\tau-1} + \frac{2}{3}\right)^{\frac{1}{1-\tau}}, \quad \rho_2 = \left(\frac{2}{3}n^{\tau-1} + \frac{1}{3}\right)^{\frac{1}{1-\tau}}, \quad (17)$$

and $\rho_1 = 1 - \rho_0 - \rho_2$, and for Metropolis:

$$\rho_0 = \frac{1}{2 + e^{-\beta/2}}, \quad \rho_1 = \frac{1}{2 + e^{-\beta/2}}, \quad \rho_2 = \frac{e^{-\beta/2}}{2 + e^{-\beta/2}}. \quad (18)$$

For EO with threshold updating, we obtain

$$\rho_0 = \frac{1}{3} - \frac{1}{3n} - \frac{s}{n} - \frac{1}{3nr}\ln\left[1 + e^{r(n-s)}\right]$$
$$+ \frac{1}{nr}\ln\left[\left(e^{nr} + e^{rs}\right)\left(1 + e^{r(1-s)}\right)^{\frac{1}{3}} + e^{\frac{r}{3}(2n+1)}\left(1 + e^{r(n-s)}\right)^{\frac{1}{3}}\right],$$

$$\rho_2 = \frac{1}{3} + \frac{2}{3n} + \frac{s}{n} - \frac{2}{3nr}\ln\left[1 + e^{r(n-s)}\right]$$
$$+ \frac{1}{nr}\ln\left[\left(e^{nr} + e^{rs}\right)\left(1 + e^{r(1-s)}\right)^{\frac{2}{3}} + e^{\frac{r}{3}(n+2)}\left(1 + e^{r(n-s)}\right)^{\frac{2}{3}}\right], (19)$$

and, assuming a threshold anywhere between $1 < s < n$, for $r \to \infty$:

$$\rho_0 = 1 - \frac{2s+1}{3n}, \quad \rho_2 = \frac{s+2}{3n}, \quad \rho_1 = \frac{s-1}{3n}. \quad (20)$$

Therefore, according to Eq. (7), Metropolis reaches its best, albeit sub-optimal, cost $\epsilon = 1/2 > 0$ at $\beta \to \infty$, due to the energetic barrier faced by the variables in ρ_1, see Fig. 3. The result for threshold updating in EO are more promising: near-optimal results are obtained, to within $O(1/n)$, for any finite threshold s. But results are best for small $s \to 1 \ll n$, in which limit we revert back to "basic" EO (only update the worst) obtained also for $\tau \to \infty$.

Finally, the result for τ-EO is most remarkable: For $n \to \infty$ at $\tau < 1$ EO remains sub-optimal, but reaches the optimal cost in the *entire* domain $\tau > 1$! This transition at $\tau = 1$ separates a (diffusive) random walk phase with too much fluctuation, and a greedy descent phase with too little fluctuation, which would trap τ-EO in problems with a complex landscape. This transition derives *generically* from the scale-free power-law in Eq. (9), as had been argued on the basis of numerical results for real NP-hard problems in Refs. [2,4]. The difference between reaching optimality in a limit only ($\tau \to \infty$ as in basic EO, $s \to 1$ for our naive threshold-EO model) or within a phase ($\tau > 1$) seems insignificant in the stationary regime, $T_{\text{run}} \to \infty$. Yet, it is the hallmark of a local search in a complex landscape that stationarity is rarely reached within any reasonable computational time $T_{\text{run}} < \infty$. At *intermediate* times, constrained variables jam each others evolution, requiring a subtle interplay between greedy descent and activated fluctuations to escape metastable states, as we will analyze in the following.

5.5 Jamming Model for τ-EO

Naturally, the range of phenomena found in a local search of NP-hard problems is not limited to energetic barriers. After all, so far we have only considered constant entries for $T_{b,a}$. Therefore, as our next AOM we want to review the case of $\mathbf{T}$ depending linearly on the ρ_i for τ-EO [41]. It highlights the importance of the fact that τ-EO attains optimality in the entire phase $\tau > 1$, instead of just an extreme limit such as basic EO ($\tau \to \infty$) or $s \to 1$ for EO with fixed threshold. From Fig. 4, we can read off $\mathbf{T}$ and obtain for Eq. (6):

$$\dot{\rho}_0 = \frac{1}{n}\left[-Q_0 + \frac{1}{2}Q_1\right],$$

$$\dot{\rho}_1 = \frac{1}{n}\left[\frac{1}{2}Q_0 - Q_1 + (\theta - \rho_1)Q_2\right], \tag{21}$$

and $\dot{\rho}_2 = -\dot{\rho}_0 - \dot{\rho}_1$ from Eq. (8). Aside from the dependence of $\mathbf{T}$ on ρ_1, we have also introduced the threshold parameter θ. In fact, if $\theta \geq 1$, the model behaves effectively like the previous model, and for $\theta \leq 0$ there can be no flow from state 2 to the lower states at all. The interesting regime is the case $0 < \theta < 1$, where further flow from state 2 into state 1 can be blocked for increasing ρ_1, providing a negative feed-back to the system. In effect, the model is capable of exhibiting a "jam" as observed in many models of glassy dynamics [40], and which is certainly an aspect of local search processes. Indeed, the emergence of such a jam is characteristic of the low-temperature properties of spin glasses and real optimization problems: After many update steps, most variables freeze into a near-perfect local arrangement and resist further change, while a finite fraction remains frustrated (temporarily in this model, permanently in real problems) in a poor local arrangement [44]. More and more of the frozen variables have to be dislodged collectively to accommodate the frustrated variables before the system as a whole can improve its state. In this highly correlated state, frozen variables block the progression of frustrated variables, and a jam emerges.

We obtain for the steady state, $\dot{\rho} = 0$:

$$0 = \frac{3}{2}(A-1) + \left[\theta - A^{1/(1-\tau)} + (3A-2)^{1/(1-\tau)}\right]\left(3A - 2 - n^{\tau-1}\right), \tag{22}$$

which can be solved for A to obtain

$$\rho_0 = 1 - A^{1/(1-\tau)}, \qquad \rho_2 = (3A-2)^{1/(1-\tau)}, \tag{23}$$

and $\rho_1 = 1 - \rho_0 - \rho_2$. Eq. (22) has a unique physical solution ($A > 2/3$) for the ρ's for all $0 \leq \tau \leq \infty$, $0 < \theta < 1$, and all n. As advertised in Sec. 5.4, in the thermodynamic limit $n \to \infty$ the generic critical point of τ-EO at $\tau = 1$ emerges: If $\tau < 1$, the sole n-dependent term in Eq. (22) vanishes, allowing A, and hence the ρ's, to take on finite values, i. e. $e > 0$ in Eq. (7). If $\tau > 1$, the n-dependent term diverges, forcing A to diverge in kind, resulting in $\rho_0 \to 1$ and $\rho_i \to 0$ for $i > 0$ in Eqs. (23), i. e. $e \to 0$.

While the steady state ($t \to \infty$) features of this model do not seem to be much different from the model in Sec. 5.4, the dynamics at intermediate times t is more subtle. In particular, as was shown in Ref. [41], a jam in the flow of variables towards better fitness may ensue under certain circumstances. The emergence of the jam depends on initial conditions, and its duration will prove to get longer for larger values of τ. If the initial conditions place a fraction $\rho_0 > 1 - \theta$ already into the lowest state, most likely no jam will emerge, since $\rho_1(t) < \theta$ for all times, and the ground state is reached in $t = O(n)$ steps. But if initially $\rho_1 + \rho_2 = 1 - \rho_0 > \theta$, and τ is sufficiently large, τ-EO will drive the system to a situation where $\rho_1 \approx \theta$ by preferentially transferring variables from ρ_2 to ρ_1, as Fig. 5 shows. Then, further evolution becomes extremely slow, delayed by the τ-dependent, small probability that a variable in state 1 is updated ahead of an *extensive* ($\propto n$) number of less-fit variables in state 2.

Clearly, this jam is *not* a stationary solution of Eq. (21). We consider initial conditions leading to a jam, $\rho_1(0) + \rho_2(0) > \theta$ and make the Ansatz

$$\rho_1(t) = \theta - \eta(t) \tag{24}$$

with $\eta \ll 1$ for $t \lesssim t_{jam}$, where t_{jam} is the time before $\rho_0 \rightarrow 1$. To determine t_{jam}, we apply Eq. (24) to the evolution equations in (21) and use the norm and $\dot{\rho}_1 = 0$ to leading-order, $\dot{\rho}_0 = -\dot{\rho}_2$, which yields an equation solely for $\rho_2(t)$,

$$-\frac{d\rho_2}{dt} \sim \frac{1}{n^\tau} \left[1 - \frac{3}{2}(\theta + \rho_2)^{1-\tau} + \frac{1}{2}\rho_2^{1-\tau} \right], \tag{25}$$

or, using the fact that ρ_2 almost instantly takes on the value of $\rho_1(0) + \rho_2(0) - \theta = 1 - \theta - \rho_0(0)$ (see Fig. 5), we solve Eq. (25) to get

$$t \sim n^\tau \int_{\rho_2(t)}^{1-\theta-\rho_0(0)} \frac{2d\xi}{2 - 3(\theta + \xi)^{1-\tau} + \xi^{1-\tau}}. \tag{26}$$

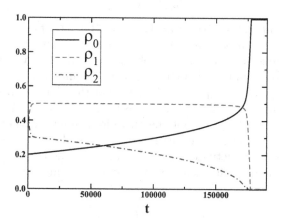

Fig. 5. Plot of the typical evolution of the system in Eqs. (21) for some generic initial condition that leads to a jam. Shown are $\rho_0(t)$, $\rho_1(t)$, and $\rho_2(t)$ for $n = 1000$, $\tau = 2$, $\theta = 0.5$ and initial conditions $\rho_0(0) = 0.2$, $\rho_1(0) = \rho_2(0) = 0.4$. Since $\rho_1(0) < \theta$, ρ_1 fills up to θ almost instantly with variables from ρ_2 while ρ_0 stays $\approx$constant. After that, $\rho_1 \approx \theta$ for a very long time ($\gg n$) while variables slowly trickle down through state 1. Eventually, after $t = O(n^\tau)$, ρ_2 vanishes and EO can empty out ρ_1 directly which leads to the ground state $\rho_0 = 1$ (i. e. $e = 0$) almost instantly.

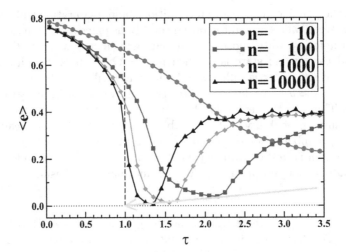

Fig. 6. Plot of the energy $\langle e \rangle$, averaged over initial conditions, vs. τ in many τ-EO runs of Eqs. (21) with $T_{\mathrm{run}} = 100n$, $n = 10$, 100, 1000, and 10000 and $\theta = 1/2$. For small values of τ, $\langle e \rangle$ closely follows the steady state solutions from Eqs. (22-23). It reaches a minimum at a value near the prediction for $\tau_{\mathrm{opt}} \approx 3.5$, 2.1, 1.6, and 1.4, approaching $\tau = 1^+$ along the yellow arrow, and rises sharply beyond that, comparable to Fig. 1.

We can estimate the duration of the jam ending at $t = t_{\mathrm{jam}}$ by setting $\rho_2(t_{\mathrm{jam}}) \approx 0$, see Fig. 5, leaving the integral as a constant to find:

$$t_{\mathrm{jam}} \sim n^\tau. \tag{27}$$

Instead of repeating the lengthy calculation in Ref. [41] for the ground state energy averaged over all possible initial conditions for finite runtime $T_{\mathrm{run}} \propto n$, we can content ourselves here with the obvious remark that a finite fraction of the initial conditions will lead to a jam, hence will require a runtime $T_{\mathrm{run}} \gg t_{\mathrm{jam}}$ to reach optimality. With $T_{\mathrm{run}} \propto n^k$, the fact that the phase transition in τ-EO provides good solutions for all $\tau > 1$ allows us to choose $1 < \tau < k$, as is apparent from Fig. 1 where $k = 3$. Fig. 6, obtained here from simulations of this jammed model in Eqs. (21), verifies the general asymptotic scaling, $\tau_{\mathrm{opt}} - 1 \sim 1/\ln(n)$, with small enough τ to fluctuate out of any jam in a time near-linear in n while still attaining optimal results as it would for *any* $\tau > 1$ at infinite runtime.

References

1. Boettcher, S., Percus, A.G.: Nature's way of optimizing. Artificial Intelligence 119, 275 (2000)
2. Boettcher, S., Percus, A.G.: Optimization with extremal dynamics. Phys. Rev. Lett. 86, 5211–5214 (2001)
3. Boettcher, S., Percus, A.G.: Extremal optimization: Methods derived from co-evolution. In: GECCO 1999: Proceedings of the Genetic and Evolutionary Computation Conference, pp. 825–832. Morgan Kaufmann, San Francisco (1999)

4. Boettcher, S., Percus, A.G.: Extremal optimization for graph partitioning. Phys. Rev. E 64, 026114 (2001)
5. Boettcher, S., Percus, A.G.: Extremal optimization at the phase transition of the 3-coloring problem. Phys. Rev. E 69, 066703 (2004)
6. Hoos, H.H., Stützle, T.: Stochastic Local Search: Foundations and Applications. Morgan Kaufmann, San Francisco (2004)
7. Boettcher, S.: Numerical results for ground states of mean-field spin glasses at low connectivities. Phys. Rev. B 67, R060403 (2003)
8. Boettcher, S.: Extremal optimization for Sherrington-Kirkpatrick spin glasses. Eur. Phys. J. B 46, 501–505 (2005)
9. Boettcher, S.: Extremal optimization and graph partitioning at the percolation threshold. J. Math. Phys. A: Math. Gen. 32, 5201–5211 (1999)
10. Dall, J., Sibani, P.: Faster Monte Carlo Simulations at Low Temperatures: The Waiting Time Method. Computer Physics Communication 141, 260–267 (2001)
11. Wang, J.S., Okabe, Y.: Comparison of Extremal Optimization with flat-histogram dynamics for finding spin-glass ground states. J. Phys. Soc. Jpn. 72, 1380 (2003)
12. Wang, J.: Transition matrix Monte Carlo and flat-histogram algorithm. In: AIP Conf. Proc. 690: The Monte Carlo Method in the Physical Sciences, pp. 344–348 (2003)
13. Boettcher, S., Sibani, P.: Comparing extremal and thermal explorations of energy landscapes. Eur. Phys. J. B 44, 317–326 (2005)
14. Boettcher, S., Frank, M.: Optimizing at the ergodic edge. Physica A 367, 220–230 (2006)
15. Shmygelska, A.: An extremal optimization search method for the protein folding problem: the go-model example. In: GECCO 2007: Proceedings of the 2007 GECCO conference companion on Genetic and evolutionary computation, pp. 2572–2579. ACM, New York (2007)
16. Mang, N.G., Zeng, C.: Reference energy extremal optimization: A stochastic search algorithm applied to computational protein design. J. Comp. Chem. 29, 1762–1771 (2008)
17. Meshoul, S., Batouche, M.: Robust point correspondence for image registration using optimization with extremal dynamics. In: Van Gool, L. (ed.) DAGM 2002. LNCS, vol. 2449, pp. 330–337. Springer, Heidelberg (2002)
18. Meshoul, S., Batouche, M.: Ant colony system with extremal dynamics for point matching and pose estimation. In: 16th International Conference on Pattern Recognition (ICPR 2002), vol. 3, p. 30823 (2002)
19. Meshoul, S., Batouche, M.: Combining Extremal Optimization with singular value decomposition for effective point matching. Int. J. Pattern Rec. and AI 17, 1111–1126 (2003)
20. Yom-Tov, E., Grossman, A., Inbar, G.F.: Movement-related potentials during the performance of a motor task i: The effect of learning and force. Biological Cybernatics 85, 395–399 (2001)
21. Svenson, P.: Extremal Optimization for sensor report pre-processing. Proc. SPIE 5429, 162–171 (2004)
22. de Sousa, F.L., Vlassov, V., Ramos, F.M.: Heat pipe design through generalized extremal optimization. Heat Transf. Eng. 25, 34–45 (2004)
23. Zhou, T., Bai, W.J., Cheng, L.J., Wang, B.H.: Continuous Extremal Optimization for Lennard-Jones clusters. Phys. Rev. E 72, 016702 (2005)
24. Menai, M.E., Batouche, M.: Extremal Optimization for Max-SAT. In: Proceedings of the International Conference on Artificial Intelligence (ICAI), pp. 954–958 (2002)

25. Menai, M.E., Batouche, M.: Efficient initial solution to Extremal Optimization algorithm for weighted MAXSAT problem. In: Chung, P.W.H., Hinde, C.J., Ali, M. (eds.) IEA/AIE 2003. LNCS, vol. 2718, pp. 592–603. Springer, Heidelberg (2003)
26. Menai, M.E., Batouche, M.: A Bose-Einstein Extremal Optimization method for solving real-world instances of maximum satisfiablility. In: Proceedings of the International Conference on Artificial Intelligence (ICAI), pp. 257–262 (2003)
27. Duch, J., Arenas, A.: Community detection in complex networks using Extremal Optimization. Phys. Rev. E 72, 027104 (2005)
28. Danon, L., Diaz-Guilera, A., Duch, J., Arenas, A.: Comparing community structure identification. J. Stat. Mech.-Theo. Exp., P09008 (2005)
29. Neda, Z., Florian, R., Ravasz, M., Libal, A., Györgyi, G.: Phase transition in an optimal clusterization model. Physica A 362, 357–368 (2006)
30. Onody, R.N., de Castro, P.A.: Optimization and self-organized criticality in a magnetic system. Physica A 322, 247–255 (2003)
31. Middleton, A.A.: Improved Extremal Optimization for the Ising spin glass. Phys. Rev. E 69, 055701(R) (2004)
32. Iwamatsu, M., Okabe, Y.: Basin hopping with occasional jumping. Chem. Phys. Lett. 399, 396–400 (2004); cond-mat/0410723
33. de Sousa, F.L., Vlassov, V., Ramos, F.M.: Generalized Extremal Optimization for solving complex optimal design problems. In: Cantú-Paz, E., Foster, J.A., Deb, K., Davis, L., Roy, R., O'Reilly, U.-M., Beyer, H.-G., Kendall, G., Wilson, S.W., Harman, M., Wegener, J., Dasgupta, D., Potter, M.A., Schultz, A., Dowsland, K.A., Jonoska, N., Miller, J., Standish, R.K. (eds.) GECCO 2003. LNCS, vol. 2723, pp. 375–376. Springer, Heidelberg (2003)
34. de Sousa, F.L., Ramos, F.M., Galski, R.L., Muraoka, I.: Generalized extremal optimization: A new meta-heuristic inspired by a model of natural evolution. Recent Developments in Biologically Inspired Computing (2004)
35. Heilmann, F., Hoffmann, K.H., Salamon, P.: Best possible probability distribution over Extremal Optimization ranks. Europhys. Lett. 66, 305–310 (2004)
36. Hoffmann, K.H., Heilmann, F., Salamon, P.: Fitness threshold accepting over Extremal Optimization ranks. Phys. Rev. E 70, 046704 (2004)
37. Hartmann, A.K., Rieger, H.: New Optimization Algorithms in Physics. Wiley-VCH, Berlin (2004)
38. Kauffman, S.A., Johnsen, S.: Coevolution to the edge of chaos: Coupled fitness landscapes, poised states, and coevolutionary avalanches. J. Theor. Biol. 149, 467–505 (1991)
39. Percus, A., Istrate, G., Moore, C.: Computational Complexity and Statistical Physics. Oxford University Press, New York (2006)
40. Fischer, K.H., Hertz, J.A.: Spin Glasses. Cambridge University Press, Cambridge (1991)
41. Boettcher, S., Grigni, M.: Jamming model for the extremal optimization heuristic. J. Phys. A: Math. Gen. 35, 1109–1123 (2002)
42. Kirkpatrick, S., Gelatt, C.D., Vecchi, M.P.: Optimization by simulated annealing. Science 220, 671–680 (1983)
43. Salamon, P., Sibani, P., Frost, R.: Facts, Conjectures, and Improvements for Simulated Annealing. Society for Industrial & Applied Mathematics (2002)
44. Palmer, R.G., Stein, D.L., Abraham, E., Anderson, P.W.: Models of hierarchically constrained dynamics for glassy relaxation. Phys. Rev. Lett. 53, 958–961 (1984)

A Variable Neighborhood Descent Search Algorithm for Delay-Constrained Least-Cost Multicast Routing

Rong Qu[1], Ying Xu[1,2], and Graham Kendall[1]

[1] The Automated Scheduling, Optimisation and Planning (ASAP) Group, School of Computer Science, The University of Nottingham, Nottingham, UK
[2] School of Computer and Communication, Hunan University, Hunan, China
{rxq,yxx,gxk}@cs.nott.ac.uk

Abstract. The rapid evolution of real-time multimedia applications requires Quality of Service (QoS) based multicast routing in underlying computer networks. The constrained Steiner Tree, as the underpinning mathematical structure, is a well-known NP-complete problem. In this paper we investigate a variable neighborhood descent (VND) search, a variant of variable neighborhood search, for the delay-constrained least-cost (DCLC) multicast routing problem. The neighborhood structures designed in the VND approaches are based on the idea of path replacement in trees. They are simple, yet effective operators, enabling a flexible search over the solution space of this complex problem with multiple constraints. A large number of simulations demonstrate that our algorithm is highly efficient in solving the DCLC multicast routing problem in terms of the tree cost and execution time. To our knowledge, this is the first study of VND algorithm on the DCLC multicast routing problem. It outperforms other existing algorithms over a range of problem instances.

1 Introduction

The general problem of multicast routing has received significant research attention in the area of computer networks and algorithmic network theory [1,2,3]. It is defined as sending messages from a source to a set of destinations that belong to the same multicast group. Many real-time multimedia applications (e.g. video conferencing, distance education) require the underlying network to satisfy certain quality of service (QoS). These QoS requirements include the cost, delay, delay variation and hop count, etc, among which the delay and cost are the most important for constructing multicast trees. The end-to-end delay is the total delay along the paths from the source to each destination. The cost of the multicast tree is the sum of costs on its edges.

To search for the minimum cost tree in the multicast routing problem is the problem of finding a Steiner Tree [4], which is known to be NP-complete [5]. The Delay-Constrained Least-Cost (DCLC) multicast routing problem is the problem of finding a Delay-Constrained Steiner tree (DCST), also known to be

T. Stützle (Ed.): LION 3, LNCS 5851, pp. 15–29, 2009.

NP-complete [6]. Surveys in the literature of multicast communication problems exist on both the early solutions [7] and recent optimization algorithms [8].

Algorithms for multicast routing problems can usually be classified as source-based and destination-based algorithms. Source-based algorithms assume that each node has all the necessary information to construct the multicast tree (e.g. [9,10,11]). Destination-based algorithms do not require that each node maintains the status information of the entire network, and multiple nodes participate in constructing the multicast tree (e.g. [6,12]).

The first DCST heuristic, Kompella-Pasquale-Polyzos (KPP) heuristic, uses Prim's algorithm [14] to obtain a minimum spanning tree. Another heuristic, Constrained Dijkstra (CDKS) heuristic, constructs delay-constrained shortest path tree for large networks by using Dijkstra's heuristic [15]. Bounded Shortest Multicast Algorithm (BSMA) [11], a well known deterministic multicast algorithm for the DCST problem, iteratively refines the tree to lower costs. Although developed in the mid 1990s, it is still being frequently compared with many multicast routing algorithms in the current literature. However, it requires excessive execution time for large networks as it uses the k Shortest Path algorithm [16] to find lower cost paths.

The second group of algorithms considers distributed multicast routing problems. The idea of Destination-Driven MultiCasting (DDMC) comes from Prim's minimum spanning tree algorithm and Dijkstra's shortest path algorithm. The QoS Dependent Multicast Routing (QDMR) algorithm extends the DDMC algorithm by using a weight function to dynamically adjust how far a node is from the delay bound and adds the node with the lowest weight to the current tree.

In recent years, metaheuristic algorithms such as simulated annealing [17,18], genetic algorithm [19,20], tabu search [21,22,23,24], GRASP [25] and path relinking [26] have been investigated for various multicast routing problems. In the tabu search algorithm in [24], initial solutions are generated based on Dijkstra's algorithm. A modified Prim's algorithm iteratively refines the initial solution by switching edges chosen from a backup path set. In the path relinking algorithm in [26], pairs of solutions in a reference set are iteratively improved. A repair procedure is used to repair any infeasible solution. Simulation results show that this path relinking algorithm outperforms other algorithms with regards to the tree cost. However, when the network size increases and many infeasible solutions need to be repaired, it is time consuming only suitable for real-time small networks.

In this paper we investigate variable neighborhood descent (VND) search, a variant of variable neighborhood search (VNS), for DCLC multicast routing problems. Although VNS algorithms have been applied to Steiner tree problems (e.g. VNS as a post-optimization procedure to the prize collecting Steiner tree problem [27], and the bounded diameter minimum spanning tree problem [28]), as far as we are aware, no research has been carried out using VND on DCST problems. Experimental results show that our VND algorithms obtained the best quality solutions when compared against the algorithms discussed above.

The rest of the paper is organized as follows. In Section 2, we present the network model and the problem formulation. Section 3 presents the proposed VND algorithms. We evaluate our algorithms by computer simulations on a range of problem instances in Section 4. Finally, Section 5 concludes this paper and presents possible directions for future work.

2   The Delay-Constrained Least-Cost Multicast Routing Problem

We consider a computer network represented by a directed graph $G = (V, E)$ with $|V| = n$ nodes and $|E| = l$ edges, where V is a set of nodes and E is a set of links. Each link $e = (i, j) \in E$ is associated with two parameters, namely the link cost $C(e)$: $E \mapsto \Re^+$ and the link delay $D(e)$: $E \mapsto \Re^+$. Due to the asymmetric nature of computer networks, it is possible that $C(e) \neq C(e')$ and $D(e) \neq D(e')$, for link $e = (i, j)$ and link $e' = (j, i)$. The nodes in V include a source node s, destination nodes which receive data stream from the source, denoted by $R \subseteq V - \{s\}$, called multicast groups, and relay nodes which are intermediate hops on the paths from the source to destinations.

We define a path from node u to node v as an ordered set of links, denoted by $P(u, v) = \{(u, i), (i, j), \ldots, (k, v)\}$. A multicast tree $T(s, R)$ is a set of paths rooted from the source s and spanning all members of R. We denote by $P_T(r_i) \subseteq T$ the set of links in T that constitute the path from s to $r_i \in R$. The total delay from s to r_i, denoted by $Delay[r_i]$, is simply the sum of the delay of all links along $P_T(r_i)$, i.e.

$$Delay[r_i] = \sum_{e \in P_T(r_i)} D(e), \ \forall r_i \in R \tag{1}$$

The delay of the tree, denoted by $Delay[T]$, is the maximum delay among all the paths from the source to each destination, i.e.

$$Delay[T] = max\{Delay[r_i] \mid \forall r_i \in R\} \tag{2}$$

The total cost of the tree, denoted by $Cost(T)$, is defined as the sum of the costs of all links in the tree, i.e.

$$Cost(T) = \sum_{e \in T} C(e) \tag{3}$$

Applications may assign different upper bounds δ_i for each destination $r_i \in R$. In this paper, we assume that the upper bound for all destinations is the same, and is denoted by $\Delta = \delta_i, r_i \in R$.

Given these definitions, we formally define the Delay-Constrained Steiner Tree (DCST) problem as follows [6]:

The Delay-Constrained Steiner Tree (DCST) Problem: Given a network G, a source node s, destination nodes set R, a link delay function $D(\cdot)$, a link cost function $C(\cdot)$, and a delay bound Δ, the objective of the DCST Problem is to construct a multicast tree $T(s, R)$ such that the delay bound is satisfied, and the tree cost $Cost(T)$ is minimized. We can define the objective function as:

$$min\{Cost(T) \mid P_T(r_i) \subseteq T(s, R), Delay[r_i] \leq \Delta, \forall r_i \in R\} \qquad (4)$$

3   The Variable Descent Neighborhood Search Algorithms

Variable neighborhood search (VNS), jointly invented by Mladenović and Hansen [29] in 1996, is a metaheuristic for solving combinatorial and global optimization problems. Unlike many standard metaheuristics where only a single neighborhood is employed, VNS systematically changes different neighborhoods within a local search. The idea is that a local optimum defined by one neighborhood structure is not necessarily the local optimum of another neighborhood structure, thus the search can systematically traverse different search spaces which are defined by different neighborhood structures. This makes the search much more flexible within the solution space of the problem, and potentially leads to better solutions which are difficult to obtain by using single neighborhood based local search algorithms [29,30,31]. The basic principles of VNS are easy to apply, parameters being kept to a minimum. Our proposed algorithm is based on basic variable neighborhood descent search (VND), a variant of VNS algorithm [29].

3.1   Initialisation

In our VND Multicast Routing (VNDMR) algorithm, let us denote N_k, $k = 1, \ldots, k_{max}$ as the set of solutions of the kth neighborhood operator upon an incumbent solution x. We first create an initial solution T_0 and then iteratively improve T_0 by employing three neighborhoods, defined in Section 3.2, until the tree cost cannot be reduced, while the delay constraint is satisfied. To investigate the effects of different initial solutions, we design two variants of the algorithm, namely VNDMR0 and VNDMR1, with the same neighborhood structures, but starting from different initial solutions:

- Initialisation by DKSLD (VNDMR0): Dijkstra's shortest path algorithm is used to construct the least delay multicast tree;
- Initialisation by DBDDSP (VNDMR1): A modified Delay-Bounded DDSP (DBDDSP) algorithm is used as the initialisation method based on the Destination-Driven Shortest Path (DDSP) algorithm, a destination-driven shortest path multicast tree algorithm with no delay constraint developed in [32].

3.2   Neighborhood Structures within the VND Algorithms

The first group of neighborhood structures within our VNDMR algorithms are designed based on an operation called path replacement, i.e. a path in a tree T_i

is replaced by another new path not in the tree T_i, resulting in a new tree T_{i+1}. Our delay-bounded path replacement operation guarantees that the tree T_{i+1} is always delay-bounded and loop free. To present the candidate paths chosen in the path replacement, we define *superpath* (based on [11], also called *key-path* in the literature [33,34]) as the longest simple path in the tree T_i, where all internal nodes, except the two end nodes of the path, are relay nodes and each relay node connects exactly two edges. The pseudo-code of VNDMR is presented in Fig.1.

- VNDMR($G = (V, E)$, S, R, Δ, k_{max}, N_k, $k = 1, \ldots, k_{max}$)
- /*S: the source node; R: the destination nodes set; $\Delta \geq 0$: the delay bound; k_{max} = 3: the number of neighborhood structures; $N_k(T)$: the set of neighborhoods by employing neighborhood N_k */
  - Create initial solution T_0; // by using DKSLD or DBDDSP, see Section 3.1
  - if $T_0 = NULL$ then return $FAILED$; // a feasible tree does not exist
  - else
    * $T_{best} = T_0$; $k = 1$;
    * while $k \leq k_{max}$
      · select the best neighbor T_i, $T_i \in N_k(T_{best})$;
      · if(T_i has lower cost or low delay) then $T_{best} = T_i$; $k = 1$;
      · else $k++$;
    * end of while loop
  - return T_{best}

Fig. 1. The Pseudo-code of the VNDMR Algorithm

The three neighborhood structures of VNDMR0 and VNDMR1 are described as below:

1. **Neighbor1:** the most expensive edges on each superpath in tree T_i are the candidates of the path replacement. At each step, one chosen edge is deleted, leading to two separate subtrees T_i^1 and T_i^2. The Dijkstra's shortest path algorithm is then used to find a new delay-bounded shortest path that connects the two subtrees and reduces the tree cost;
2. **Neighbor2:** this operator operates on all superpaths in the tree T_i (either connecting or not connecting to a destination node). At each step, one superpath is replaced by a cheaper delay-bounded path using the same path replacement strategy in **Neighbor1**;
3. **Neighbor3:** all the superpaths connected to destination nodes in T_i are the candidate paths to be replaced. At each step, the deletion of a superpath divides the tree T_i into a subtree T_i' and a destination node r_i. Then the same path replacement strategy is used to search for a new delay-bounded shortest path reconnecting r_i to T_i'.

To test how different neighborhood structures will affect the performance of the VND algorithm with the same initial solution, another VND algorithm, named VNDMR2, is developed with an extended new node-based neighborhood structure. The three neighborhood structures of VNDMR2 are as follows:

1. **Neighbor1':** one neighborhood tree is defined by deleting a non-source and non-destination node from the current multicast tree and creating a minimum spanning tree which spans the remaining nodes by using Prim's spanning tree algorithm. Once a better tree is found, the current tree is updated. These steps are repeated until no better tree can be found for 3 times;
2. **Neighbor2':** the same as **Neighbor2** in VNDMR0 and VNDMR1;
3. **Neighbor3':** the same as **Neighbor3** in VNDMR0 and VNDMR1.

3.3   Time Complexity of the VNDMR Algorithm

Proof of the probability of transition from a spanning tree s_i to s_j (see [35]):

According to Cayley's theorem [36], for a n node network, there are n^{n-2} possible spanning trees. Thus, the number of Steiner trees is bounded by n^{n-2}. Let us consider a Markov chain of n^{n-2} states, where each state corresponds to a spanning tree. We sort these states in a decreasing order with respect to the cost of the Steiner tree. Replacing each state in the sorted list with n copies of itself results into a total number of n^{n-1} states. In the Markov chain, transition edges from a state s_i go only to a right state s_j of s_i. Assume that each possible transition is equally likely. Thus the probability of a transition from s_i to s_j is:

$$p_{ij} = \frac{1}{i-1} \ (1 \leq j < i, P_{11} = 1) \tag{5}$$

We prove the time complexity of VNDMR based on the method used in [35]. Let m_i be the number of transitions needed to go from state s_i to s_1, the expected value $E[m_i] = \log(i)$. Therefore, if the VNDMR algorithm starts from the most expensive state, i.e. n^{n-1}, the expected number of transitions is $O(\log(n^{n-1})) = O(n\log(n))$. So the expected maximum number of iterations of the neighborhood structures in VNDMR is $O(n\log(n))$. The VNDMR algorithm includes three neighborhood structures (N_1, N_2, N_3), then the time complexity of VNDMR is:

$$O(nlog(n)(O(N1) + O(N2) + O(N3))) \tag{6}$$

For example, the three neighborhoods of VNDMR0 and VNDMR1 use the same path replacement strategy. A path-replacement operation is dominated by Dijkstra's shortest path algorithm which takes $O(l\log(n))$, where $l = |E|$ is the total links in the network. In the worst case, each neighborhood requires replacing at most $O(l)$ superpaths. Thus the time complexity of VNDMR0 and VNDMR1 is:

$$O(nlog(n)(3 * l * llog(n))) = O(l^2 nlog^2(n)) \tag{7}$$

4   Performance Evaluation

To evaluate the efficiency of our VNDMR algorithm, we use a multicast routing simulator (MRSIM) implemented in C++ based on Salama's generator [1]. MRSIM generates random network topologies using a graph generation algorithm described in [37]. The positions of the nodes are fixed in a rectangle of size $4000 \times 4000 km^2$. The simulator defines the link delay function $D(e)$ as the propagation delay of the link (queuing and transmission delays are negligible) and the link cost function $C(e)$ as the current total bandwidth reserved on the link in the network. The Euclidean metric is used to determine the distance $l(u, v)$ between pairs of nodes (u, v). Edges connect nodes (u, v), with a probability

$$P(u, v) = \beta exp(-l(u, v)/\alpha L) \quad \alpha, \beta \in (0, 1] \tag{8}$$

where parameters α and β can be set to obtain desired characteristics in the graph. A large β gives nodes a high average degree, and a small α gives long connections. L is the maximum distance between two nodes. In our simulations, we set $\alpha = 0.25$, $\beta = 0.40$, the average degree $= 4$ and the capacity of each link $= 155Mb/s$ (in this paper we set the capacity to a large enough value so that such constraint is not considered in the problem). All simulations were run on a Windows XP computer with Pentium VI 3.4GHZ, 1G RAM.

To encourage scientific comparisons, we have put the problem details of all instances tested at http://www.cs.nott.ac.uk/~yxx/resource.html, with some example solutions obtained by the proposed algorithms.

4.1   VNDMR with Different Initialisations

In the first set of experiments, we randomly generate 20 different network topologies for each size of 20, 50, 100, 200 and 300 nodes in the networks. For each network topology, the source node and the destination nodes are randomly selected. The delay bound in our experiments for each network topology is set as 2 times the tree delay of the DKSLD algorithm, i.e. $\Delta = 2 \times Delay(T_{DKSLD})$. For each network topology, the simulation was run 50 times, where the average tree costs and execution times were reported. We investigate the performance of two variants of VNDMR with different initializations, e.g. VNDMR0 with DKSLD and VNDMR1 with DBDDSP. Both variants employ the same neighborhood structures as defined in Section 3.2.

Fig.2 presents the tree cost and execution time of VNDMR0 and VNDMR1 for problems of different network sizes with a group size (number of destinations) of 10. We can see that the tree cost of the initial solutions obtained from DBDDSP and DKSLD can both be improved by the VNDMR algorithms. The paired t-test value of the average tree cost between VNDMR0 and VNDMR1 is 3.85, meaning VNDMR1 is significantly better than VNDMR0. We conclude that VNDMR1 performs better than VNDMR0 in terms of both tree cost and computational time.

Fig.3.(a) presents the tree costs of the two VNDMR algorithms with different initial solutions for networks of 50 nodes with different group sizes. In the table, the above observations still hold. The initial solutions from DBDDSP for VNDMR1 are better than that of DKSLD for VNDMR0. Both VNDMR algorithms can further reduce the tree cost, and VNDMR1 performs slightly better than VNDMR0. Fig.3.(b) also shows that VNDMR1 requires less execution time than that of VNDMR0.

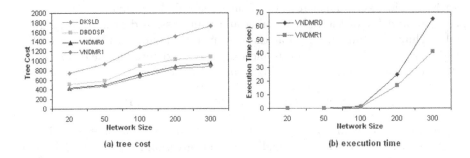

Fig. 2. Results of VNDMR with different initialisations, group size = 10

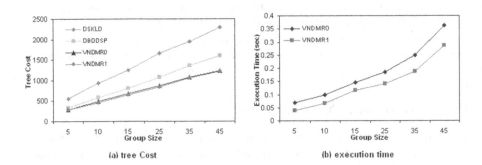

Fig. 3. Execution time by VNDMR with different initialisations, network size = 50

This group of experiments show that our VNDMR algorithms can always improve the initial solutions when constructing the DCLC multicast trees. The quality of initial solutions affects the performance of the VND algorithm. It is shown that better initial solutions from more intelligent heuristics lead to better final results, and also reduce the execution time of the VND algorithm.

4.2   VNDMR with Different Neighborhood Structures

In the second group of experiments, we test VNDMR1 and VNDMR2 on the same randomly generated network topologies in the same manner as mentioned in Section 4.1. Both VNDMR1 and VNDMR2 start from the same initial solution (DBDDSP), whereas they apply the different neighborhood structures described in Section 3.1 and 3.2, respectively.

The average tree cost and execution time of VNDMR1 and VNDMR2 on 5 different network sizes with group sizes equal to 10 are shown in Table 1. VNDMR2 always gets better average tree cost than VNDMR1 on these different network sizes. The paired t-test value of the average tree cost between VNDMR1 and VNDMR2 is 4.84, indicating a significantly difference between them. It is also observed that VNDMR2 spends longer computing time than VNDMR1.

Table 1. Average tree cost and execution time vs. network size, group size = 10

Network	Initial Solution	VNDMR1		VNDMR2	
Size	DBDDSP	Cost	Time(s)	Cost	Time(s)
20	513.85	416.25	0.008	**407.9**	0.053
50	583.1	466.5	0.067	**456.85**	0.509
100	892.75	667.85	0.924	**650.2**	4.969
200	1029.15	840.55	16.474	**829.55**	29.891
300	1084.55	875.05	41.379	**851.25**	86.365

Table 2. Average tree cost and execution time vs. group size, network size = 50

Group	Initial Solution	VNDMR1		VNDMR2	
Size	DBDDSP	Cost	Time(s)	Cost	Time(s)
5	330.05	280.75	0.038	**280.15**	0.075
10	583.1	466.5	0.067	**456.85**	0.214
15	809.1	682.95	0.117	**643.65**	0.373
25	1077.75	**840.25**	0.141	845.15	0.473
35	1359.95	**1055.75**	0.187	1063.45	0.583
45	1591.45	**1214.75**	0.287	1224.95	0.595

The average tree cost and execution time of VNDMR1 and VNDMR2 on the same group of 50-node networks with different group sizes are shown in Table 2. We can see that VNDMR2 gets better tree costs on the networks with small group sizes (5, 10, 15), while VNDMR1 performs better than VNDMR2 when the group size increases (25, 35, 45). It means the design of the neighborhood structures affects the performance of the VND algorithm. The *Neighbor1'* of VNDMR2 is based on an operation on the nodes in the multicast tree. With the increasing group size, i.e. the number of destination nodes, the amount of nodes which can be deleted from the current tree decreases. Since the possible neighborhood trees of the current tree that can be explored are reduced, *Neighbor1'* plays not much role when exploring the solution space. However, the edge-based VNDMR1 still performs well even on the networks with large group sizes. On the other hand, VNDMR1 spends less computing time than VNDMR2 on the tested problems.

4.3   Comparisons with Existing Algorithms

In the second set of experiments, we compare VNDMR1 with four existing multicast routing algorithms in terms of both the solution quality and the computational time on the same network topologies in Section 4.1. The four algorithms include BSMA, CDKS, QDMR, which are DCLC multicast routing algorithms, and DKSLC, which uses Dijkstra's algorithm to construct the least cost multicast trees without the delay constraint. These algorithms have already been integrated in the MRSIM simulator and reviewed in Section 1.

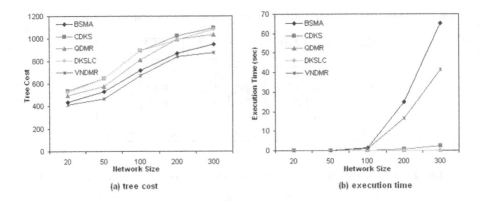

Fig. 4. Tree cost and execution time with group size $= 10$ from different approaches

Fig.4 presents the tree cost and execution time of these four algorithms and our VNDMR1 algorithm. It can be clearly seen in Fig.4.(a) that VNDMR1 out-performs the other four algorithms in terms of the tree cost. CDKS and DKSLD have the worst and similar tree cost; BSMA is better than QDMR but worse on the tree cost than VNDMR. In addition, Fig.4.(b) shows that VNDMR1 requires less execution time than BSMA. The other three algorithm CDKS, QDMR and DKSLC require lower computational time. However, the solution quality is of much lower quality than both BSMA and VNDMR1.

Fig.5 presents the results of our VNDMR1 algorithm and other algorithms in terms of the tree cost and execution time for problems of different network sizes, where the group size is 10% of the overall network size. Again, it can be seen in Fig.5.(a) that VNDMR1 outperforms the other four algorithms upon the solution quality. Fig.5.(b) shows that VNDMR1 requires less execution time than BSMA. This is due to that the time complexity of VNDMR1 is $O(l^2 n log^2(n))$, while BSMA's time complexity is $O(kn^3 log(n))$ (n: the number of nodes, l: the number of edges, k: the kth shortest path between source and a destination).

In [26], Ghaboosi and Haghighat develop a path relinking algorithm and show that it outperforms a number of existing algorithms including KPP, BSMA, GA-based algorithms [19,20], tabu search based algorithms [23,24,22,21] and another path relinking algorithm [38]. In order to compare our VNDMR algorithms with these algorithms in the literature, we generate a group of random graphs with different network sizes (10, 20, 30, 40, 50, 60, 70, 80, 90, 100 nodes). For a fair comparison, three random topologies are generated for each network size, which are the same as the simulations designed in [26]. In these graphs, the link cost depends on the link length, all the link delays are set to 1, the group size is set to 30% of the network size, the delay bounds are set to different values depending on the network sizes ($\Delta = 7$ for network size 10-30, $\Delta = 8$ for network size 40-60, $\Delta = 10$ for network size 70-80 and $\Delta = 12$ for network size 90-100).

We test two variants of VND, VNDMR1 and VNDMR2, with the same ini-tial solution DBDDSP but different neighborhood structures as described in Section 3.2. The simulation results are reported in Tables 3 and 4.

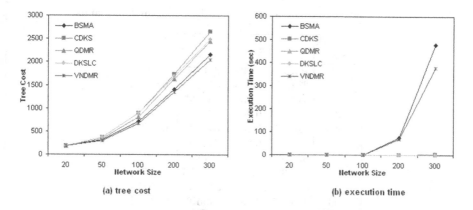

Fig. 5. Results of different approaches, group size = 10% of network size

Table 3. Average tree costs of existing algorithms on random graphs

	Algorithms	Average Tree Costs
Heuristics	KPP1 [9]	905.581
	KPP2 [9]	911.684
	BSMA [11]	872.681
GA-based Algorithms	Wang et al. [19]	815.969
	Haghighat et al. [20]	808.406
TS-based Algorithms	Skorin-Kapov and Kos [22]	897.578
	Youssef et al. [21]	854.839
	Wang et al. [23]	1214.75
	Ghaboosi and Haghighat [24]	739.095
Path relinking	Ghaboosi and Haghighat [26]	691.434
VNS Algorithms	VNDMR1	**680.067**
	VNDMR2	**658.967**

As only the average tree cost over all problem instances of different sizes are reported in [26], we report the same in Table 3. It shows that the VNDMR2 performs the best in terms of the average tree cost from 10 runs for each graph. Details of the average tree cost and the execution time of VNDMR1 and VNDMR2 on each network size are given in Table 4, showing that VNDMR2 obtains the best solutions on 9 out of 10 network sizes, while VNDMR1 gets 1 best result. We also observe that VNDMR2 spends longer computing time than VNDMR1 to get the better results. The standard deviations of both VNDMR1 and VNDMR2 for each graph are 0, due to that the order of the nodes changed in the search is fixed for comparisons, i.e. there is no random factor in VNDMR1 and VNDMR2. For the 3 graphs of each size, results vary in VNDMR1 and VNDMR2. For example, for the largest graph, VNSMR2 obtained solution of 1097, 922 and 998 (average 1005.67), compared with those of 1130, 916 and 1076 (average 1040.67) from VNSMR1.

Table 4. Average results of our VNDMR on the random graphs

| Network | VNDMR1 | | VNDMR2 | |
Size	Cost	Time(s)	Cost	Time(s)
10	**94.67**	0.005	**94.67**	0.003
20	282.33	0.015	**275.33**	0.032
30	415.67	0.036	**399.67**	0.17
40	518	0.063	**514**	0.362
50	726.67	0.151	**674.67**	0.859
60	812.33	0.292	**777.67**	1.392
70	805.33	0.682	**805**	2.571
80	922.33	1.286	**905.33**	5.127
90	1182.67	3.151	**1137.67**	11.705
100	1040.67	4.292	**1005.67**	15.332

We re-implemented the path relinking algorithm in [26]. Fig.6 presents the execution time of the path relinking algorithm, VNDMR1 and VNDMR2 tested on the same computer. Our VNDMR algorithms can obtain better results in a very short time compared with that of the path relinking algorithm.

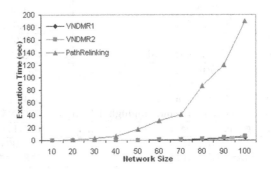

Fig. 6. Average execution time of VNDMR and the Path Relinking [26]

In summary, over a large number of simulations on instances of different characteristics, we have demonstrated that the proposed VND algorithms outperform other existing algorithms with regard to both the average tree cost and computational time. Our VNDMR2 obtains the best average tree cost on the random graphs so far.

5   Conclusions

In this paper, we have investigated variable neighborhood descent (VND) search algorithms for solving multicast network routing problems, where delay-constrained least-cost multicast trees are constructed. The problem is a Delay-Constrained Steiner tree problem and has been proved to be NP-complete. The main characteristic of our VND algorithms is that of using three simple, yet effective, neighborhood structures. Each neighborhood is designed to reduce the tree cost in different ways and at the same time satisfy the delay constraint. This enables a much more flexible search over the search space. A large number of experimental results demonstrate that our VND algorithms are the best performing algorithms in comparison with other existing algorithms in terms of both the total tree cost and the execution time.

Many promising directions of future work are possible. Real world network scenarios are mostly dynamic with some nodes leaving and joining the multicast groups at various times. Additionally, our VND algorithm can be easily adapted for solving a variety of network routing problems with different constraints.

Acknowledgements. This research is supported by Hunan University, China, and the School of Computer Science at The University of Nottingham, UK.

References

1. Salama, H.F., Reeves, D.S., Viniotis, Y.: Evaluation of multicast routing algorithms for realtime communication on high-speed networks. IEEE Journal on Selected Areas in Communications 15, 332–345 (1997)
2. Yeo, C.K., Lee, B.S., Er, M.H.: A survey of application level multicast techniques. Computer Communications 27, 1547–1568 (2004)
3. Masip-Bruin, X., Yannuzzi, M., Domingo-Pascual, J., Fonte, A., Curado, M., Monteiro, E., Kuipers, F., Van Mieghem, P., Avallone, S., Ventre, G., Aranda-Gutierrez, P., Hollick, M., Steinmetz, R., Iannone, L., Salamatian, K.: Research challenges in QoS routing. Computer Communications 29, 563–581 (2006)
4. Hwang, F.K., Richards, D.S.: Steiner tree problems. IEEE/ACM Trans. Networking 22, 55–89 (1992)
5. Garey, M.R., Johnson, D.S.: Computers and Intractability: A Guide to the Theory of NP-Completeness. W.H. Freeman and Company, New York (1979)
6. Guo, L., Matta, I.: QDMR: An efficient QoS dependent multicast routing algorithm. In: Proceedings of the 5th IEEE Real Time Technology and Applications Symposium, pp. 213–222 (1999)
7. Diot, C., Dabbous, W., Crowcroft, J.: Multicast communication: a survey of protocols, functions, and mechanisms. IEEE Journal on Selected Areas in Communications 15, 277–290 (1997)
8. Oliveira, C.A.S., Pardalos, P.M.: A survey of combinatorial optimization problems in multicast routing. Computers & Operations Research 32(8), 1953–1981 (2005)
9. Kompella, V.P., Pasquale, J.V., Polyzos, G.C.: Multicast routing for multimedia communication. IEEE/ACM Transactions on Networking 1, 286–292 (1993)
10. Sun, Q., Langendoerfer, H.: Efficient multicast routing for delay-sensitive applications. In: Proceedings of the 2nd Workshop on Protocols for Multimedia Systems, pp. 452–458 (1995)
11. Zhu, Q., Parsa, M., Garcia-Luna-Aceves, J.J.: A source-based algorithm for delay-constrained minimum-cost multicasting. In: Proceedings of the 14th Annual Joint Conference of the IEEE Computer and Communication (INFOCOM 1995), pp. 377–385. IEEE Computer Society Press, Boston (1995)

12. Shaikh, A., Shin, K.: Destination-driven routing for low-cost multicast. IEEE Journal on Selected Areas in Communications 15, 373–381 (1997)
13. Kou, L., Markowsky, G., Berman, L.: A fast algorithm for Steiner trees. Acta Informatica 15, 141–145 (1981)
14. Cormen, T.H., Leiserson, C.E., Revest, R.L.: Introduction to Algorithms. MIT Press, Cambridge (1997)
15. Betsekas, D., Gallager, R.: Data Networks, 2nd edn. Prentice-Hall, Englewood Cliffs (1992)
16. Eppstein, D.: Finding the k shortest paths. SIAM Journal of Computing 28, 652–673 (1998)
17. Wang, X.L., Jiang, Z.: QoS multicast routing based on simulated annealing algorithm. In: Proceedings of International Society for Optical Engineering on Network Architectures, Management, and Applications, pp. 511–516 (2004)
18. Zhang, K., Wang, H., Liu, F.Y.: Distributed multicast routing for delay variation-bounded Steiner tree using simulated annealing. Computer Communications 28, 1356–1370 (2005)
19. Wang, Z., Shi, B., Zhao, E.: Bandwidth-delay-constrained least-cost multicast routing based on heuristic genetic algorithm. Computer communications 24, 685–692 (2001)
20. Haghighat, A.T., Faez, K., Dehghan, M., Mowlaei, A., Ghahremani, Y.: GA-based heuristic algorithms for bandwidth-delay-constrained least-cost multicast routing. Computer Communications 27, 111–127 (2004)
21. Youssef, H., Sait, M., Adiche, H.: Evolutionary algorithms, simulated annealing and tabu search: a comparative study. Engineering Applications of Artificial Intelligence 14, 167–181 (2001)
22. Skorin-Kapov, N., Kos, M.: The application of steiner trees to delay constrained multicast routing: a tabu search approach. In: Proceedings of the seventh international Conference on Telecommunications, Zagreb, Croatia, pp. 443–448 (2003)
23. Wang, H., Fang, J., Wang, H., Sun, Y.M.: TSDLMRA: an efficient multicast routing algorithm based on tabu search. Journal of Network and Computer Applications 27, 77–90 (2004)
24. Ghaboosi, N., Haghighat, A.T.: A tabu search based algorithm for multicast routing with QoS constraints. In: 9th International Conference on Information Technology, pp. 18–21 (2006)
25. Skorin-Kapov, N., Kos, M.: A GRASP heuristic for the delay-constrained multicast routing problem. Telecommunication Systems 32, 55–69 (2006)
26. Ghaboosi, N., Haghighat, A.T.: A path relinking approach for Delay-Constrained Least-Cost Multicast routing problem. In: 19th International Conference on Tools with Artificial Intelligence, pp. 383–390 (2007)
27. Canuto, S.A., Resende, M.G.C., Ribeiro, C.C.: Local search with perturbations for the prize collecting Steiner tree problem in graphs. Networks 38, 50–58 (2001)
28. Gruber, M., Raidl, G.R.: Variable neighborhood search for the bounded diameter minimum spanning tree problem. In: Hansen, P., Mladenović, N., Pérez, J.A.M., Batista, B.M., Moreno-Vega, J.M. (eds.) Proceedings of the 18th Mini Euro Conference on Variable Neighborhood Search, Tenerife, Spain (2005)
29. Mladenovic, N., Hansen, P.: Variable neighborhood search. Computers & Operations Research 24, 1097–1100 (1997)
30. Jari, K., Teemu, N., Olli, B., Michel, G.: An efficient variable neighborhood search heuristic for very large scale vehicle routing problems. Computers & Operations Research 34, 2743–2757 (2007)

31. Burke, E.K., Curtois, T.E., Post, G., Qu, R., Veltman, B.: A hybrid heuristic ordering and variable neighbourhood search for the nurse rostering problem. European Journal of Operational Research 2, 330–341 (2008)
32. Zhang, B., Mouftah, H.T.: A destination-driven shortest path tree algorithm. In: IEEE International Conference on Communications, pp. 2258–2262 (2002)
33. Martins, S.L., Resende, M.G.C., Ribeiro, C.C., Pardalos, P.M.: A parallel GRASP for the Steiner tree problem in graphs using a hybrid local search strategy. Journal of Global Optimization 17(1-4), 267–283 (2000)
34. Leitner, M., Raidl, G.R.: Lagrangian Decomposition, Metaheuristics, and Hybrid Approaches for the Design of the Last Mile in Fiber Optic Networks. In: Blesa, M.J., Blum, C., Cotta, C., Fernández, A.J., Gallardo, J.E., Roli, A., Sampels, M. (eds.) HM 2008. LNCS, vol. 5296, pp. 158–174. Springer, Heidelberg (2008)
35. Sun, Q., Langendoerfer, H.: An efficient delay-constrained multicast routing algorithm. Technical Report, Internal Report, Institute of Operating Systems and Computer Networks. TU Braunschweig, Germany (1997)
36. Cayley, A.: A theorem on trees. Journal of Math. 23, 376–378 (1989)
37. Waxman, B.M.: Routing of multipoint connections. IEEE Journal on Selected Areas in Communications 6, 1617–1622 (1988)
38. Bastos, M.P., Ribeiro, C.C.: Reactive tabu search with path relinking for the Steiner problem in graphs. In: Proceedings of the third Metaheuristics International Conference, Angra dos Reis, Brazil (1999)

Expeditive Extensions of Evolutionary Bayesian Probabilistic Neural Networks

Vasileios L. Georgiou[1], Sonia Malefaki[2], Konstantinos E. Parsopoulos[1], Philipos D. Alevizos[1], and Michael N. Vrahatis[1]

[1] Department of Mathematics, University of Patras, Patras, Greece
{vlg,kostasp,philipos,vrahatis}@math.upatras.gr
[2] Department of Statistics and Insurance Science,
University of Piraeus, Piraeus, Greece
smalefak@unipi.gr

Abstract. Probabilistic Neural Networks (PNNs) constitute a promising methodology for classification and prediction tasks. Their performance depends heavily on several factors, such as their spread parameters, kernels, and prior probabilities. Recently, Evolutionary Bayesian PNNs were proposed to address this problem by incorporating Bayesian models for estimation of spread parameters, as well as Particle Swarm Optimization (PSO) as a means to select prior probabilities. We further extend this class of models by introducing new features, such as the Epanechnikov kernels as an alternative to the Gaussian ones, and PSO for parameter configuration of the Bayesian model. Experimental results of five extended models on widely used benchmark problems suggest that the proposed approaches are significantly faster than the established ones, while exhibiting competitive classification accuracy.

1 Introduction

Classification models exhibit a rapid development in the past few years, due to their wide applicability in modern scientific and engineering applications. Probabilistic Neural Networks (PNNs) [1] is a widely used classification methodology, which has been used in several applications in bioinformatics [2, 3, 4, 5], as well as in different scientific fields [6, 7] with promising results. PNNs constitute a variant of the well–known Discriminant Analysis [8], presented in the framework of artificial neural networks. Their main task is the classification of unknown feature vectors into predefined classes [1], where the Probability Density Function (PDF) of each class is estimated by kernel functions. For this purpose, the Gaussian kernel function is usually employed.

The type of kernels and their spread parameters, as well as the prior probability of each class affect the performance of PNNs, significantly [9, 10]. Evolutionary Bayesian PNNs (EBPNNs) [11] were proposed as variants of the standard PNNs, where the spread parameters are estimated by Bayesian models, while the prior probabilities are determined by the Particle Swarm Optimization (PSO) algorithm. However, the employed Bayesian models included also several parameters, configured through a time consuming exhaustive search procedure.

T. Stützle (Ed.): LION 3, LNCS 5851, pp. 30–44, 2009.

The present work aims at extending the EBPNN model in order to reduce the required execution time. For this purpose, new features are introduced, such as the Epanechnikov kernels. Also, besides the prior probabilities, PSO is used for determining the constants of the Bayesian model's prior distributions. The new class of models is called Extended EBPNN (EEBPNN), and five models are compared with different established EBPNN and PNN approaches on four widely used classification problems from the UCI repository, with promising results.

The paper is organized as follows: Section 2 contains the necessary background information on PNNs and PSO. The proposed EEBPNN model is described in Section 3, and experimental results are reported in Section 4. The paper concludes in Section 5.

2   Background Material

PNNs and PSO are briefly described in this section for presentation completeness.

2.1   Probabilistic Neural Networks

PNNs are supervised neural network models, closely related to the Bayes classification rule [7, 12] and Parzen nonparametric probability density function estimation theory [1, 13]. Their training procedure consists of a single pass over all training patterns [1], thereby rendering PNNs faster to train, compared to the Feedforward Neural Networks (FNNs).

Consider a p–dimensional classification task and let K be the number of classes. Let also $\mathcal{T}_{tr}$ be the training data set with a total of N_{tr} feature vectors, while N_k be the number of training vectors that belong to the k–th class, $k = 1, 2, \ldots, K$. The i–th feature vector of the k–th class is denoted as $X_{ik} \in \mathbb{R}^p$, where $i = 1, 2, \ldots, N_k$, $k = 1, 2, \ldots, K$. Then, a PNN consists of four layers: the *input, pattern, summation* , and *output layer*, as depicted in Fig. 1 [1, 9].

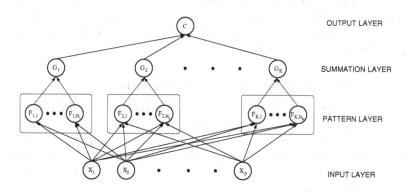

Fig. 1. The probabilistic neural network model

An input feature vector, $X \in \mathbb{R}^p$, is applied to the p input neurons that comprise the input layer, and it is passed to the pattern layer. The latter is fully interconnected with the input layer and organized into K groups of neurons. Each group of neurons in the pattern layer consists of N_k neurons, and the i–th neuron in the k–th group computes its own output by using a kernel function. The kernel function is typically a multivariate Gaussian of the form,

$$f_{ik}(X) = \frac{1}{(2\pi)^{p/2}|\Sigma|^{1/2}} \exp\left(-\frac{1}{2}(X - X_{ik})^{\top} \Sigma^{-1}(X - X_{ik})\right), \qquad (1)$$

where $X_{ik} \in \mathbb{R}^p$ is the center of the kernel, and Σ is the matrix of spread (smoothing) parameters. PNNs that exploit a global smoothing parameter are called *homoscedastic*, while the use of a different parameter per class is referred to as *heteroscedastic* PNN [14].

The summation layer consists of K neurons and each one estimates the conditional probability of the corresponding class given the input feature vector, X, according to the Bayes decision rule:

$$G_k(X) = \sum_{i=1}^{N_k} \pi_k f_{ik}(X), \quad k \in \{1, 2, \ldots, K\}, \qquad (2)$$

where π_k is the prior probability of the k–th class, and $\sum_{k=1}^{K} \pi_k = 1$. Thus, X is classified in the class that achieves the maximum output of the summation neurons.

A limitation of PNNs is the curse of dimensionality. When the dimension of the data set is large, PNNs usually do not yield good results. A faster version of the PNN can be obtained by using only a part, instead of the whole training data set. Such a training set can be obtained either by randomly sampling from the available data or by finding "representative" training vectors for each class through a clustering technique.

For this purpose, the widely used K–medoids clustering algorithm [15] can be applied on the training data of each class. The extracted medoids are then used as centers for the PNN's kernels, instead of using all the available training data. The resulted PNNs are significantly smaller with respect to the number of neurons in the pattern layer, although there is no sound procedure for estimating the optimal required number of medoids.

2.2   The Particle Swarm Optimization Algorithm

Particle Swarm Optimization (PSO) was introduced in 1995 by Eberhart and Kennedy [16, 17], drawing inspiration from the dynamics of socially organized groups. PSO is a stochastic, population–based optimization algorithm that exploits a population of individuals to synchronously probe the search space. In this context, the population is called a *swarm* and the individuals (i.e., the search points) are called the *particles*.

Each particle moves with an adaptable velocity within the search space and retains in memory the best position it has ever encountered. This position is also shared with other particles in the swarm. In the *global* PSO variant, the best position ever attained by all individuals of the swarm is communicated to every particle at each iteration. On the other hand, in *local* PSO, best positions are communicated only within strict neighborhoods of each particle.

Assume a d–dimensional search space, $\mathcal{S} \subset \mathbb{R}^d$, and a swarm consisting of N particles. Let

$$Z_i = (z_{i1}, z_{i2}, \ldots, z_{id})^\top \in \mathcal{S},$$

be the i-th particle and

$$V_i = (v_{i1}, v_{i2}, \ldots, v_{id})^\top, \quad B_i = (b_{i1}, b_{i2}, \ldots, b_{id})^\top \in \mathcal{S},$$

be its velocity and best position, respectively. Assume g to be the index of the particle that attained the best previous position among all particles, and t be the iteration counter. Then, the swarm is manipulated by the equations:

$$V_i(t+1) = \chi \left[V_i(t) + c_1\, r_1\big(B_i(t) - Z_i(t)\big) + c_2\, r_2\big(B_g(t) - Z_i(t)\big) \right], \quad (3)$$
$$Z_i(t+1) = Z_i(t) + V_i(t+1), \quad (4)$$

where $i = 1, 2, \ldots, N$; χ is a parameter called the *constriction coefficient*; c_1 and c_2 are two positive constants called the *cognitive* and *social* parameter, respectively; and r_1, r_2, are random vectors uniformly distributed within $[0, 1]^d$ [18]. All vector operations in Eqs. (3) and (4) are performed componentwise.

The best positions are then updated according to the equation:

$$B_i(t+1) = \begin{cases} Z_i(t+1), & \text{if } f\big(Z_i(t+1)\big) < f\big(B_i(t)\big), \\ B_i(t), & \text{otherwise.} \end{cases}$$

The particles are bounded within the search space $\mathcal{S}$, while the constriction coefficient is derived analytically through the formula:

$$\chi = \frac{2\kappa}{\left|2 - \varphi - \sqrt{\varphi^2 - 4\varphi}\right|},$$

for $\varphi > 4$, where $\varphi = c_1 + c_2$ and $\kappa = 1$, based on the stability analysis due to Clerc and Kennedy [18].

3   The Proposed Extended Model

EBPNN models were proposed as a new variant of PNNs that estimate the spread parameters through Bayesian models. More specifically, a different diagonal matrix of spread parameters,

$$\Sigma_k = \text{diag}\left(\sigma_{1k}^2, \sigma_{2k}^2, \ldots, \sigma_{pk}^2\right), \quad k = 1, 2, \ldots, K,$$

for each one of the K classes is used to increase model flexibility [10, 11].

The centered data per class, received from the preprocessing phase with the K-medoids algorithm as described in Section 2.1, are modeled by:

$$X_{ik} \overset{iid}{\sim} \mathcal{N}_p(\mu_k, \Sigma_k), \quad i = 1, 2, \ldots, N_k, \quad k = 1, 2, \ldots, K.$$

The prior distributions of the model parameters are defined as:

$$\mu_{jk} \sim \mathcal{N}(0, \nu^2),$$
$$\tau_{jk} \sim \mathcal{G}(\alpha, \beta), \quad j = 1, 2, \ldots, p,$$

where $\tau_{jk} = \sigma_{jk}^{-2}$ and $\alpha, \beta, \nu > 0$.

In EBPNNs, it is assumed that the class centers, X_{ik}, are conditionally independent given μ_k and τ_{jk}. Also, μ_k and τ_{jk} are also considered independent, with joint posterior distribution:

$$\pi(\mu_{jk}, \tau_{jk}|X._{k,j}) \propto \tau_{jk}^{\frac{N_k}{2}+\alpha-1} \times \exp\left\{ -\tau_{jk}\left(\frac{\sum_{i=1}^{N_k}(X_{ik,j}-\mu_{jk})^2}{2} + \beta \right) - \frac{\mu_{jk}^2}{2\nu^2} \right\},$$

where $X_{ik,j}$ stands for the j-th component of the p-dimensional vector X_{ik}.

Simulation from the posterior distribution of (μ_{jk}, τ_{jk}), for $j = 1, 2, \ldots, p$, $k = 1, 2, \ldots, K$, requires the application of an indirect method, such as Gibbs sampler, since direct simulation is not feasible. The Gibbs sampler [19] produces a Markov chain by an iterative, recursive sampling from the conditional distributions that converges in distribution to the joint distribution. The full conditional distributions are given as follows:

$$\mu_{jk}|\tau_{jk}, X._{k,j} \sim \mathcal{N}\left(\frac{\tau_{jk}\sum_{i=1}^{N_k} X_{ik,j}}{\tau_{jk}N_k + \frac{1}{\nu^2}}, \frac{1}{\tau_{jk}N_k + \frac{1}{\nu^2}} \right), \tag{5}$$

$$\tau_{jk}|\mu_{jk}, X._{k,j} \sim \mathcal{G}\left(\frac{N_k}{2} + \alpha, \frac{\sum_{i=1}^{N_k}(X_{ik,j}-\mu_{jk})^2}{2} + \beta \right). \tag{6}$$

Starting from any point in the support of the joined distribution, we draw successively from the conditional distributions of μ_{jk} and τ_{jk}, each in turn, using the previously drawn value of the other parameter, and the obtained sequence converges to the joint distribution.

Conjugated prior distributions were chosen in EBPNNs, such that closed forms are available for the full conditional distributions. The choice of conjugated prior distributions has minor importance, since any distribution can be used as prior. In such cases, we can use a hybrid Gibbs sampler (Gibbs sampler embedding a Metropolis Hastings step) or different Monte Carlo or Markov Chain Monte Carlo simulation methods, such as Importance Sampling and Metropolis Hastings [20].

Based on the aforementioned Bayesian model, EBPNNs estimate the spread parameters of their kernels. Thus, instead of estimating $p \times K$ spread parameters, only the values of α, β, and ν, need to be determined. In recent implementations [10, 11], an exhaustive search was carried out in the range $[10^{-4}, 10]$, using

Table 1. Pseudocode of the sampling procedure and the determination of the prior probabilities with PSO in EBPNNs

---

Create the clustered training set $\mathcal{T}_{\text{tr}}^{\text{c}}$ from the original training set $\mathcal{T}_{\text{tr}}$.
// *Estimation of the spread parameters using the Gibbs sampler* //
Do $(k = 1, 2, \ldots, K)$
  Do $(j = 1, 2, \ldots, p)$
    Select initial value for μ_{jk}.
    Do $(m = 1, 2, \ldots, G_{\max})$
      Draw from Eq. (6) a new τ_{jk}^{new}, using μ_{jk}.
      Draw from Eq. (5) a new μ_{jk}^{new}, using τ_{jk}^{new}.
      Set $\mu_{jk} \leftarrow \mu_{jk}^{\text{new}}$ and $\tau_{jk}^{m} \leftarrow \tau_{jk}^{\text{new}}$.
    End Do
    Compute mean value, τ_{jk}, of τ_{jk}^{m}, $m = 1, 2, \ldots, G_{\max}$.
  End Do
  Set the spread matrix Σ_k of class k by using the relation $\tau_{jk} = \sigma_{jk}^{-2}$.
End Do
// *Estimation of the prior probabilities by PSO* //
Initialize a swarm of particles Z_i, $i = 1, 2, \ldots, N$, within the range $[0, 1]^K$.
Initialize best positions, B_i, and velocities, V_i, $i = 1, 2, \ldots, N$.
While (stopping condition not met)
  Update swarm using Eqs. (3) and (4).
  Constrain particles within $[0, 1]^K$.
  Evaluate particles based on the classification accuracy on $\mathcal{T}_{\text{tr}}$.
  Update best positions.
End While
Write the obtained spread matrices Σ_k, $k = 1, 2, \ldots, K$, and prior weights.

---

variable step size, for the selection of α and β. Furthermore, the value of ν was set arbitrarily to $\nu = 1$ [10, 11].

In standard PNNs, the prior probabilities of Eq. (2) are either estimated from the available data or set randomly. In contrast to this procedure, EBPNNs employ the PSO algorithm to determine the most promising values for the prior probabilities with respect to classification accuracy. Thus, a swarm of weights is randomly generated and probes the search space of weights to find the most promising values. The underlying objective function utilized by PSO is the classification accuracy of the PNN over the whole training data set [11]. The pseudocode of the Gibbs sampling phase as well as the determination of the priors with PSO, is presented in Table 1.

The proposed *Extended Evolutionary Bayesian Probabilistic Neural Network* (EEBPNN) model extends the aforementioned EBPNN models, as follows:

(1). The *Epanechnikov kernel*, which is defined as:

$$f_{ik}(X) = \max\left\{0, 1 - \frac{1}{2\kappa^2}(X - X_{ik})^{\top} \Sigma_k^{-1}(X - X_{ik})\right\}, \qquad (7)$$

where κ is the kernel's parameter [21], is used instead of the typical Gaussian kernels. This choice is based on the fact that the Epanechnikov kernel has the smallest asymptotic mean integrated squared error (AMISE) and it is considered as optimal kernel [22]. The expected gain is significantly faster execution time, since there is no need to calculate the time–consuming exponential functions included in the Gaussian kernel. The parameter κ can be set arbitrarily by the user or, alternatively, determined by using the PSO algorithm.

(2). The PSO algorithm is employed for the selection of the most appropriate values of α and β in the Bayesian model. The constant, ν, is set to the value 0.2, which is adequate to cover the range $[-0.5, 0.5]$ of the data in the considered problems.

The described EEBPNN model introduces several new features in different aspects of the standard PNN and EBPNN model. Generally, it is not necessary to use all the new features concurrently in the same model. Thus, one can define EBPNN models with Gaussian kernels, using PSO for determining the constants of the prior distributions in the Bayesian model, as well as the prior probabilities. Alternatively, Epanechnikov kernel can be used with the established EBPNN and BPNN models, where Bayesian constants are determined through exhaustive search. In the next section, we define several alternative models, and report their performance on widely used classification tasks.

4   Experimental Settings and Results

We considered four widely used benchmark problems from the Proben1 benchmark data sets [23] of the UCI repository [24]. Specifically, we used the following data sets:

1. *Wisconsin Breast Cancer Database* (WBCD): the aim is to predict whether a breast tumor is benign or malignant [25]. There are 9 continuous attributes based on cell descriptions gathered by microscopic examination and 699 instances.

2. *Card Data Set*: the aim is to predict the approval or non–approval of a credit card to a customer [26]. There are 51 attributes (not explicitly reported for confidential reasons) and the number of observations is 690.

3. *The Pima Indians Diabetes Data Set*: the aim is to predict the onset of diabetes, therefore, there are two classes [27]. The input features are the diastolic blood pressure; triceps skin fold thickness; plasma glucose concentration in a glucose tolerance test; and diabetes pedigree function. These 8 inputs are all continuous without missing values and there are 768 instances.

4. *Heart Disease Data Set*: the aim is to predict whether at least one of the four major vessels of the heart is reduced in diameter by more than 50%, so there are two classes [28]. The 35 attributes of the 920 patients are age, sex, smoking habits, subjective patient pain descriptions and results of various medical examinations such as blood pressure and cardiogram.

Table 2. Characteristics of the four benchmark data sets

	Cancer	Card	Diabetes	Heart
Number of Instances	699	690	768	920
Variables	9	51	8	35
Classes	2	2	2	2

The characteristics of the four data sets are summarized in Table 2. In our experiments, the number of medoids extracted from each class was only the 5% of the class size. This reduces the size of the pattern layer by a factor of 20, compared to the standard PNN that utilizes the whole training data set. The choice of 5% was based on numerous trials with different fractions of the class size. Also, for the Gibbs sampler, a number of $G_{\max} = 10^4$ draws was used.

The following new models of the EEBPNN class were considered in our experiments:

M1. **Epan.BPNN:** BPNN that uses Epanechnikov kernels with $\kappa = 1$, exhaustive search for the selection of the prior distributions' constants of the Bayesian model, and the prior probabilities are set explicitly based on the fraction of each class in the data set.

M2. **Epan.EBPNN:** EBPNN that uses Epanechnikov kernels with $\kappa = 1$, exhaustive search for the selection of the prior distributions' constants of the Bayesian model, and the prior probabilities are computed with PSO.

M3. **Gauss.MCPNN:** EBPNN with Gaussian kernels, prior distributions' constants of the Bayesian model estimated by PSO, and prior probabilities are set explicitly based on the fraction of each class in the data set.

M4. **Gauss.PMCPNN:** EBPNN with Gaussian kernels, prior distributions' constants of the Bayesian model as well as the prior probabilities are estimated by PSO.

M5. **Epan.EEBPNN:** EEBPNN with Epanechnikov kernels, where the prior distributions' constants of the Bayesian model, the prior probabilities and Epanechnikov's κ are all estimated by PSO.

Moreover, the following established PNN–based models were used for comparison purposes:

M6. **PNN:** Standard PNN with exhaustive search for the selection of the spread parameter σ.

M7. **CL.PNN:** Standard PNN that uses the clustered instead of the whole training set.

M8. **GGEE.PNN:** A variant of the standard PNN, proposed by Gorunescu et al. [29], which incorporates a Monte Carlo search technique.

M9. **Hom.EPNN:** Homoscedastic EPNN [9] that utilizes the whole training data set for the construction of the PNN's pattern layer.

M10. **Het.EPNN:** Heteroscedastic EPNN [9] that utilizes the whole training set.

M11. **CL.Hom.EPNN:** Same with the Hom.EPNN, where only the clustered training set was used for PNN's construction.

M12. **CL.Het.EPNN:** Same with the Het.EPNN, where only the clustered training set was used.

M13. **Bag.EPNN:** Bagging EPNN that incorporates the bagging technique for the prior weighting, clustered training set and generalized spread parameters' matrix [30].

M14. **Bag.P.EPNN:** Bagging EPNN with bagging, clustered training set, generalized spread parameters' matrix and prior probabilities estimation by PSO.

M15. **Gaus.BPNN:** BPNN with Gaussian kernels and prior distributions' constants of the Bayesian model are selected by an exhaustive search.

M16. **Gaus.EBPNN:** EBPNN with Gaussian kernels, prior distributions' constants of the Bayesian model are selected by an exhaustive search, and prior probabilities estimated by PSO.

Regarding PSO, we used the default parameter values, $c_1 = c_2 = 2.05$, and $\chi = 0.729$ [18]. The number of particles was set to 10, and a maximum number of 50 generations was allowed for the detection of the prior probabilities. In the Hom.EPNN (M9) model, the number of particles was set to 5, while in the Het.EPNN (M10), Bag.EPNN (M13), and Bag.P.EPNN (M14) models, 10 particles were used and a maximum number of 100 iterations was allowed. For the bagging EPNNs, 11 bootstrap samples were drawn from each clustered training data set. Based on these samples, an ensemble of 11 EPNNs was constructed, and the final classification was obtained by a majority voting procedure. In the proposed EBPNN variants (M1, M2), a swarm of 10 particles was used for 50 iterations, while for the MCPNN and EEBPNN variants (M3–M5), 5 particles were used for 30 iterations.

Every benchmark data set was applied 10 times using 10–fold cross–validation, where every time the folds were randomly selected. The mean, median, standard deviation, minimum and maximum of the obtained classification accuracies and CPU times were recorded for all models and they are reported in Tables 3–6. Each table consists of two parts divided by a horizontal line. The upper part contains all statistics for the five proposed models M1–M5, while the lower part contains the statistics for the rest of the models. The best classification performance for the proposed models, as well as for the rest of the models, is boldfaced. Thus, there is a boldfaced line in each of the two parts of the table, which corresponds to the best performing model of the corresponding part of the table, with respect to its classification accuracy.

In Table 3, which corresponds to the Cancer data set, the M14 model, i.e., EPNN model with bagging, clustered training set, generalized spread parameters' matrix and prior probabilities estimation by PSO, exhibited the highest classification accuracy of 97.17% and a CPU time of 90.01 seconds. On the other hand, the most promising of the proposed models was M3, i.e., EBPNN

Table 3. Test set classification accuracy percentages and CPU times for the Cancer data set

Model	Classification Accuracy (%)					CPU time (sec.)				
	Mean	Median	St.D.	Min	Max	Mean	Median	St.D.	Min	Max
M1	96.39	96.35	0.18	96.14	96.71	21.40	21.42	0.06	21.32	21.47
M2	96.53	96.56	0.22	96.14	96.85	24.39	24.42	0.07	24.28	24.48
M3	**96.75**	**96.71**	**0.22**	**96.42**	**97.14**	**41.12**	**41.11**	**0.69**	**40.04**	**42.72**
M4	96.75	96.71	0.17	96.42	97.00	65.04	65.65	2.91	57.23	67.78
M5	96.55	96.49	0.24	96.28	97.13	62.36	64.27	3.59	57.04	65.31
M6	95.79	95.85	0.25	95.27	96.14	42.09	42.42	0.66	40.66	42.69
M7	91.91	92.06	0.84	90.42	92.99	0.08	0.08	0.00	0.08	0.09
M8	96.39	96.42	0.20	95.99	96.71	1.52	1.61	0.17	1.22	1.65
M9	95.82	95.85	0.28	95.28	96.28	89.12	88.82	1.07	88.12	91.73
M10	95.32	95.21	0.57	94.42	96.14	171.78	171.75	1.07	170.21	174.04
M11	90.50	90.84	1.58	87.85	92.56	0.16	0.15	0.02	0.14	0.20
M12	87.89	87.78	1.74	85.27	90.56	0.32	0.33	0.06	0.24	0.43
M13	96.85	96.78	0.46	96.14	97.85	82.78	78.07	8.86	76.22	99.75
M14	**97.17**	**97.14**	**0.16**	**96.86**	**97.43**	**90.01**	**89.86**	**0.92**	**88.97**	**92.12**
M15	96.36	96.35	0.22	96.13	96.85	27.74	28.08	1.08	24.67	28.17
M16	96.51	96.49	0.14	96.28	96.71	31.62	32.02	1.33	27.84	32.16

Table 4. Test set classification accuracy percentages and CPU times for the Card data set

Model	Classification Accuracy (%)					CPU time (sec.)				
	Mean	Median	St.D.	Min	Max	Mean	Median	St.D.	Min	Max
M1	80.58	80.94	1.03	78.55	81.59	193.86	193.69	1.37	191.86	195.82
M2	82.83	83.04	0.89	81.45	84.06	203.71	203.47	1.42	201.77	205.95
M3	84.84	84.57	0.76	84.06	86.23	223.64	221.09	20.04	199.17	262.71
M4	84.64	84.64	0.66	83.77	85.66	350.22	347.19	45.94	268.26	408.51
M5	**85.90**	**85.87**	**0.57**	**84.78**	**86.96**	**354.45**	**351.22**	**34.93**	**310.75**	**399.69**
M6	82.10	81.96	0.76	80.87	83.48	182.01	186.37	7.88	169.82	187.93
M7	80.49	80.58	0.66	79.13	81.45	0.23	0.23	0.00	0.22	0.24
M8	84.31	84.28	0.63	83.48	85.51	5.46	5.45	0.06	5.38	5.53
M9	85.35	85.22	0.38	84.93	86.09	266.10	274.39	74.56	168.72	342.27
M10	**87.67**	**87.76**	**0.51**	**86.96**	**88.55**	**521.60**	**510.24**	**142.74**	**327.08**	**671.83**
M11	82.02	81.81	1.15	80.73	84.49	0.49	0.47	0.06	0.46	0.66
M12	85.20	85.36	0.97	83.34	86.52	0.66	0.70	0.14	0.42	0.86
M13	86.64	86.67	0.51	85.80	87.39	309.85	309.36	1.88	307.58	314.33
M14	86.83	86.81	0.34	86.38	87.39	309.73	309.84	2.62	305.26	314.95
M15	84.93	85.00	0.25	84.49	85.22	215.39	214.92	1.24	214.14	217.49
M16	86.21	86.02	0.54	85.66	87.54	229.49	228.98	1.37	228.09	231.70

Table 5. Test set classification accuracy percentages and CPU times for the Diabetes data set

Model	Classification Accuracy (%)					CPU time (sec.)				
	Mean	Median	St.D.	Min	Max	Mean	Median	St.D.	Min	Max
M1	73.90	73.93	1.16	71.89	75.91	25.18	25.38	0.64	23.37	25.44
M2	71.68	71.79	1.08	69.92	73.55	28.61	28.86	0.81	26.31	28.93
M3	66.79	66.72	0.56	66.05	67.93	37.82	37.87	0.76	36.59	39.19
M4	73.88	73.64	0.53	73.35	74.49	49.92	49.52	1.18	48.80	51.59
M5	**74.64**	**74.47**	**1.18**	**72.80**	**76.69**	**56.29**	**56.53**	**1.42**	**54.15**	**58.18**
M6	65.08	65.08	0.05	64.99	65.15	49.58	49.64	0.38	49.06	50.09
M7	65.08	65.08	0.05	64.99	65.15	0.10	0.10	0.00	0.10	0.11
M8	69.43	69.24	0.68	68.53	70.38	1.87	1.87	0.03	1.83	1.90
M9	67.67	67.58	0.88	66.03	68.80	101.17	101.13	0.48	100.40	102.01
M10	69.37	69.46	0.80	67.73	70.54	195.27	195.66	0.92	193.82	196.62
M11	65.35	65.14	0.48	64.99	66.35	0.18	0.18	0.00	0.17	0.18
M12	69.30	69.18	1.59	67.08	72.36	0.36	0.36	0.01	0.35	0.38
M13	71.00	71.16	1.02	68.90	72.09	106.42	106.53	0.92	104.25	107.73
M14	71.22	71.39	1.00	69.75	72.54	106.24	106.26	0.81	105.31	108.06
M15	**74.21**	**74.35**	**0.93**	**72.43**	**75.91**	**25.18**	**25.63**	**1.09**	**22.48**	**25.79**
M16	72.93	73.26	1.50	69.92	75.06	29.62	30.27	1.51	25.94	30.43

Table 6. Test set classification accuracy percentages and CPU times for the Heart data set

Model	Classification Accuracy (%)					CPU time (sec.)				
	Mean	Median	St.D.	Min	Max	Mean	Median	St.D.	Min	Max
M1	72.26	72.12	0.48	71.52	72.94	88.26	88.17	0.47	87.52	89.27
M2	73.32	73.31	0.48	72.72	74.02	104.54	104.28	0.72	103.55	106.12
M3	**82.11**	**82.17**	**0.66**	**80.54**	**83.04**	**158.79**	**152.13**	**14.25**	**145.29**	**182.52**
M4	81.82	81.90	1.06	79.78	83.37	160.80	163.87	9.40	147.96	174.22
M5	81.82	81.90	1.06	79.78	83.37	151.42	150.62	8.70	143.03	173.56
M6	79.23	79.13	0.48	78.59	80.00	207.99	223.48	45.27	125.62	241.32
M7	79.84	79.78	0.71	78.48	80.98	0.32	0.32	0.02	0.30	0.35
M8	80.68	80.65	0.52	79.89	81.41	6.47	6.94	0.92	4.95	7.18
M9	81.50	81.52	0.27	80.87	81.74	223.28	224.35	4.28	215.15	228.97
M10	**82.60**	**82.45**	**0.40**	**82.07**	**83.26**	**438.10**	**440.29**	**6.82**	**422.45**	**449.24**
M11	79.96	79.95	0.56	79.24	81.09	0.67	0.63	0.08	0.61	0.83
M12	77.62	77.66	1.16	75.98	79.35	1.37	1.31	0.16	1.19	1.70
M13	82.28	82.34	0.62	81.20	83.15	394.49	392.36	5.93	387.13	404.55
M14	82.35	82.50	1.05	80.43	84.13	393.22	391.47	4.95	388.02	401.03
M15	80.46	80.43	0.69	79.13	81.52	88.55	88.54	0.38	87.92	89.03
M16	81.60	81.68	0.65	80.44	82.61	106.71	106.79	0.56	105.79	107.53

Table 7. The gain and loss percentages for classification accuracy and CPU between the best performing of the proposed and the rest of the models, for each benchmark problem. Negative values denote loss instead of gain.

	Best proposed model	Best of the rest models	Gain in classification accuracy	Gain in CPU time
Cancer	M3	M14	−0.4%	54.3%
Card	M5	M10	−2.1%	32.1%
Diabetes	M5	M15	0.5%	−55.2%
Heart	M3	M10	−0.5%	63.7%

with Gaussian kernels, prior distributions' constants of the Bayesian model are selected by PSO, and prior probabilities that are set explicitly based on the fraction of each class in the data set, with a classification accuracy of 96.75% and CPU time equal to 41.12 seconds. Thus, the proposed model provides a satisfactory performance that is almost 0.4% worst with respect to its classification accuracy than the best performing model, but at a 54.3% gain in CPU time.

In the results for the Card data set, reported in Table 4, the M10 model, i.e., Heteroscedastic EPNN trained with the whole training set, had the best performance, 87.67%, among all models, with a CPU time of 521.60 seconds. The most promising from the proposed models, was M5, i.e., EEBPNN with Epanechnikov kernels, where the prior distributions' constants of the Bayesian model, prior probabilities and Epanechnikov's κ are all estimated by PSO. M5 had a classification accuracy of 85.90% at the cost of a CPU time equal to 354.45 seconds. Thus, M5 had a competitive performance that is about 2% worst than the best model, although requiring 32% less CPU time.

In the Diabetes data set, reported in Table 5, the proposed M5 model had the best performance, 74.64%, among all models. However, this came with an increased CPU time of 56.29 seconds, compared to M15, which was the best performing among the rest of the models, with classification accuracy of 74.21% and CPU time 25.18 seconds. M15 consists of a BPNN with Gaussian kernels and prior distributions' parameters of the Bayesian model estimated by an exhaustive search.

In the Heart data set, reported in Table 6, M10 was again the best performing model as for the Card data set, with a classification accuracy of 82.60% and CPU time equal to 438.10 seconds. On the other hand, M3 was the best performing from the proposed models, with an accuracy of 82,11%, which is 0.5% worse than M10, but at a computational cost of 158.79 seconds, i.e., it was 63.7% faster than M10.

In Table 7, we summarize all the gain and loss in classification accuracy and CPU time between the the best performing of the proposed and the rest of the models, for each benchmark problem, with negative values denote loss instead of gain. As we observe, the proposed models M3 and M5 have the best performance among the five proposed models M1–M5. They were able to achieve highly competitive classification accuracies but at significantly lower computational times, rendering them promising variants that can be useful especially in time–critical applications.

5   Conclusions

We proposed a class of Extended EBPNN models that incorporate several new features compared to the standard EBPNNs. These features include the use of the Epanechnikov kernel instead of the standard Gaussian kernels, as well as the selection of the prior distributions' constants of the Bayesian model by using the PSO algorithm. Five models are proposed that incorporate alternatively the aforementioned features, and four widely used benchmark classification problems from the UCI repository are employed to investigate their performance against several established PNN–based classification models.

The obtained results show that the proposed models can be competitive to the best performing of the rest models, while achieving significantly faster computation times in most cases. Thus, the proposed model can be considered as a promising alternative in time–critical PNN applications, although further research is needed to fully reveal the potential of EEBPNNs in such applications. However, in one of the test problems, the proposed model had the best performance overall, but at a higher computational cost.

Acknowledgements. The work of K.E. Parsopoulos was supported by the State Scholarship Foundation of Greece (I.K.Y.).

References

1. Specht, D.F.: Probabilistic neural networks. Neural Networks 2, 109–118 (1990)
2. Guo, J., Lin, Y., Sun, Z.: A novel method for protein subcellular localization based on boosting and probabilistic neural network. In: 2nd Asia-Pacific Bioinf. Conf. (APBC 2004), Dunedin, New Zealand, pp. 20–27 (2004)
3. Holmes, E., Nicholson, J.K., Tranter, G.: Metabonomic characterization of genetic variations in toxicological and metabolic responses using probabilistic neural networks. Chem. Res. Toxicol. 14(2), 182–191 (2001)
4. Huang, C.J.: A performance analysis of cancer classification using feature extraction and probabilistic neural networks. In: 7th Conference on Artificial Intelligence and Applications, Wufon, Taiwan, pp. 374–378 (2002)
5. Wang, Y., Adali, T., Kung, S., Szabo, Z.: Quantification and segmentation of brain tissues from mr images: A probabilistic neural network approach. IEEE Transactions on Image Processing 7(8), 1165–1181 (1998)
6. Ganchev, T., Tasoulis, D.K., Vrahatis, M.N., Fakotakis, N.: Generalized locally recurrent probabilistic neural networks with application to text-independent speaker verification. Neurocomputing 70(7-9), 1424–1438 (2007)
7. Romero, R., Touretzky, D., Thibadeau, R.: Optical chinese character recognition using probabilistic neural network. Pattern Recognition 8(30), 1279–1292 (1997)

8. Hand, J.D.: Kernel Discriminant Analysis. Research Studies Press (1982)
9. Georgiou, V.L., Pavlidis, N.G., Parsopoulos, K.E., Alevizos, P.D., Vrahatis, M.N.: New self-adaptive probabilistic neural networks in bioinformatic and medical tasks. International Journal on Artificial Intelligence Tools 15(3), 371–396 (2006)
10. Georgiou, V.L., Malefaki, S.N.: Incorporating Bayesian models for the estimation of the spread parameters of probabilistic neural networks with application in biomedical tasks. In: Int. Conf. on Statistical Methods for Biomedical and Technical Systems, Limassol, Cyprus, pp. 305–310 (2006)
11. Georgiou, V.L., Malefaki, S.N., Alevizos, P.D., Vrahatis, M.N.: Evolutionary Bayesian probabilistic neural networks. In: Int. Conf. on Numerical Analysis and Applied Mathematics (ICNAAM 2006), pp. 393–396. Wiley-VCH, Chichester (2006)
12. Raghu, P.P., Yegnanarayana, B.: Supervised texture classification using a probabilistic neural network and constraint satisfaction model. IEEE Transactions on Neural Networks 9(3), 516–522 (1998)
13. Parzen, E.: On the estimation of a probability density function and mode. Annals of Mathematical Statistics 3, 1065–1076 (1962)
14. Specht, D.F., Romsdahl, H.: Experience with adaptive probabilistic neural network and adaptive general regression neural network. In: Proc. IEEE Int. Conf. Neural Networks, vol. 2, pp. 1203–1208 (1994)
15. Kaufman, L., Rousseeuw, P.J.: Finding Groups in Data: An Introduction to Cluster Analysis. John Wiley and Sons, New York (1990)
16. Eberhart, R.C., Kennedy, J.: A new optimizer using particle swarm theory. In: Proceedings Sixth Symposium on Micro Machine and Human Science, Piscataway, NJ, pp. 39–43. IEEE Service Center, Los Alamitos (1995)
17. Kennedy, J., Eberhart, R.C.: Particle swarm optimization. In: Proc. IEEE Int. Conf. Neural Networks, vol. IV, pp. 1942–1948 (1995)
18. Clerc, M., Kennedy, J.: The particle swarm–explosion, stability, and convergence in a multidimensional complex space. IEEE Tr. Ev. Comp. 6(1), 58–73 (2002)
19. Geman, S., Geman, D.: Stochastic relaxation, gibbs distributions and the Bayesian restoration of images. IEEE Trans. Pattn. Anal. Mach. Intel. 6, 721–741 (1984)
20. Gilks, W.R., Richardson, S., Spiegelhalter, D.J.: Markov Chain Monte Carlo in Practice. Chapman and Hall, Boca Raton (1996)
21. Looney, C.G.: A fuzzy classifier network with ellipsoidal epanechnikov functions. Neurocomputing 48, 489–509 (2002)
22. Herrmann, E.: Asymptotic distribution of bandwidth selectors in kernel regression estimation. Statistical Papers 35, 17–26 (1994)
23. Prechelt, L.: Proben1: A set of neural network benchmark problems and benchmarking rules. Technical Report 21/94, Fak. Informatik, Univ. Karlsruhe (1994)
24. Asuncion, A., Newman, D.: UCI machine learning repository (2007)
25. Mangasarian, O.L., Wolberg, W.H.: Cancer diagnosis via linear programming. SIAM News 23(5), 1–18 (1990)
26. Quinlan, J.: Simplifying decision trees. International Journal of Man-Machine Studies 27(3), 221–234 (1987)
27. Smith, J.W., Everhart, J.E., Dickson, W.C., Knowler, W.C., Johannes, R.S.: Using the adap learning algorithm to forecast the onset of diabetes mellitus. In: Proc. Symp. Comp. Appl. & Med. Care, pp. 261–265 (1988)

28. Detrano, R., Janosi, A., Steinbrunn, W., Pfisterer, M., Schmid, J., Sandhu, S., Guppy, K., Lee, S., Froelicher, V.: International application of a new probability algorithm for the diagnosis of coronary artery disease. American Journal of Cardiology 64, 304–310 (1989)
29. Gorunescu, F., Gorunescu, M., Revett, K., Ene, M.: A hybrid incremental/monte carlo searching technique for the "smoothing" parameter of probabilistic neural networks. In: Int. Conf. Knowl. Eng., Princ. and Techn., KEPT 2007, Cluj-Napoca, Romania, pp. 107–113 (2007)
30. Georgiou, V.L., Alevizos, P.D., Vrahatis, M.N.: Novel approaches to probabilistic neural networks through bagging and evolutionary estimating of prior probabilities. Neural Processing Letters 27, 153–162 (2008)

New Bounds on the Clique Number of Graphs Based on Spectral Hypergraph Theory

Samuel Rota Bulò and Marcello Pelillo

Dipartimento di Informatica, Università Ca' Foscari di Venezia, Venice, Italy
{srotabul,pelillo}@dsi.unive.it

Abstract. This work introduces new bounds on the clique number of graphs derived from a result due to Sós and Straus, which generalizes the Motzkin-Straus Theorem to a specific class of hypergraphs. In particular, we generalize and improve the spectral bounds introduced by Wilf in 1967 and 1986 establishing an interesting link between the clique number and the emerging spectral hypergraph theory field. In order to compute the bounds we face the problem of extracting the leading H-eigenpair of supersymmetric tensors, which is still uncovered in the literature. To this end, we provide two approaches to serve the purpose. Finally, we present some preliminary experimental results.

1 Introduction

Let $G = (V, E)$ be a (undirected) graph, where $V = \{1, \ldots, n\}$ is the vertex set and $E \subseteq \binom{V}{2}$ is the edge set, with $\binom{V}{k}$ denoting the set of all k-element subsets of V. A *clique* of G is a subset of mutually adjacent vertices in V. A clique is called *maximal* if it is not contained in any other clique. A clique is called *maximum* if it has maximum cardinality. The maximum size of a clique in G is called the *clique number* of G and is denoted by $\omega(G)$.

The problem of finding the clique number of a graph is one of the most famous NP-complete problems, and turns out to be even intractable to approximate [1]. An interesting field of research consists in trying to bound the clique number of a graph. In the literature we find several bounds on the clique number, but, in this paper, our attention will be on bounds that employ spectral graph theory, since our new bounds are obtained using spectral hypergraph theory. However, it is worth noting that there is another very promising class of bounds, which will not be covered in this work, based on semidefinite programming [2,3].

In 1967, Wilf [4] used for the first time spectral graph theory for computing bounds on the clique number of graphs. His result was inspired by a theorem due to Motzkin and Straus [5], which establishes a link between the problem of finding the clique number of a graph G and the problem of optimizing the Lagrangian of G over the simplex Δ, where the Lagrangian of a graph $G = (V, E)$ is the function $L_G : \mathbb{R}^n \to \mathbb{R}$ defined as

$$L_G(\mathbf{x}) = \sum_{\{i,j\} \in E} x_i x_j,$$

T. Stützle (Ed.): LION 3, LNCS 5851, pp. 45–58, 2009.

and the *standard simplex* Δ is the set of nonnegative n-dimensional real vectors that sum up to 1, i.e., $\Delta = \{\mathbf{x} \in \mathbb{R}^n_+ : \sum_{i=1}^n x_i = 1\}$.

Theorem 1 (Motzkin-Straus). *Let G be a graph with clique number $\omega(G)$, and $\mathbf{x}^*$ a maximizer of L_G over Δ then*

$$L_G(\mathbf{x}^*) = \frac{1}{2}\left[1 - \frac{1}{\omega(G)}\right].$$

Assuming S a maximum clique of G, Motzkin and Straus additionally proved that the *characteristic vector* $\mathbf{x}^S$ of S defined as

$$\mathbf{x}_i^S = \begin{cases} \frac{1}{|S|} & i \in S \\ 0 & i \notin S \end{cases}$$

is a global maximizer of L_G over Δ.

Before introducing the bounds of Wilf, we introduce some notions from spectral graph theory, namely, a discipline that studies the properties of graphs in relationship to the eigenvalues and eigenvectors of its adjacency matrix or Laplacian matrix. The *spectral radius* $\rho(G)$ of a graph G is the largest eigenvalue of the adjacency matrix of G. An eigenvector of unit length having $\rho(G)$ as eigenvalue is called *Perron eigenvector* of G. The Perron eigenvector is always nonnegative and it may not be unique unless the multiplicity of the largest eigenvalue is exactly 1. By definition, the spectral radius ρ and an associated Perron eigenvector $\mathbf{x}_P$ of a graph G satisfy the eigenvalue equation

$$A_G \mathbf{x}_P = \rho \mathbf{x}_P,$$

which can be equivalently expressed in terms of the graph Lagrangian L_G as

$$\nabla L_G(\mathbf{x}_P) = \rho \mathbf{x}_P,$$

where ∇ is the standard gradient operator. Since G is undirected and hence, A_G is symmetric, a useful variational characterization of ρ and $\mathbf{x}_P$ is given by the following constrained program,

$$\rho = \max_{\mathbf{x} \in S_2} \mathbf{x}^T A_G \mathbf{x} = 2 \max_{\mathbf{x} \in S_2} L_G(\mathbf{x}), \tag{1}$$

where $S_k = \{\mathbf{x} \in \mathbb{R}^n : \|\mathbf{x}\|_k^k = 1\}$. Note that the eigenvectors of A_G are the critical points of this maximization problem. A further alternative characterization of the spectral radius and Perron eigenvector, that will be useful in the sequel, consists in maximizing the *Rayleigh quotient*, i.e.,

$$\rho = \max_{\mathbf{x} \in \mathbb{R}^n} \frac{\mathbf{x}^T A_G \mathbf{x}}{\mathbf{x}^T \mathbf{x}} = 2 \max_{\mathbf{x} \in \mathbb{R}^n} \frac{L_G(\mathbf{x})}{\mathbf{x}^T \mathbf{x}}. \tag{2}$$

Note that every eigenvector associated to ρ is a maximizer in (2), whereas in (1) only a Perron eigenvector is a global maximizer.

As mentioned, in 1967 Wilf [4] obtained a spectral upper bound to the clique number exploiting both the Motzkin-Straus theorem and (2).

Theorem 2. *Let G be an undirected graph with clique number $\omega(G)$ and spectral radius ρ. Then*

$$\omega(G) \leq \rho + 1 \,.$$

Proof. Let $\mathbf{x}_\omega$ be the characteristic vector of a maximum clique of G, then $\mathbf{x}_\omega^T \mathbf{x}_\omega = 1/\omega(G)$ and by the Motzkin-Straus theorem $\mathbf{x}_\omega^T A_G \mathbf{x}_\omega = 1 - 1/\omega(G)$. By (2) we have that

$$\frac{\mathbf{x}_\omega^T A_G \mathbf{x}_\omega}{\mathbf{x}_\omega^T \mathbf{x}_\omega} = \frac{1 - \frac{1}{\omega(G)}}{\frac{1}{\omega(G)}} = \omega(G) - 1 \leq \rho \,,$$

from which the property derives.

Later in 1986, Wilf [6] introduced also a lower bound again combining the Motzkin-Straus theorem and (2), but this time also the Perron eigenvector is involved in the result.

Theorem 3. *Let G be an undirected graph with spectral radius ρ and Perron eigenvector $\mathbf{x}_P$. Then*

$$\omega(G) \geq \frac{s_P^2}{s_P^2 - \rho} \,,$$

where $s_P = \|\mathbf{x}_P\|_1$.

Proof. Let $\mathbf{y} = \mathbf{x}_P/s_P$. Clearly $\mathbf{y} \in \Delta$. By the Motzkin-Straus theorem we have that

$$\mathbf{y}^T A_G \mathbf{y} = \frac{\mathbf{x}_P^T A_G \mathbf{x}_P}{s_P^2} = \frac{\rho}{s_P^2} \leq 1 - \frac{1}{\omega(G)} \,,$$

from which the result derives.

For a review of further spectral bounds we refer to [7]. We also refer to [8] for bounds that employ spectral graph theory based on the Laplacian of the graph.

In this paper, we introduce new classes of upper and lower bounds on the clique number of graphs, generalizing the Wilf's ones. More precisely, we achieve our goal combining a reformulation of a theorem due to Sós and Straus in terms of hypregraphs, and the emerging spectral hypergraph theory field. Further, we tackle the computational aspects of our new bounds and, finally, we present some preliminary experiments on random graphs.

2   A Reformulation of the Sós and Straus Theorem

In this section, we provide a reformulation of a generalization of the Motzkin-Straus theorem due to Sós and Straus [9], by explicitly establishing a connection to hypergraph theory, which will form the basis of our new bounds. To this end we start introducing hypergraphs.

A k-*uniform hypergraph*, or simply a k-*graph*, is a pair $G = (V, E)$, where $V = \{1, \ldots, n\}$ is a finite set of *vertices* and $E \subseteq \binom{V}{k}$ is a set of k-subsets of V, each of which is called a *hyperedge*. 2-graphs are typically called *graphs*. The *complement* of a k-graph G is given by $\bar{G} = (V, \bar{E})$ where $\bar{E} = \binom{V}{k} \setminus E$. A subset of vertices $C \subseteq V$ is called a *hyperclique* if $\binom{C}{k} \subseteq E$. To improve readability, in the sequel we will drop the prefix "hyper" when referring to edges and cliques of a k-graph. A clique is said to be *maximal* if it is not contained in any other clique, while it is called *maximum* if it has maximum cardinality. The *clique number* of a k-graph G, denoted by $\omega(G)$, is defined as the cardinality of a maximum clique. The *Lagrangian* of a k-graph G with n vertices is the following homogeneous multilinear polynomial in n variables:

$$L_G(\mathbf{x}) = \sum_{e \in E} \prod_{i \in e} x_i \,. \tag{3}$$

Given an undirected graph G and a positive integer k not exceeding the clique number $\omega(G)$, we can build a hypergraph H, that we call the k-*clique* $(k+1)$-*graph* of G having k-cliques of G as vertices and $(k+1)$-cliques of G as edges. Note that each $(k+1)$-clique of G has exactly $(k+1)$ different k-cliques as subsets. By shrinking each k-clique of G into a vertex in H, each $(k+1)$-clique of G can be transformed into an edge of H containing exactly $k+1$ vertices. Before addressing a more formal definition we provide an example. Figure 1 shows an undirected graph G on the left and the related 3-clique 4-graph H on the right. Each vertex in H is associated to the 3-clique of G reported in red over it. For each 4-clique C in G, there exists an edge in H containing the vertices associated to 3-subsets of C. For example, the clique $\{1, 2, 3, 4\}$ in G is associated to the edge $\{1, 2, 3, 4\}$ in H, and the vertices $1, 2, 3, 4$ of H are associated to the 3-cliques $\{1, 2, 3\}, \{2, 3, 4\}, \{1, 2, 4\}, \{1, 3, 4\}$ of G respectively. It is worth noting that the construction of the k-clique $(k+1)$-graph of a graph G depends on how we label the vertices in H, i.e., an enumeration of the set of k-cliques of G. Therefore, we provide also a more formal definition. Let G be an undirected graph G, $\mathcal{C}_k(G)$ the set of all k-cliques of G, and let Φ_k^G be the set of possible enumerations of $\mathcal{C}_k(G)$, i.e., one-to-one mappings from $\{1, \ldots, |\mathcal{C}_k(G)|\}$ to $\mathcal{C}_k(G)$. The k-*clique* $(k+1)$-*graph* of G with respect to $\phi \in \Phi_k^G$ is the $(k+1)$-graph $H = (V, E)$, where $V = \{1, \ldots, |\mathcal{C}_k(G)|\}$, and $E = \left\{ e \in \binom{V}{k+1} : \bigcup_{i \in e} \phi(i) \in \mathcal{C}_{k+1}(G) \right\}$.

Let G be an undirected graph with clique number ω and let H be its k-clique $(k+1)$-graph with respect to an enumeration $\phi \in \Phi_k^G$. Sós and Straus provided a characterization of the clique number of a graph G in terms of the maximum of the Lagrangian of H over Δ_k, where

$$\Delta_k = \{\mathbf{x} \in \mathbb{R}_+^n : \|\mathbf{x}\|_k^k = 1\} \,,$$

leading to the following result.

Theorem 4. *Let G be an undirected graph with clique number ω and let H be its k-clique $(k+1)$-graph with respect to any $\phi \in \Phi_k^G$. The Lagrangian of H attains its maximum over Δ_k at $\binom{\omega}{k+1} / \binom{\omega}{k}^{(k+1)/k}$.*

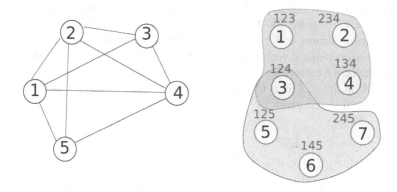

Fig. 1. Example of a 3-clique 4-graph (right) of an undirected graph (left)

Moreover, a subset of vertices C is a maximum clique of G if and only if the vector $\mathbf{x} \in \Delta_k$ defined as

$$x_i = \begin{cases} \binom{\omega}{k}^{-1/k} & \text{if } \phi(i) \subset C \\ 0 & \text{otherwise}, \end{cases} \tag{4}$$

is a global maximizer of L_H over Δ_k.

Proof. See [9].

If we consider $k = 1$, then Theorem 4 is the Motzkin-Straus theorem.

As the Motzkin-Straus theorem inspired new bounds on the clique number of graphs, Theorem 4 prompted our new bounds by combining it with spectral hypergraph theory.

3   Spectral Hypergraph Theory

Opposed to spectral graph theory, which has a long history, spectral hypergraph theory is a novel field roughed out by Drineas and Lim [10]. Clearly, spectral hypergraph theory includes spectral graph theory as a special case, but this generality introduces some ambiguities. For example, the adjacency matrix for graphs becomes an adjacency tensor for hypergraphs, for which there are several possible notions of eigenvalues and eigenvectors that can be taken into account for studying the properties of the hypergraph [10,11].

For our purposes, we will employ spectral hypergraph theory in a transversal way, as we will use the spectral properties of the adjacency tensor of the k-clique $(k + 1)$-graph of a graph G to study properties of G. This means that we do not want to study the properties of the hypergraph by itself, but those of the graph it is constructed from.

A real kth-order n-dimensional *tensor* A consists of n^k real entries, $A_{i_1,\ldots,i_k} \in \mathbb{R}$, where $i_j = 1, \ldots, n$ for $j = 1, \ldots, k$. The tensor A is called *supersymmetric* if its entries are invariant under any permutation of their indices.

Given a n-dimensional vector $\mathbf{x}$, we denote by X^k the kth-order n-dimensional rank-one tensor with entries $x_{i_1} \cdots x_{i_k}$, and by $\mathbf{x}^k$ the n-dimensional vector with entries x_i^k. Finally, if A is a kth-order n-dimensional tensor and $\mathbf{x} \in \mathbb{R}^n$, then AX^k is the real scalar given by

$$AX^k = \sum_{i_1,\ldots,i_k=1}^{n} a_{i_1,\ldots,i_k} x_{i_1} \cdots x_{i_k},$$

and AX^{k-1} is the n-dimensional vector, whose ith entry is given by

$$\left(AX^{k-1}\right)_i = \sum_{i_2,\ldots,i_k=1}^{n} a_{i,i_2,\ldots,i_k} x_{i_2} \cdots x_{i_k}.$$

The *adjacency tensor* of a k-graph $H = (V, E)$ with n vertices is the kth-order n-dimensional tensor A_H, whose entry indexed by $(i_1, \ldots, i_k)$ is 1, if $\{i_1, \ldots, i_k\} \in E$, 0 otherwise. Clearly, A_H is supersymmetric.

As mentioned before, there are several notions of eigenvalue/eigenvector for tensors, but only one fits our needs. In 2005, Lim [10] and Qi [11] independently introduced the same notion of eigenvalue, which they called ℓ^p-eigenvalue and H-eigenvalue respectively. Given a kth-order n-dimensional tensor A, a real value λ and a vector $\mathbf{x} \in \mathbb{R}^n$ are an *eigenvalue* and an *eigenvector* of A respectively, if they satisfy the following equation

$$AX^{k-1} = \lambda \mathbf{x}^{k-1}. \tag{5}$$

The eigenvalues derived from (5) reflect many properties of the eigenvalues of matrices [11]. In fact, Drineas and Lim [10] successfully generalize some results from spectral graph theory to hypergraphs employing this notion of eigenvalue.

The *spectral radius* $\rho(H)$ of a k-graph H is the largest eigenvalue of the adjacency tensor of H. An eigenvector of unit k-norm having $\rho(H)$ as eigenvalue is called *Perron eigenvector* of H. As it happens for graphs, the Perron eigenvector may not be unique unless the multiplicity of the largest eigenvalue is 1. In the sequel, a *leading eigenpair* of a k-graph will be a pair comoposed by the spectral radius and a Perron eigenvector.

The eigenvalue equation for hypergraphs can be rewritten in terms of the Lagrangian as follows,

$$(k-1)! \nabla L_H(\mathbf{x}) = \lambda \mathbf{x}^{k-1}, \tag{6}$$

and there is a variational characterization of the spectral radius and a related Perron eigenvector of H derived from the following constrained program,

$$\rho(H) = \max_{\mathbf{x} \in S_k} A_H X^k = k! \max_{\mathbf{x} \in S_k} L_H(\mathbf{x}), \tag{7}$$

which has the Perron eigenvector as maximizer. Note that all the eigenvalues of A_H are the critical points of (7).

By a generalization of the Perron-Frobenius theory to nonnegative tensors [12], it turns out that a Perron eigenvector of H is always nonnegative. This allows us to add a non negativity constraint to (7) without affecting the solution, i.e. we can replace S_k with Δ_k, yielding

$$\rho(H) = \max_{\mathbf{x} \in \Delta_k} A_H X^k = k! \max_{\mathbf{x} \in \Delta_k} L_H(\mathbf{x}), \tag{8}$$

Alternatively, a further characterization is obtained by maximizing a generalization of the Rayleigh quotient, namely

$$\rho(H) = \max_{\mathbf{x} \in \mathbb{R}^n} \frac{A_H X^k}{\|\mathbf{x}\|_k^k} = k! \max_{\mathbf{x} \in \mathbb{R}^n} \frac{L_H(\mathbf{x})}{\|\mathbf{x}\|_k^k}, \tag{9}$$

where all eigenvectors associated to ρ are global maximizers.

4   New Bounds Based on Spectral Hypergraph Theory

In this section, we provide new classes of upper and lower bounds that generalize those introduced by Wilf [4,6] for graphs. The basic idea is to combine the Sós and Straus theorem with spectral hypergraph theory.

Theorem 5 (New Upper Bound). *Let G be an undirected graph with clique number $\omega(G)$ and H a k-clique $(k+1)$-graph of G with spectral radius $\rho(H)$. Then*

$$\omega(G) \leq \frac{\rho(H)}{k!} + k.$$

Proof. Let C be a maximum clique of G and $\mathbf{x}_\omega$ the vector defined as (4). We write ω for $\omega(G)$. By (9) and Theorem 4,

$$(k+1)! \frac{L_H(\mathbf{x}_\omega)}{\|\mathbf{x}_\omega\|_{k+1}^{k+1}} = (k+1)! \frac{\binom{\omega}{k+1}\binom{\omega}{k}^{-(k+1)/k}}{\binom{\omega}{k}^{-1/k}} =$$

$$= (k+1)! \frac{\binom{\omega}{k+1}}{\binom{\omega}{k}} = k!(\omega - k) \leq \rho(H),$$

from which the result derives.

Note that for each choice of $0 < k \leq \omega(G)$, we have a new bound. In particular, by taking $k = 1$, we have $H = G$ obtaining Wilf's upper bound (Theorem 2). Note also that if we take $k = \omega(G)$, then H is a hypergraph consisting only of vertices (one per maximum clique of G) and no edges. In this case, the bound gives trivially $\omega(G)$.

Theorem 6 (New Lower Bound). *Let G be an undirected graph with clique number $\omega(G)$ and H a k-clique $(k+1)$-graph of G with spectral radius $\rho(H)$ and Perron eigenvector $\mathbf{x}_P$. Then*

$$\omega(G) \geq \psi_k^{-1}\left(\frac{\rho(H)}{k!\|\mathbf{x}_P\|_k^{k+1}}\right) \qquad \psi_k(x) = (x-k)\binom{x}{k}^{-\frac{1}{k}}.$$

Before proving this result we prove that $\psi_k(x)$ is monotonically increasing for $x \geq k$.

Lemma 1. *$\psi_k(x)$ is monotonically increasing for $x \geq k$.*

Proof. For all $x \geq k$,

$$\psi_k'(x) = \binom{x}{k}^{-\frac{1}{k}}\left[1 - \frac{1}{k}\sum_{j=0}^{k-1}\frac{x-k}{x-j}\right],$$

which is positive for $x \geq k$. Hence, $\psi_k(x)$ is monotonically increasing for $x \geq k$.

Proof (Theorem 6). Let $\mathbf{y} = \mathbf{x}_P/\|\mathbf{x}_P\|_k$. Clearly, $\mathbf{y} \in \Delta_k$. We write ω for $\omega(G)$. By (7) and Theorem 4,

$$L_H(\mathbf{y}) = \frac{L_H(\mathbf{x}_P)}{\|\mathbf{x}_P\|_k^{k+1}} = \frac{\rho(H)}{(k+1)!\|\mathbf{x}_P\|_k^{k+1}} \leq$$

$$\leq \binom{\omega}{k+1}\bigg/\binom{\omega}{k}^{\frac{k+1}{k}} = \frac{\omega-k}{k+1}\binom{\omega}{k}^{-\frac{1}{k}} = \frac{\psi_k(\omega)}{k+1}.$$

Since by Lemma 1, $\psi_k(x)$ is monotonically increasing and $\omega \geq k$, the result follows.

Theorem 6 introduces a class of lower bounds, one for each choice of $0 < k \leq \omega(G)$. Again, by taking $k = 1$, we obtain the Wilf's lower bound (Theorem 3). Moreover, by taking $k = \omega(G)$, the bound trivially returns $\omega(G)$.

By Lemma 1, $\psi_k(x)$ is invertible for $x \geq k$, however, it is difficult, and maybe not possible, to find a general analytical inverse. It is straightforward to find the inverse for $k = 1, 2$ analytically. In fact we obtain,

$$\psi_1^{-1}(y) = \frac{1}{1-y} \qquad \psi_2^{-1}(y) = \frac{8 - y^2 + y\sqrt{y^2 + 16}}{2(2 - y^2)},$$

but, in general, the best way for computing the inverse is numerically through some kind of section search, like the dicotomic search.

The computational complexity of our bounds is dominated by the construction of the k-clique $(k+1)$-graph. This, in fact, increases exponentially with the parameter k, that can be chosen between 1 and the clique number. This fact

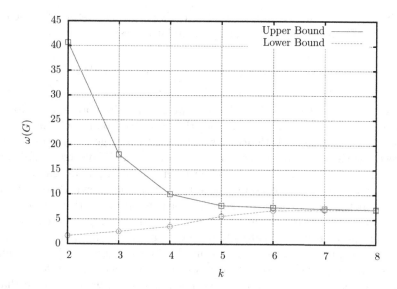

Fig. 2. Application of our classes of upper and lower bounds to a random graph [$n = 100; \delta = 0.4$]. We conjecture that, by increasing k, the upper bound is monotonically non increasing and our lower bound is monotonically non decreasing.

intuitively suggests that also the tightness of the bound should increase with k, as we require more computational effort for its calculation. Although we have no prove by now, we conjecture that, by increasing k, our new upper and lower bounds are monotonically non increasing and non decreasing respectively. This fact is also supported by all the experiments that we conducted. As an example of this monotonicity, Figure 2 plots our bounds on the clique number of a random graph [$n = 100; \delta = 0.4$] at varying values of k.

Although our new bounds are intuitively simple, their computation is not obvious, in particular we refer to the extraction of the Perron vector and spectral radius of a k-graph. Therefore the next section is devoted to the computational aspects of our new spectral bounds.

5   Computing the Bounds

As mentioned, the computational complexity of our approach is dominated by the construction of the k-clique $(k + 1)$-graph, which increases exponentially with k. Despite the complexity, the construction of the hypergraph is a fairly simple task. Indeed, the computation of the bounds involves more interesting problems such as the computation of the spectral radius and Perron eigenvector of a k-graph.

We have seen that a useful characterization of the leading eigenpair of a k-graph derives from the maximization in (7), whose critical points are those satisfying Equation (6). From this, we can derive a primal method for the optimization of (7) consisting in the following update rules:

$$\mathbf{y}^{(t+1)} = \nabla L_H\left(\mathbf{x}^{(t)}\right)^{\frac{1}{k-1}} \quad \mathbf{x}^{(t+1)} = \frac{\mathbf{y}^{(t+1)}}{\|\mathbf{y}^{(t+1)}\|_k} \quad \rho^{(t+1)} = k! L_H\left(\mathbf{x}^{(t+1)}\right), \quad (10)$$

where $\mathbf{x}^{(0)} \in \Delta_k$ and $\mathbf{x}^{(0)} > 0$ and where we remind the notation $\mathbf{x}^k$, for representing a vector with entries x_i^k. Note that the fixed points of this iterative process satisfy equation (6). Hence, the solution at convergence is a nonnegative eigenvector with the related eigenvalue. This approach can be straightforwardly extended to the extraction of the leading eigenpair of any nonnegative tensor, yielding a Generalization to nonnegative tensors of the known Power Method, which is a method for extracting the leading eigenpair of matrices (case $k = 2$). Thereby, we call GPM the method governed by (10). Although, we achieved this method autonomously, it was already proposed in [12,13], in a more general setting, for optimizing nonnegative generalized polynomials under ℓ^p constraints. There is still no prove of convergence for this approach, except for the case $k = 2$, however, supported by the experimental results, we conjecture that it always converges.

Another interesting method (called GBE) that we developed for extracting the leading eigenpair of k-graphs, and more in general of nonnegative tensors, derives from a Generalization of the Baum-Eagon Theorem [14] to nonnegative generalized homogeneous polynomials.

A *generalized polynomial* is a function $P : \mathbb{R}^n \to \mathbb{R}$ of the form:

$$P(\mathbf{x}) = \sum_{\boldsymbol{\alpha}} c_{\boldsymbol{\alpha}} \prod_{i=1}^{n} x_i^{\alpha_i}$$

where $\boldsymbol{\alpha}$ ranges over a finite set of $\mathbb{R}_+^n$. By definition, the degree of P is $h = \max_{\boldsymbol{\alpha}} h_{\boldsymbol{\alpha}}$, where $h_{\boldsymbol{\alpha}} = \sum_i \alpha_i$. If $h_{\boldsymbol{\alpha}} = h$ for all $\boldsymbol{\alpha}$, we call P a *homogeneous* generalized polynomial of degree h. We say that P is *nonnegative* if $c_{\boldsymbol{\alpha}} \geq 0$ for all $\boldsymbol{\alpha}$.

Theorem 7. *Let $P(\mathbf{x})$ be a nonnegative generalized homogeneous polynomial in the variables x_i, and let $\mathbf{x} \in \Delta$. Define the mapping $\mathbf{z} = \mathcal{M}(\mathbf{x})$ as follows:*

$$z_i = x_i \partial_i P(\mathbf{x}) \Big/ \sum_{j=1}^{n} x_j \partial_j P(\mathbf{x}), \quad i = 1, \dots, n.$$

Then $P(\mathcal{M}(\mathbf{x})) > P(\mathbf{x})$, unless $\mathcal{M}(\mathbf{x}) = \mathbf{x}$. In other words, $\mathcal{M}$ is a growth transformation for the polynomial P.

Proof. See [15].

In order to apply this result, we cast (8) into an equivalent optimization problem, which satisfies the conditions of Theorem 7, obtaining in this way a new optimization approach. We define the diffeomorphism $\varphi_k : \mathbb{R}_+^n \to \mathbb{R}_+^n$ by putting $\varphi_k(\mathbf{x}) = \mathbf{x}^{\frac{1}{k}}$. By setting $\mathbf{y} = \varphi_k^{-1}(\mathbf{x})$, we have that

$$\rho(H) = k! \max_{\mathbf{x} \in \Delta_k} L_H(\mathbf{x}) = k! \max_{\mathbf{y} \in \Delta} (L_H \circ \varphi_k)(\mathbf{y}) . \tag{11}$$

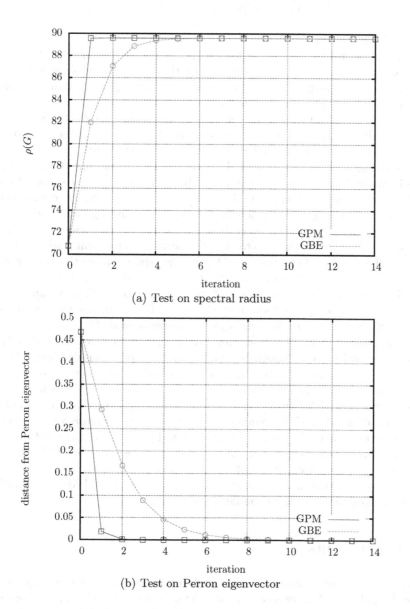

(a) Test on spectral radius

(b) Test on Perron eigenvector

Fig. 3. Comparison of GPM and GBE on the extraction of (a) the spectral radius and (b) the Perron eigenvector, of a random graph $[n = 100; \delta = 0.9]$

Here $L_H \circ \varphi_k$ is a generalized nonnegative homogeneous polynomial of degree 1 and by applying Theorem 7, we obtain the following update rules:

$$x_i^{(t+1)} = \left[\frac{x_i^{(t)} \partial_i L_H(\mathbf{x}^{(t)})}{\sum_j x_j^{(t)} \partial_j L_H(\mathbf{x}^{(t)})} \right]^{\frac{1}{k}} \qquad \rho^{(t+1)} = k! L_H \left(\mathbf{x}^{(t+1)} \right) , \qquad (12)$$

where $\mathbf{x}^{(0)} \in \Delta_k$ and $\mathbf{x}^{(0)} > \mathbf{0}$.

It is worth noting that this idea can be extended to compute the leading eigenpair of any nonnegative tensor. In particular, it provides us with a new approach for computing the leading eigenpair of nonnegative matrices, which is different to the standard known techniques.

Differently from GPM, it is not obvious that the fixed points of GBE are only those satisfying equation (7). Experimentally, this was always the case and we aim at proving this result in future developments of this work.

The nonnegative eigenvector of A_H that we obtain from one of the two proposed approaches, may unfortunately not be a Perron eigenvector, because our methods are basically local optimizers. However, we can easily overcome this problem, yielding a global solution. In fact, by employing a generalization of the Perron-Frobenious theory to generalized polynomials [12], it can be shown [15] that every nonnegative eigenvector has the positive components indexed by the vertices of a connected component of the hypergraph (or more than one in some special cases). This suggests a simple approach to achieve a global solution. We first find the connected components of the hypergraph H. Then we apply GPE or GBE on each component and keep the best solution. In this way we are able to extract the spectral radius and a Perron eigenvector of H.

There is experimental evidence that GPM converges faster than GBE. As an example, we calculated through GPM and GBE the spectral radius and a Perron eigenvector of a random graph with $n = 100$ vertices and density $\delta = 0.9$. In Figure 3(a) we plot the evolution of $2L_G(\mathbf{x}^{(t)})$, which has $\rho(G)$ as limit point. It is evident that GPM with just one step reaches a good approximation of the spectral radius, while GBE manifests a smoother curve. Figure 3(b) focuses on the approximation of the Perron eigenvector and plots the distance between $\mathbf{x}^{(t)}$ and the Perron eigenvector, i.e., $\|\mathbf{x}^{(t)} - \mathbf{x}_P\|$. Here, we see that GPM needs 2 steps for a good approximation of the Perron eigenvector, whereas GBE needs about 9 steps. Therefore for the experiments we will use GPM for computing the leading eigenpair of the k-cliques $(k+1)$-graphs.

6   Experiments

In this section, we evaluate only the performances of our 3th-order upper and lower bounds, which have a complexity $O(\gamma n^3)$, where γ is the number of iterations of GPM and can be assumed constant. Clearly, provided that our conjecture holds, by increasing k, we can only improve the results obtained with $k = 3$, but also the computational effort will increase.

Table 1. Experiments on random graphs. The columns n, δ and ω are the order, density and average clique number of the random graphs, respectively. The results, expecting the last row, are expressed in terms of relative error.

Random graphs			Upper bound errors			Lower bound errors		
n	δ	ω	Wilf	Amin	Order 3	Wilf	Budin.	Order 3
100	0.05	3.12	1.25	10.58	**0.13**	0.648	0.641	**0.119**
	0.10	3.96	1.99	9.26	**0.15**	0.714	0.707	**0.363**
	0.20	5.00	3.33	7.84	**0.59**	0.745	0.738	**0.512**
	0.30	6.13	4.17	6.52	**0.91**	0.761	0.753	**0.616**
	0.40	7.51	4.49	5.24	**1.34**	0.775	0.765	**0.664**
	0.50	9.11	4.58	4.19	**2.00**	0.779	0.768	**0.685**
	0.60	11.51	4.28	3.16	**2.29**	0.782	0.769	**0.711**
	0.70	14.55	3.85	**2.33**	2.52	0.772	0.756	**0.714**
	0.80	19.99	3.03	**1.45**	2.30	0.754	0.734	**0.708**
	0.90	30.69	1.94	**0.61**	1.70	0.695	0.662	**0.646**
	0.95	43.50	1.19	**0.16**	1.08	0.606	**0.562**	0.587
200	0.10	4.17	4.25	19.97	**0.29**	0.728	0.725	**0.463**
	0.50	11.00	8.19	7.71	**3.62**	0.817	0.811	**0.746**
	0.90	?	180.10	**99.08**	164.19	9.646	10.330	**10.851**

We compare our 3rd-order bounds against other state-of-the-art spectral bounds, which were the best performing approaches reviewed in the work of Budinich [7]. For the upper bound, we compare against Wilf's upper bound [4] (which is our bound with $k = 2$), that will never perform better than our 3rd order bound according to our conjecture, and we compare also against the Amin's bound [16]. For the lower bound, we compare against Wilf's lower bound [6] and Budinich's lower bound [7].

Table 1 reports the obtained results. The columns n, δ and ω are the order, density and average clique number of the random graphs, respectively. The results, expecting the last row, are expressed in terms of relative error, i.e. if $\bar{\omega}$ is the value of the bound then the relative error for the upper and lower bounds are $(\bar{\omega} - \omega)/\omega$ and $(\omega - \bar{\omega})/\omega$, respectively. In the last row, where the average clique number could not be computed, we reported the absolute value of the bounds. It is clear that, as expected, our 3th-order bounds strictly improve the Wilf's one. It is also evident that our lower bound outperforms the competitors, excepting very dense graphs ($\delta = 0.95$), whereas our upper bound outperforms Amin's one on low and medium densities ($\delta \leq 0.6$). The decrease of the performances with respect to Amin and Budinich on high densities is due to the fact that the advantage of knowing the triangles of the graph becomes irrelevant when approaching the complete graph. However, also the relative error slowly approaches zero, since for the complete graph all our bounds are exact.

7 Conclusions

In this work, we introduce a new class of bounds on the clique number of graphs by employing, for the first time to our knowledge, spectral hypergraph theory.

The bounds are derived from a result due to Sós and Straus and generalize the classic spectral upper and lower bounds of Wilf.

The computation of our new bounds introduces the side problem of establishing the spectral radius and a Perron eigenvector of a k-graph, which is still uncovered. To this end, we introduce two dynamics that serve our purposes. The first is a generalization of the known Power Method, while the second derives from our generalization of the Baum-Eagon result to generalized polynomials.

Finally, we test our 3th-order bounds comparing them against state-of-the-art spectral approaches on random graphs. The results show the superiority of our bounds on all graphs excepting the dense ones, where anyway we achieve low relative errors.

References

1. Hastad, J.: Clique is hard to approximate within $n^{1-\varepsilon}$. Ann. Symp. Found. Comput. Sci. 37, 627–636 (1996)
2. Lovás, L.: On the Shannon capacity of a graph. IEEE Trans. Inform. Theory IT-25, 1–7 (1979)
3. Schrijver, A.: A comparison of the Delsarte and Lovász bounds. IEEE Trans. Inform. Theory 25, 425–429 (1979)
4. Wilf, H.S.: The eigenvalues of a graph and its chromatic number. J. London Math. Soc. 42, 330–332 (1967)
5. Motzkin, T.S., Straus, E.G.: Maxima for graphs and a new proof of a theorem of Turán. Canad. J. Math. 17, 533–540 (1965)
6. Wilf, H.S.: Spectral bounds for the clique and independence numbers of graphs. J. Combin. Theory Series B 40, 113–117 (1986)
7. Budinich, M.: Exact bounds on the order of the maximum clique of a graph. Discr. Appl. Math. 127, 535–543 (2003)
8. Lu, M., Liu, H., Tian, F.: Laplacian spectral bounds for clique and independence numbers of graphs. J. Combin. Theory Series B 97(5), 726–732 (2007)
9. Sós, V., Straus, E.G.: Extremal of functions on graphs with applications to graphs and hypergraphs. J. Combin. Theory Series B 63, 189–207 (1982)
10. Lim, L.H.: Singular values and eigenvalues of tensors: a variational approach. In: IEEE Int. Workshop on Comput. Adv. in Multi-Sensor Adapt. Processing, vol. 1, pp. 129–132 (2005)
11. Qi, L.: Eigenvalues of a real supersymmetric tensor. J. of Symb. Comp. 40, 1302–1324 (2005)
12. Baratchart, L., Berthod, M., Pottier, L.: Optimization of positive generalized polynomials under ℓ^p constraints. J. of Convex Analysis 5(2), 353–379 (1998)
13. Berthod, M.: Definition of a consistent labeling as a global extremum. In: Int. Conf. Patt. Recogn., pp. 339–341 (1982)
14. Baum, L.E., Eagon, J.A.: An inequality with applications to statistical estimation for probabilistic functions of Markov processes and to a model for ecology. Bull. Amer. Math. Soc. 73, 360–363 (1967)
15. Rota Bulò, S.: A game-theoretic framework for similarity-based data clustering. PhD Thesis. Work in progress (2009)
16. Amin, A.T., Hakimi, S.L.: Upper bounds of the order of a clique of a graph. SIAM J. on Appl. Math. 22(4), 569–573 (1972)

Beam-ACO Based on Stochastic Sampling: A Case Study on the TSP with Time Windows

Manuel López-Ibáñez and Christian Blum

Dept. Llenguatges i Sistemes Informàtics,
Universitat Politècnica de Catalunya, Barcelona, Spain
{m.lopez-ibanez,cblum}@lsi.upc.edu

Abstract. Beam-ACO algorithms are hybrid methods that combine the metaheuristic ant colony optimization with beam search. They heavily rely on accurate and computationally inexpensive bounding information for choosing between different partial solutions during the solution construction process. In this work we present the use of stochastic sampling as a useful alternative to bounding information in cases were computing accurate bounding information is too expensive. As a case study we choose the well-known travelling salesman problem with time windows. Our results clearly demonstrate that Beam-ACO, even when bounding information is replaced by stochastic sampling, may have important advantages over standard ACO algorithms.

1 Introduction

Ant colony optimization (ACO) is a metaheuristic that is based on the probabilistic construction of solutions [1]. At each algorithm iteration, n solutions are constructed independently from each other. A recently proposed ACO hybrid, known as Beam-ACO [2,3], employs at each iteration a probabilistic beam search procedure that constructs n solutions non-independently in parallel. A crucial component of beam search is bounding information for choosing between different partial solutions at each step of the solution construction process [4]. At each step, beam search keeps a certain number of the best partial solutions available for further extension, and excludes the rest from further examination. A problem arises when bounding information is either misleading (that is, the wrong partial solutions are kept for further examination) or when bounding information is computationally expensive.

Browsing the relevant artificial intelligence literature, we came across a different method for the evaluation of partial solutions in the context of tree search procedures: *probing* or *stochastic sampling* [5,6]. Hereby, each given partial solution is completed a number of N^s times in a stochastic way. The information that is obtained is used to differentiate between different partial solutions.

In this work we propose to replace the use of bounding information in Beam-ACO with a stochastic sampling procedure. For this case study we choose the travelling salesman problem with time windows (TSPTW), due to the fact that accurate bounding information that is computationally inexpensive is—to our

T. Stützle (Ed.): LION 3, LNCS 5851, pp. 59–73, 2009.

knowledge—not available for this problem. Finally, we want to state clearly at this point that the primary goal of this research is not to obtain state-of-the-art results for the TSPTW, which is left for future work. Our aim is to show that Beam-ACO based on stochastic sampling may have significant advantages over standard ACO algorithms.

The remainder of this work is organized as follows. In Section 2 we give a technical description of the TSPTW. Furthermore, in Sections 3 and 4 we first introduce our standard (in the sense of *non-hybrid*) ACO algorithm for the TSPTW and then we introduce the probabilistic beam search procedure that is needed for Beam-ACO. Finally, in Section 5 we present an experimental evaluation, and in Section 6 we offer conclusions and an outlook to future work.

2   The TSPTW

The traveling salesman problem with time windows (TSPTW) is the problem of finding an efficient route to visit a number of customers, starting and ending at a depot, with the added difficulty that each customer may only be visited within a certain time window. In practice, the TSPTW is an important problem in logistics.

The TSPTW is proven to be NP-hard, and even finding a feasible solution is an NP-complete problem [7]. The problem is closely related to a number of important problems. For example, the well-known traveling salesman problem (TSP) is a special case of the TSPTW. The TSPTW itself can be seen as a special case with a single vehicle of the vehicle routing problem with time windows (VRPTW). The state of the art in solving the TSPTW is a simulated annealing approach by Ohlmann and Thomas [8].

2.1   Formal Problem Definition

The TSPTW is formally defined as follows. Let $G = (N, A)$ be a finite graph, where $N = \{0, 1, \ldots, n\}$ consists of a set of nodes representing the depot (node 0) and n customers, and $A = N \times N$ is the set of arcs connecting the nodes.

For every arc $a_{ij} \in A$ between two nodes i and j, there is an associated cost $c(a_{ij})$. This cost typically represents the travel time between customers i and j, plus a service time at customer i.

For every node $i \in N$, there is an associated time window, $[e_i, l_i]$, where e_i represents the earliest service start time and l_i is the latest service start time.

A solution to the problem is a tour visiting each node once, starting and ending at the depot. Hence, a tour is represented as $P = (p_0 = 0, p_1, \ldots, p_n, p_{n+1} = 0)$, where the sub-sequence $(p_1, \ldots, p_k, \ldots, p_n)$ is a permutation of the nodes in $N \setminus \{0\}$ and p_k denotes the index of the customer at the kth position of the tour. Two additional elements, $p_0 = 0$ and $p_{n+1} = 0$, represent the depot.

It is assumed that waiting times are permitted, that is, a node i can be reached before the start of its time window e_i, but cannot be left before e_i. Therefore, the departure time from customer p_k is calculated as $D_{p_k} = \max(A_{p_k}, e_{p_k})$, where $A_{p_k} = D_{p_{k-1}} + c(a_{p_{k-1}, p_k})$ is the arrival time at customer p_k in the tour.

The literature defines two related but different objectives for this problem. One is the minimization of the cost of the arcs traversed along the tour $\sum_{k=0}^{n} c(a_{p_k,p_{k+1}})$. The other alternative is to minimise $A_{p_{n+1}}$, the arrival time at the depot. In this work, we focus on the former, and, therefore, we formally defined the TSPTW as:

$$\text{minimise:} \quad F(P) = \sum_{k=0}^{n} c(a_{p_k,p_{k+1}})$$

subject to:

$$\Omega(P) = \sum_{k=0}^{n+1} \omega(p_k) = 0 \tag{1}$$

where:

$$\omega(p_k) = \begin{cases} 1 & \text{if } A_{p_k} > l_{p_k} , \\ 0 & \text{otherwise;} \end{cases}$$

$$A_{p_{k+1}} = \max(A_{p_k}, e_{p_k}) + c(a_{p_k,p_{k+1}}) .$$

In the above definition, $\Omega(P)$ denotes the number of time window constraints that are violated by tour P, which must be zero for feasible solutions.

3   The ACO Algorithm

The application of the ACO framework to any problem implies the definition of a solution construction mechanism and the specification of appropriate pheromone information $\mathcal{T}$. In the case of the TSPTW, ants construct a complete tour by starting at the depot (node 0) and iteratively adding customers to the tour. Once all customers have been added to the tour, it is completed by adding node 0.

As for the pheromone information, $\forall a_{ij} \in A, \exists \tau_{ij} \in \mathcal{T}, 0 \leq \tau_{ij} \leq 1$, where τ_{ij} represents the desirability of visiting customer j after customer i in the tour: the greater the pheromone value τ_{ij}, the greater is the desirability of choosing j as the next customer to visit in the current tour.

The particular ACO algorithm proposed in this paper for the TSPTW combines ideas from both $\mathcal{MMAS}$ and ACS algorithms implemented in the hyper-cube framework (HCF) as proposed by Blum and Dorigo [9]. A high level description of the algorithm is given in Algorithm 1. The data structures used, in addition to counters and to the pheromone values, are: (1) the best-so-far solution P^{bf}, i.e., the best solution generated since the start of the algorithm; (2) the restart-best solution P^{rb}, that is, the best solution generated since the last restart of the algorithm; (3) the convergence factor (cf), $0 \leq cf \leq 1$, which is a measure of how far the algorithm is from convergence; and (4) the Boolean variable bs_update, which becomes true when the algorithm reaches convergence.

Roughly, the algorithm works as follows. Initially, all variables are initialized. In particular, the pheromone values are set to their initial value 0.5. Then, a

---

Algorithm 1. ACO algorithm for the TSPTW

1: **input:** $N^{\mathrm{a}} \in \mathbb{Z}^+$, $q_0 \in [0,1] \subset \mathbb{R}$
2: $P^{\mathrm{bf}} := \mathrm{NULL}$, $P^{\mathrm{rb}} := \mathrm{NULL}$, $cf := 0$, $bs\_update := \mathrm{FALSE}$
3: $\tau_{ij} := 0.5 \quad \forall \tau_{ij} \in \mathcal{T}$
4: **while** CPU time limit not reached **do**
5:     **for each** ant $a \in \{1, \dots, N^{\mathrm{a}}\}$ **do**
6:         $P_a := (0)$ {start at the depot}
7:         **repeat**
8:            choose next customer $j \in \mathcal{N}(P_a)$ following Eq. (2)
9:            add j as last element of partial solution P_a
10:        **until** all n customers are visited
11:     **end for**
12:    add 0 as last element of P_a {finish at the depot}
13:    $P^{\mathrm{ib}} := \min_{\mathrm{lex}}\{P_1, \dots, P_{N^{\mathrm{a}}}\}$ {identify *iteration-best*}
14:    **if** $P^{\mathrm{ib}} <_{\mathrm{lex}} P^{\mathrm{rb}}$ **then** $P^{\mathrm{rb}} := P^{\mathrm{ib}}$
15:    **if** $P^{\mathrm{ib}} <_{\mathrm{lex}} P^{\mathrm{bf}}$ **then** $P^{\mathrm{bf}} := P^{\mathrm{ib}}$
16:    $cf := \mathsf{ComputeConvergenceFactor}(\mathcal{T})$
17:    **if** $bs\_update = \mathrm{TRUE}$ **and** $cf > 0.99$ **then**
18:       $\tau_{ij} := 0.5 \quad \forall \tau_{ij} \in \mathcal{T}$
19:       $P^{\mathrm{rb}} := \mathrm{NULL}$
20:       $bs\_update := \mathrm{FALSE}$
21:    **else**
22:       **if** $cf > 0.99$ **then**
23:          $bs\_update := \mathrm{TRUE}$
24:       **end if**
25:       $\mathsf{ApplyPheromoneUpdate}(cf, bs\_update, \mathcal{T}, P^{\mathrm{ib}}, P^{\mathrm{rb}}, P^{\mathrm{bf}})$
26:    **end if**
27: **end while**
28: **output:** P^{bf}

---

main loop is repeated until a termination criteria, such as a CPU time limit, is met. Each algorithm iteration consists of the following steps.

First, a number of ants (N^{a}) construct complete tours by following the state transition rule defined in Eq. (2). Each ant a constructs a single tour P_a by iteratively adding customers to its partial tour. At each construction step, ant a chooses one customer j among the set $\mathcal{N}(P_a)$ of customers not visited yet by the current partial tour P_a. The decision is made by firstly generating a random number q uniformly distributed within $[0,1]$ and comparing this value with a parameter q_0 called the determinism rate. If $q \leq q_0$, j is chosen deterministically as the value with the highest product of pheromone and heuristic information. Otherwise, j is stochastically chosen from a distribution of probabilities. This rule is described by the following equation:

$$\begin{cases} j = \arg\max_{k \in \mathcal{N}(P_a)}\{\tau_{ik} \cdot \eta_{ik}\} & \text{if } q \leq q_0, \\[2mm] j \sim \{\mathbf{p}_i(k) \mid k \in \mathcal{N}(P_a)\} & \text{otherwise.} \end{cases} \qquad (2)$$

where i is the last customer added to the tour P_a, η_{ij} is a heuristic value that represents an estimation of the benefit of visiting customer j after customer i, and the symbol $\sim$ denotes drawing a random number from a probability distribution defined by the probabilities $\mathbf{p}_i(k)$. These probabilities depend on the pheromone and heuristic information associated to each choice and are defined by the following probabilistic rule:

$$\mathbf{p}_i(j) = \frac{\tau_{ij} \cdot \eta_{ij}}{\sum_{k \in \mathcal{N}(P_a)} \tau_{ik} \cdot \eta_{ik}} \quad \text{if } j \in \mathcal{N}(P_a) \tag{3}$$

For the TSPTW, we define a heuristic information that combines the travel cost between customers (c_{ij}) and the latest service time (l_j). The values are first normalized to $[0, 1]$, with the maximum value corresponding to 0 and the minimum to 1, and then combined with equal weight:

$$\eta_{ij} = \frac{1}{2} \left(\frac{c^{\max} - c_{ij}}{c^{\max} - c^{\min}} + \frac{l^{\max} - l_j}{l^{\max} - l^{\min}} \right) \tag{4}$$

After all ants have completed their tours, the tours are compared to identify the *iteration-best* solution (P^{ib}), i.e., the best solution among the ones constructed in the current iteration, denoted as $\min_{\mathrm{lex}} = \{P_1, \ldots, P_{N^a}\}$. To identify the best solution, tours are compared lexicographically ($<_{lex}$) by first minimising the number of constraint violations (Ω) and, if they have equal number of constraint violations, comparing their tour cost (F). More formally, we compare two different tours P and P' as follows:

$$P <_{\mathrm{lex}} P' \iff \Omega(P) < \Omega(P') \vee (\Omega(P) = \Omega(P') \wedge F(P) < F(P')) \tag{5}$$

Next, a new value for the convergence factor *cf* is computed. Depending on this value, as well as on the value of the Boolean variable *bs_update*, a decision on whether to restart the algorithm or not is made. If the algorithm is restarted, all the pheromone values are reset to their initial value (0.5). The algorithm is iterated until the CPU time limit is reached. Once terminated, the algorithm returns the best solution found which corresponds to P^{bf}. In the following we describe the two remaining procedures of Algorithm 1 in more detail.

Procedure ComputeConvergenceFactor($\mathcal{T}$) computes the convergence factor *cf*, which is a function of the current pheromone values, as follows:

$$cf = 2 \left(\frac{\sum_{\tau_{ij} \in \mathcal{T}} \max\{\tau^{\max} - \tau_{ij}, \tau_{ij} - \tau^{\min}\}}{|\mathcal{T}| \cdot (\tau^{\max} - \tau^{\min})} - 0.5 \right) \tag{6}$$

where $\tau^{\max}$ and $\tau^{\min}$ are, respectively, the maximum and minimum pheromone values allowed. Hence, $cf = 0$ when the algorithm is initialized (or reset), that is, when all pheromone values are set to 0.5. In contrast, when the algorithm has converged and all pheromone values have either the value $\tau^{\max}$ or the value $\tau^{\min}$, then $cf = 1$. In all other cases, *cf* has a value within $(0, 1)$.

The next step of the algorithm updates the pheromone information by means of the procedure ApplyPheromoneUpdate(*cf*, *bs_update*, $\mathcal{T}$, P^{ib}, P^{rb}, P^{bf}). In general, three solutions are used for updating the pheromone values. These are

Table 1. Setting of κ^{ib}, κ^{rb} and κ^{bf} depending on the convergence factor cf and the Boolean control variable bs_update

bs_update	FALSE				TRUE
cf	$[0, 0.4)$	$[0.4, 0.6)$	$[0.6, 0.8)$	$[0.8, 1]$	—
κ^{ib}	1	2/3	1/3	0	0
κ^{rb}	0	1/3	2/3	1	0
κ^{bf}	0	0	0	0	1

the *iteration-best* solution P^{ib}, the *restart-best* solution P^{rb}, and the *best-so-far* solution P^{bf}. The influence of each solution on the pheromone update depends on the state of convergence of the algorithm as measured by the convergence factor cf. Hence, each pheromone value $\tau_{ij} \in \mathcal{T}$ is updated as follows:

$$\tau_{ij} = \tau_{ij} + \rho \cdot (\xi_{ij} - \tau_{ij}) \ , \tag{7}$$

with

$$\xi_{ij} = \kappa^{ib} \cdot P_{ij}^{ib} + \kappa^{rb} \cdot P_{ij}^{rb} + \kappa^{bf} \cdot P_{ij}^{bf} \ , \tag{8}$$

where ρ is a parameter that determines the learning rate, P_{ij}^* is 1 if customer j is visited after customer i in solution P^* and 0 otherwise, κ^{ib} is the weight (that is, the influence) of solution P^{ib}, κ^{rb} is the weight of solution P^{rb}, κ^{bf} is the weight of solution P^{bf}, and $\kappa^{ib} + \kappa^{rb} + \kappa^{bf} = 1$. Equation (8) allows to choose how to schedule the relative influence of the three solutions used for updating the pheromone values. For our application we used a standard update schedule as shown in Table 1 and a value of $\rho = 0.1$.

After the pheromone update rule in Eq. (7) is applied, pheromone values that exceed $\tau^{max} = 0.999$ are set back to τ^{max} (similarly for $\tau^{min} = 0.001$). This is done in order to avoid a complete convergence of the algorithm, which is a situation that should be avoided. This completes the description of our ACO approach for the TSPTW problem.

4   Beam-ACO with Stochastic Sampling

As mentioned before, a Beam-ACO algorithm is obtained from a standard ACO algorithm by the replacement of the independent construction of solutions with a probabilistic beam search procedure. The probabilistic beam search that we invented for the TSPTW is described in Algorithm 2. The algorithm requires three input parameters: $k_{bw} \in \mathbb{Z}^+$ is the so-called *beam width*, $\mu \in \mathbb{R}^+ \geq 1$ is a parameter that determines the number of children that can be chosen at each step, and N^s is the number of *stochastic samples* taken for evaluating a partial solution. Moreover, B_t denotes a set of partial tours called the *beam*. Hereby, index t denotes the current iteration of the beam search. At any time it holds that $|B_t| \leq k_{bw}$, that is, the beam is smaller or equal to the beam width.

Algorithm 2. Probabilistic Beam search (PBS) for the TSPTW

1: $B_0 := \{(0)\}$
2: **for** $t := 0$ to n **do**
3:    $C := \mathcal{C}(B_t)$
4:    **for** $k = 1, \ldots, \min\{\lfloor \mu \cdot k_{\mathrm{bw}} \rfloor, |C|\}$ **do**
5:       $\langle P, j \rangle := \mathsf{ChooseFrom}(C)$
6:       $C := C \setminus \langle P, j \rangle$
7:       $B_{t+1} := B_{t+1} \cup \langle P, j \rangle$
8:    **end for**
9:    $B_{t+1} := \mathsf{Reduce}(B_{t+1}, k_{\mathrm{bw}})$
10: **end for**
11: **output:** $\arg\min_{\mathrm{lex}} \{T \mid T \in B_n\}$

A problem-dependent greedy function $\nu()$ is used to assign a weight to partial solutions.

At the start of the algorithm the beam only contains one partial tour starting at the depot, that is, $B_0 := \{(0)\}$. Let $C := \mathcal{C}(B_t)$ denote the set of all possible extensions of the partial tours in B_t. A partial tour P may be extended by adding a customer j not yet visited by that tour. Such a candidate extension of a partial tour is henceforth denoted by $\langle P, j \rangle$. At each iteration, at most $\lfloor \mu \cdot k_{\mathrm{bw}} \rfloor$ candidate extensions are selected from C by means of the procedure $\mathsf{ChooseFrom}(C)$ to form the new beam B_{t+1}. At the end of each step, the new beam B_{t+1} is reduced by means of the procedure Reduce in case it contains more than k_{bw} partial solutions. When the current iteration is equal to the number of customers $(t = n)$, all elements in B_n are completed by adding the depot, and finally the best solution is returned.

The procedure $\mathsf{ChooseFrom}(C)$ chooses a candidate extension $\langle P, j \rangle$ from C, either deterministically or probabilistically according to a parameter q_0 called *determinism rate* (see also the description of ACO). More precisely, for each call to $\mathsf{ChooseFrom}(C)$, a random number q is generated and if $q \leq q_0$, the decision is taken deterministically by choosing the candidate extension that maximises the product of the pheromone information $\mathcal{T}$ and the greedy function $\nu()$:

$$\langle P, j \rangle = \arg\max_{\langle P', k \rangle \in C} \tau(\langle P', k \rangle) \cdot \nu(\langle P', k \rangle)^{-1} \tag{9}$$

where $\tau(\langle P', k \rangle)$ corresponds to the pheromone value $\tau_{ik} \in \mathcal{T}$, supposing that i is the last customer visited in tour P'.

Otherwise, if $q > q_0$, the decision is taken stochastically according to the following probabilities:

$$\mathbf{p}(\langle P, j \rangle) = \frac{\tau(\langle P, j \rangle) \cdot \nu(\langle P, j \rangle)^{-1}}{\sum\limits_{\langle P', k \rangle \in C} \tau(\langle P', k \rangle) \cdot \nu(\langle P', k \rangle)^{-1}} \tag{10}$$

The greedy function $\nu(\langle P, j \rangle)$ assigns a heuristic value to each candidate extension $\langle P, j \rangle$. In principle, for this purpose we could use the heuristic η given by

Eq. (4), that is, $\nu(\langle P, j \rangle) = \eta(\langle P, j \rangle)$. As in the case of the pheromone information, the notation $\eta(\langle P, j \rangle)$ refers to the value of η_{ik} as defined in Eq. (4), supposing that i was the last customer visited in tour P. However, when comparing two extensions $\langle P, j \rangle \in C$ and $\langle P', k \rangle \in C$, the value of η might be misleading in case $P \neq P'$. We solved this problem by defining the greedy function $\nu()$ as follows.

Firstly, instead of using the value of η directly, we rank the extensions with respect to their value of η and use the corresponding ranks for comparison. More specifically, the extension with the highest value of η for all candidate extensions of the same tour receives rank 1. Formally, $r(\langle P, j \rangle) = 1$ where $\langle P, j \rangle = \arg\max_{k \in \mathcal{N}(P)} \eta(\langle P, k \rangle)$. The extension with the second highest value of η receives rank 2, and so on and so forth. Secondly, the value of the greedy function of an extension $\nu(\langle P, j \rangle)$ is calculated as the sum of the ranks that correspond to the sequence of extensions generated during the construction of partial tour $P = \{p_0, p_1, \ldots, p_{|P|}\}$. Formally:

$$\nu(\langle P, j \rangle) = r(\langle 0, p_1 \rangle) + \left(\sum_{i=2}^{|P|} r(\langle (p_0, \ldots, p_{i-1}), p_i \rangle) \right) + r(\langle P, j \rangle) \, , \qquad (11)$$

where $p_0 = 0$ is the depot, and p_i denotes the index of the customer visited in the ith position of the tour. This definition of $\nu()$ allows us to compare extensions of different partial tours by giving more priority to those extensions maximising $\nu()^{-1}$.

Finally, the application of procedure Reduce(B_t) removes the worst $\max\{|B_t| - k_{\mathrm{bw}}, 0\}$ partial solutions from B_t. As mentioned before, in standard applications of beam search, the *worst* solutions are determined by applying—in the case of minimization—a lower bound to each partial solution. The solutions removed from B_t are then the ones with the greatest lower bound value. However, as the literature for the TSPTW does not offer accurate and at the same time computationally inexpensive lower bounds, we use *stochastic sampling* for evaluating partial solutions. More specifically, a *sample* of a partial solution is obtained by using an ant (from the standard ACO algorithm) to complete the tour by iteratively adding unvisited customers following Eq. (2). For each partial solution, a number N^{s} of complete solutions is sampled. The value of the best of these samples (with respect to Eq. 5) is used for evaluating the corresponding partial solution. Only the k_{bw} best partial solutions (with respect to their corresponding best samples) are kept in B_t and the others are discarded. The PBS algorithm keeps track of the best solution among the ones sampled. A partial solution which is already worse than the best solution sampled can only become even worse when further extended. Hence, such partial solutions are removed from B_t without further sampling. If no solution in B_t is better than the best solution sampled, PBS returns the latter.

The above procedure defines a probabilistic beam search algorithm, henceforth denoted by PBS(k_{bw}, μ, N^{s}). In Beam-ACO, this PBS algorithm replaces the construction loop performed by the ants in Algorithm 1 (lines 5–13). Instead, a single call to PBS generates the *iteration-best* solution.

5   Experimental Evaluation

We implemented ACO and Beam-ACO in C++. The algorithms were tested on 30 instances provided by Potvin and Bengio [10] and derived from Solomon's RC2 VRPTW instances [11]. These instances are known to contain a mix of randomly-spaced and clustered customers. First, we performed a set of initial experiments in order to find appropriate values for various parameters of ACO and Beam-ACO. Next, we compared the results obtained by the two algorithms for the 30 instances mentioned above. All experiments were run on a AMD Opteron 8218 processor, with 2.6 GHz CPU and 1 MB of cache size running GNU/Linux 2.6.24.

Comparing different algorithms for the TSPTW is not a trivial task. In the literature it can sometimes be observed that algorithms are compared with respect to their average number of constraint violations and their average cost, where *average* refers to the average over several runs. However, in general one is not interested in trading a lower tour cost for a higher number of constraint violations. Comparing the two averages mentioned above, it is difficult to assess if such a trade-off has indeed occurred. As an alternative, one might focus on the median tour cost of those runs that achieved the minimum number of constraint violations. Yet, this information does not summarise the typical behaviour of an algorithm, and does therefore not provide a means for a fair comparison. (Imagine, for example, a situation in which one of the algorithms has achieved its minimum number of constraint violations in only one run.)

Instead, we decided for the following mechanism for the comparison of two or more algorithms. More specifically, we calculate a score for each algorithm,[1] measuring the quality of its solutions relative to the quality of the solutions obtained by the competing approaches. The score is given by the percentage of times that the outcome of one algorithm was better than the outcomes obtained by the alternatives minus the percentage of times that the outcome of the same algorithm was worse than the outcomes obtained by the alternatives. For example, let us compare three different algorithms X, Y, Z. Let us assume that each algorithm is applied 5 times to a problem instance. Then, we calculate the score of, for example, algorithm X as follows. First, each of the 5 solutions obtained by X are compared with each of the 10 solutions obtained by Y and Z. The comparison is done lexicographically, following the order defined in Eq. (5), by considering first the number of constraint violations (Ω) and next the tour cost (F). Hence, for each comparison, a tour may be better, equal, or worse than another. We count the number of times that an outcome of X was better minus the number of times it was worse than the solutions produced by competing algorithms. Finally, we calculate the percentage with respect to the total number of pairwise comparisons, 50 in our example. A positive score indicates that the solutions obtained by X were more often better than the solutions obtained by the alternatives Y and Z. A negative score indicates that the alternative algorithms obtained more often better outcomes than X. A value close to zero either

---

[1] Here, the term *algorithm* may refer to ACO or Beam-ACO, or to different configurations of ACO and Beam-ACO.

indicates that the outcomes were most of the times equal or that X obtained as many better outcomes as worse outcomes than the rest.

We applied the ACO algorithm to each instance with various values for the number of ants, that is, $N^a = \{10, 20, 50, 100, 200\}$. The rest of the parameters of ACO were set as follows: $\tau^{max} = 0.999$, $\tau^{min} = 0.001$, $q_0 = 0.9$, and $\rho = 0.1$. Each run of ACO was stopped after 15 CPU seconds and we repeated each experiment 25 times with different random seeds. Table 2 gives the scores obtained by ACO for each value of N^a. As discussed above, at each row of the table, each entry represents a percentage score obtained by the difference of two values: the number of times that a tour obtained by ACO using the number of ants given in the column heading was better than a tour obtained when using a different number of ants, minus the number of times that the former was worse than the latter. Therefore, larger positive values indicate a higher (relative) quality of the results in comparison with the other values of N^a, while negative values indicate a worse quality of the tours obtained. The results in Table 2 show that the best setting of N^a varies depending on the particular instance. Although higher values of N^a lead to a better overall performance, they result in significantly worse results for a few instances, such as rc203.1 and rc204.3. This suggests that instances have important structural differences that are not reflected by their corresponding number of customers n.

In a similar manner, Table 3 shows the scores obtained by Beam-ACO when using different settings of k_{bw} and N^s. We decided to study all combinations between $k_{bw} = \{10, 20, 30, 40, 50\}$ and $N^s = \{1, 5, 10, 20\}$. The remaining Beam-ACO parameters were set in the same way as for ACO, except for the beam-search parameter $\mu = 1.5$. We applied each configuration of Beam-ACO 25 times for 15 CPU seconds to each test instance. Each table cell in Table 3 gives the score obtained by Beam-ACO using the k_{bw} and N^s settings given by the column with respect to the results obtained by all the other configurations of Beam-ACO in the same row. Again, the best configuration per instance is indicated in bold-face. Note that, for example, for instance rc201.1 all configurations obtained the optimal solution in all runs. In such cases, we indicate in boldface the configuration that required less median CPU time.

The best settings of k_{bw} and N^s depend strongly on the particular instance, as shown in Table 3. In most cases, a small beam-width ($k_{bw} = 10$) and number of samples $N^s \in \{1, 5\}$ obtained the best solutions. However, in certain cases, these settings produced notably worse results and higher values of k_{bw} and N^s are required, as for example when instances rc204.3 and rc205.1 are concerned.

Finally, a comparison between ACO and Beam-ACO is presented in Table 4. In this comparison, we used for each instance and for each algorithm the configuration that obtained the highest scores in Tables 2 and 3, that is, the ones marked in boldface. For both ACO and Beam-ACO, Table 4 gives the median ($\tilde{\Omega}$), standard deviation (sd), and minimum (Ω_{min}) number of constraint violations obtained in 25 runs, then the number of runs ([#]) that obtained that minimum number of constraint violations. For those runs that obtained Ω_{min} constraint violations, $\tilde{F}$ and sd give the median and standard deviation tour

Table 2. Relative scores obtained by ACO with different settings of N^a

Instance		N^a				
Problem	n	10	20	50	100	200
rc201.1	19	0.0	0.0	**0.0**	0.0	0.0
rc201.2	25	11.3	**27.8**	-14.1	0.9	-25.9
rc201.3	31	-45.9	-7.3	19.1	**23.6**	10.6
rc201.4	25	-50.2	-48.6	-0.6	45.3	**54.0**
rc202.1	32	-37.1	-37.3	15.4	26.0	**32.9**
rc202.2	13	12.9	-3.2	-19.3	-5.5	**15.2**
rc202.3	28	-27.5	-19.3	13.1	**22.3**	11.4
rc202.4	27	-30.4	-29.0	0.8	**38.5**	20.0
rc203.1	18	-0.8	**50.2**	37.4	-3.7	-83.0
rc203.2	32	-27.2	-26.1	17.0	12.7	**23.6**
rc203.3	36	-20.7	-20.0	-4.0	14.0	**30.7**
rc203.4	14	-74.2	9.0	1.0	**34.4**	29.8
rc204.1	44	-21.0	-10.2	7.0	**18.6**	5.8
rc204.2	32	-12.4	1.4	**19.6**	-1.9	-6.8
rc204.3	33	**65.2**	32.4	-21.2	-40.9	-35.6
rc205.1	13	-3.0	-3.0	2.0	2.0	**2.0**
rc205.2	26	-41.6	-52.2	16.4	27.0	**50.4**
rc205.3	34	18.0	-2.2	-53.2	0.0	**37.4**
rc205.4	27	-38.2	-17.0	**34.2**	4.8	16.1
rc206.1	3	**0.0**	0.0	0.0	0.0	0.0
rc206.2	36	6.8	-8.5	-24.0	5.8	**20.0**
rc206.3	24	-56.9	-20.8	19.8	**37.8**	20.0
rc206.4	37	-8.4	-35.6	10.7	9.3	**24.0**
rc207.1	33	-26.2	-13.4	**22.3**	0.0	17.3
rc207.2	30	5.2	5.4	-27.5	6.7	**10.2**
rc207.3	32	-46.8	14.0	4.9	**17.6**	10.2
rc207.4	5	**0.0**	0.0	0.0	0.0	0.0
rc208.1	37	-26.3	4.6	2.0	-1.1	**20.9**
rc208.2	28	-66.7	-13.0	**37.0**	15.3	27.4
rc208.3	35	-26.4	-10.8	5.0	2.3	**29.9**

cost and $\tilde{T}_{cpu}$ and sd are the median and standard deviation CPU time (in seconds). Finally, we calculate the score (column "Score") of Beam-ACO with respect to ACO as described earlier. In other words, we compare each of the 25 solutions obtained by Beam-ACO with each of the 25 solutions generated by ACO, for each instance. Next, for the resulting 625 pairwise comparisons,

Table 3. Relative scores obtained by Beam-ACO for different values of k_{bw} and N^s

Instance		$k_{bw}=10$				$k_{bw}=20$				$k_{bw}=30$				$k_{bw}=40$				$k_{bw}=50$			
Problem	n	$N^s=1$	5	10	20	$N^s=1$	5	10	20	$N^s=1$	5	10	20	$N^s=1$	5	10	20	$N^s=1$	5	10	20
rc201.1	19	0.0	**0.0**	0.0	0.0	0.0	0.0	0.0	0.0	0.0	0.0	0.0	0.0	0.0	0.0	0.0	0.0	0.0	0.0	0.0	0.0
rc201.2	25	**50.7**	27.4	18.2	34.7	12.5	28.7	-1.3	-9.2	38.9	-47.2	-30.7	-31.2	15.1	-32.1	-23.2	-25.1	13.7	-4.0	-11.9	-24.0
rc201.3	31	-48.4	15.1	1.1	19.8	-41.6	14.4	8.3	-8.0	-10.4	15.5	7.0	-14.7	-18.6	13.6	13.1	0.6	-20.3	22.5	4.0	**27.0**
rc201.4	25	-9.5	14.8	26.1	26.1	-45.0	14.8	22.3	26.1	-29.5	3.6	18.6	14.8	-62.9	-1.3	14.8	26.1	-57.3	-5.0	7.3	-5.0
rc202.1	32	-33.8	20.2	**43.1**	27.7	-26.3	9.6	30.0	6.5	-25.6	3.6	4.3	18.4	-33.6	-14.0	-12.9	-10.8	-33.3	-6.0	10.4	22.5
rc202.2	13	**64.8**	55.7	33.0	7.5	25.9	-57.9	-68.3	-73.2	26.8	-7.7	-53.4	-55.7	51.8	43.9	2.3	-28.1	38.1	10.4	13.0	-29.2
rc202.3	28	-2.9	**43.4**	19.2	40.6	10.2	21.4	22.6	-5.0	-17.2	4.3	-4.2	-5.7	-8.7	8.1	-4.8	-16.5	-41.8	-14.9	-15.7	-32.4
rc202.4	27	-50.7	-47.5	-38.6	-45.8	-0.2	35.1	27.9	3.1	-2.3	1.5	14.6	2.2	19.2	2.5	12.5	-17.9	21.8	**51.1**	5.3	6.3
rc203.1	18	**34.7**	34.7	34.7	28.7	34.7	4.8	-22.9	-35.9	34.7	28.7	-28.9	-60.9	34.7	5.9	-40.6	-45.5	31.1	18.0	-25.7	-65.3
rc203.2	32	31.8	51.1	34.3	18.1	**65.6**	33.3	20.6	-4.4	33.6	-17.8	7.2	-42.0	-2.2	-26.5	-20.5	-43.2	-13.3	-25.9	-35.6	-64.2
rc203.3	36	-45.9	-17.0	6.8	0.9	25.8	**36.6**	35.3	25.5	-18.3	19.6	17.8	13.8	-32.1	23.1	-1.7	-5.5	-35.6	1.2	-17.5	-32.9
rc203.4	14	-5.7	2.0	-11.6	20.4	3.1	19.8	-0.3	2.9	32.5	15.5	-3.3	-10.7	**51.5**	-7.8	-29.3	-31.4	8.3	-5.7	-20.7	-29.6
rc204.1	44	45.8	**50.0**	28.6	42.6	11.7	2.7	12.9	15.7	-33.4	10.6	-3.9	10.3	-30.2	-1.7	-14.6	-25.8	-52.1	-8.2	-28.6	-32.6
rc204.2	32	44.0	**54.1**	34.3	20.9	37.6	23.4	31.2	4.6	9.6	6.3	-33.2	-42.5	1.3	-2.0	-28.4	-41.2	-13.1	-15.2	-44.5	-47.3
rc204.3	33	-32.4	-48.2	-15.1	13.2	-13.1	-34.8	-50.7	27.8	-45.7	-8.1	27.9	**54.0**	-41.1	31.7	45.3	42.5	-34.2	27.7	22.9	30.3
rc205.1	13	-88.5	-78.8	-59.7	-60.3	-38.0	-34.8	-10.2	-12.0	35.4	35.4	**35.4**	31.9	35.4	35.4	35.4	29.1	24.9	31.9	35.4	17.0
rc205.2	26	-69.5	-70.3	-36.6	-44.5	-21.8	-1.6	-5.1	-11.8	-1.3	10.1	6.2	-5.7	17.9	**59.7**	15.6	24.9	37.0	43.0	36.2	17.5
rc205.3	34	22.1	12.0	8.1	14.9	31.0	**35.0**	-2.2	20.1	-2.7	17.3	-10.4	-9.1	14.5	-22.5	-9.5	-3.4	17.6	-36.3	-54.4	-42.2
rc205.4	27	72.5	**77.7**	47.2	25.1	52.6	12.5	6.9	-25.8	2.3	-9.6	4.6	-56.2	-13.0	-7.8	-13.1	-39.7	-13.6	-33.6	-33.8	-55.3
rc206.1	3	**0.0**	0.0	0.0	0.0	0.0	0.0	0.0	0.0	0.0	0.0	0.0	0.0	0.0	0.0	0.0	0.0	0.0	0.0	0.0	0.0
rc206.2	36	**41.3**	31.6	-4.4	-33.2	33.2	26.6	3.7	-16.9	27.1	-7.1	8.0	13.8	21.7	-8.1	-19.7	-24.1	2.4	-30.7	-22.7	-42.6
rc206.3	24	5.7	1.4	5.7	-7.0	6.9	11.2	14.2	-11.3	1.4	14.2	-2.8	9.9	18.4	-2.8	-15.5	-28.3	**22.7**	14.2	-21.5	-36.7
rc206.4	37	16.2	**40.7**	12.9	-47.9	-6.6	38.8	-21.3	-40.5	2.3	17.7	-34.2	-26.1	-14.5	37.4	-6.7	-29.3	-15.5	30.5	22.1	24.1
rc207.1	33	**50.8**	23.7	27.2	22.1	41.8	6.2	-1.8	-38.0	45.2	-15.0	-27.5	-34.7	31.2	-15.0	-19.2	-46.4	41.8	-32.0	-29.7	-30.6
rc207.2	30	26.0	32.3	30.6	**53.8**	-3.6	-6.5	7.4	3.5	-11.0	-10.4	8.8	-8.0	-15.2	-8.9	-14.7	-28.5	-23.9	-23.9	18.9	-26.6
rc207.3	32	54.6	32.5	**66.0**	26.1	-14.6	-6.0	-3.8	35.7	-26.9	19.1	-4.8	16.7	-26.8	-26.9	1.3	42.0	-34.3	-36.8	-30.0	0.9
rc207.4	5	**0.0**	0.0	0.0	0.0	0.0	0.0	0.0	0.0	0.0	0.0	0.0	0.0	0.0	0.0	0.0	0.0	0.0	0.0	0.0	0.0
rc208.1	37	-0.8	23.7	46.9	**50.5**	-58.4	20.1	17.5	32.7	-57.3	18.0	14.7	22.7	-65.3	1.4	3.6	6.5	-70.9	-8.1	8.5	-6.0
rc208.2	28	43.5	**62.9**	34.9	41.9	12.1	17.5	-9.5	2.2	-10.5	-20.8	8.1	-7.5	-15.7	-11.4	-11.8	-44.9	-17.7	-22.0	-28.5	-22.9
rc208.3	35	-40.1	**44.4**	31.8	37.2	-52.2	16.7	24.0	11.2	-48.2	15.6	13.4	8.7	-79.2	9.1	30.2	15.3	-65.1	15.4	12.4	-0.6

Table 4. Comparison of the best results of ACO and Beam-ACO for each instance

Instance		Beam-ACO								ACO								Score
Problem	n	$\tilde{\Omega}$	sd	Ω_{min}	[#]	$\bar{F}$	sd	$\tilde{T}_{cpu}$	sd	$\tilde{\Omega}$	sd	Ω_{min}	[#]	$\bar{F}$	sd	$\tilde{T}_{cpu}$	sd	
rc201.1	19	0	0.00	0	25	444.54	0.00	0.01	0.00	0	0.00	0	25	444.54	0.00	0.27	0.38	0
rc201.2	25	0	0.00	0	25	712.91	2.48	8.28	3.36	1	0.58	0	12	723.14	7.92	10.18	2.59	93.8
rc201.3	31	0	0.00	0	25	797.25	2.36	5.40	4.37	0	0.41	0	20	806.26	3.57	10.71	3.03	88.5
rc201.4	25	0	0.00	0	25	793.64	0.00	0.75	1.68	0	0.00	0	25	793.64	0.00	3.74	3.17	0
rc202.1	32	0	0.00	0	25	782.77	4.39	11.44	2.82	0	0.51	0	13	788.47	11.45	13.27	0.93	78.2
rc202.2	13	0	0.00	0	25	304.14	0.00	4.27	3.54	0	0.00	0	25	304.14	0.25	6.45	3.38	12
rc202.3	28	0	0.00	0	25	854.35	11.03	8.07	3.91	0	0.33	0	22	846.70	12.78	7.53	4.21	7.8
rc202.4	27	0	0.59	0	14	854.74	16.02	11.32	3.31	1	0.20	0	1	860.75	—	4.89	—	31.5
rc203.1	18	0	0.00	0	25	453.48	0.00	3.02	2.39	0	0.00	0	25	453.48	0.57	6.32	4.34	20
rc203.2	32	0	0.33	0	22	844.13	17.82	11.52	2.72	0	0.41	0	20	837.40	29.36	13.67	3.70	3.7
rc203.3	36	1	0.20	1	24	857.43	7.98	10.49	2.72	1	0.55	0	1	852.86	—	11.13	—	78.6
rc203.4	14	0	0.00	0	25	321.35	2.6	5.09	4.50	0	0.00	0	25	319.45	2.1	4.58	3.91	-45.9
rc204.1	44	1	0.99	0	7	940.17	9.83	13.81	2.03	1	0.84	0	7	938.20	0.95	13.02	2.35	-3.7
rc204.2	32	0	0.00	0	25	684.13	8.65	11.85	2.77	0	0.00	0	25	692.17	9.44	10.78	3.25	46.4
rc204.3	33	0	0.00	0	25	461.93	2.10	5.15	3.82	0	0.00	0	25	463.04	3.55	6.03	3.57	53.0
rc205.1	13	0	0.00	0	25	343.21	0.00	1.82	3.23	0	0.00	0	25	343.21	0.00	0.84	0.79	0
rc205.2	26	0	0.00	0	25	755.93	4.33	10.22	3.68	0	0.56	0	18	768.78	9.58	8.33	4.05	72.6
rc205.3	34	2	0.00	2	25	816.25	2.75	8.38	4.03	2	0.00	2	25	816.55	3.98	7.31	4.50	23.4
rc205.4	27	1	0.51	0	12	762.41	4.72	10.73	2.29	1	0.44	1	1	764.72	—	9.01	—	50.6
rc206.1	3	0	0.00	0	25	117.85	0.00	0.00	0.00	0	0.00	0	25	117.85	0.00	0.00	0.00	0
rc206.2	36	0	0.00	0	25	830.95	1.46	7.90	3.25	0	0.00	0	25	845.81	1.28	7.45	4.13	100
rc206.3	24	0	0.00	0	25	575.55	0.70	3.18	3.03	0	0.00	0	25	574.42	2.62	4.33	3.60	-52
rc206.4	37	0	0.00	0	25	851.02	8.56	9.28	2.61	0	0.00	0	25	890.81	15.92	11.48	2.53	94.9
rc207.1	33	0	0.00	0	25	757.30	12.66	10.84	3.57	0	0.20	0	24	772.93	16.41	12.42	2.96	55.5
rc207.2	30	1	0.00	1	25	683.99	20.17	10.26	4.48	1	0.00	1	25	682.35	34.29	12.45	2.80	2.7
rc207.3	32	0	0.00	0	25	730.04	10.02	11.33	2.21	0	0.00	0	25	722.58	10.13	12.40	3.09	-35.4
rc207.4	5	0	0.00	0	25	119.64	0.00	0.00	0.00	0	0.00	0	25	119.64	0.00	0.00	0.00	0
rc208.1	37	0	0.00	0	25	842.04	12.79	10.90	3.65	0	0.00	0	25	867.05	22.55	11.55	3.20	52.3
rc208.2	28	0	0.00	0	25	550.80	7.03	12.41	2.66	0	0.00	0	25	542.51	8.10	11.01	2.88	-32.2
rc208.3	35	0	0.00	0	25	664.42	5.27	13.37	2.47	0	0.00	0	25	673.92	8.88	9.92	3.93	63.2

we count the percentage of how many times Beam-ACO was better minus how many times it was worse than ACO. Hence, a positive score indicates that the solutions generated by Beam-ACO were more frequently better than those of ACO.

Table 4 shows, first, that Beam-ACO obtains a positive score in 20 out of 30 cases. Of the remaining 10 instances, 5 appear to be excessively easy for both algorithms, since both reach an optimal solution in all runs. Only in 5 instances ACO obtains a better score than Beam-ACO. Nevertheless, comparing the median tour costs ($\tilde{F}$) reveals that the advantage of ACO is quite small. In summary, we can conclude that Beam-ACO based on stochastic sampling provides an evident advantage over the non-hybrid ACO algorithm when applied to the 30 test instances used in this study. This becomes also clear when studying the columns with heading [#]: Beam-ACO is generally more robust in finding solutions with a low number of constraint violations.

6   Conclusions

In this paper, we have proposed a Beam-ACO approach for the TSPTW. Beam-ACO is a hybrid between ant colony optimization and beam search that relies heavily on bounding information that is accurate and computationally inexpensive. We studied a new version of Beam-ACO in which the bounding information is replaced by stochastic sampling. We performed experiments on a set of standard benchmark instances for the TSPTW, comparing a pure ACO algorithm with Beam-ACO based on stochastic sampling. The results showed that Beam-ACO obtains generally better results in most instances. In a few instances, ACO achieved slightly better results than Beam-ACO. In those instances, the heuristic information is probably quite deceptive. Nonetheless, the overall positive performance of Beam-ACO based on stochastic sampling in comparison to ACO shows that Beam-ACO can be useful even when no accurate and computationally inexpensive bounding information is available.

In the future we plan to improve the performance of our Beam-ACO approach further, for example, by the inclusion of local search and by the study of different types of heuristic information.

Acknowledgements. We wish to thank Professor Ohlmann for making widely available the benchmark instances used in his work.

References

1. Dorigo, M., Stützle, T.: Ant Colony Optimization. MIT Press, Cambridge (2004)
2. Blum, C.: Beam-ACO–hybridizing ant colony optimization with beam search: an application to open shop scheduling. Computers & Operations Research 32, 1565–1591 (2005)
3. Blum, C.: Beam-ACO for simple assembly line balancing. INFORMS Journal on Computing 20(4), 618–627 (2008)

4. Ow, P.S., Morton, T.E.: Filtered beam search in scheduling. International Journal of Production Research 26, 297–307 (1988)
5. Juillé, H., Pollack, J.B.: A sampling-based heuristic for tree search applied to grammar induction. In: Proceedings of AAAI 1998 – Fifteenth National Conference on Artificial Intelligence, pp. 776–783. MIT Press, Cambridge (1998)
6. Ruml, W.: Incomplete tree search using adaptive probing. In: Proceedings of IJCAI 2001 – Seventeenth International Joint Conference on Artificial Intelligence, pp. 235–241. IEEE Press, Los Alamitos (2001)
7. Savelsbergh, M.W.P.: Local search in routing problems with time windows. Annals of Operations Research 4(1), 285–305 (1985)
8. Ohlmann, J.W., Thomas, B.W.: A compressed-annealing heuristic for the traveling salesman problem with time windows. INFORMS Journal on Computing 19(1), 80–90 (2007)
9. Blum, C., Dorigo, M.: The hyper-cube framework for ant colony optimization. IEEE Transactions on Systems, Man, and Cybernetics – Part B 34(2), 1161–1172 (2004)
10. Potvin, J.Y., Bengio, S.: The vehicle routing problem with time windows part II: Genetic search. INFORMS Journal on Computing 8, 165–172 (1996)
11. Solomon, M.M.: Algorithms for the vehicle routing and scheduling problems with time windows. Operations Research 35, 254–265 (1987)

Flexible Stochastic Local Search
for Haplotype Inference

Luca Di Gaspero[1] and Andrea Roli[2]

[1] DIEGM, University of Udine, Udine, Italy
l.digaspero@uniud.it
[2] DEIS, University of Bologna, Cesena, Italy
andrea.roli@unibo.it

Abstract. Haplotype Inference is a challenging problem in bioinformatics that consists in inferring the basic genetic constitution of diploid organisms on the basis of their genotype. This information allows researchers to perform association studies for the genetic variants involved in diseases and the individual responses to therapeutic agents. A notable approach to the problem is to encode it as a combinatorial problem (under certain hypotheses, such as the *pure parsimony* of the *entropy minimization* criteria) and to solve it using off-the-shelf combinatorial optimization techniques.

In this paper, we present and discuss an approach based on local search metaheuristics. A flexible solver is designed to tackle the Haplotype Inference under the criterion of choice, that could be defined by the user. We test our approach by solving instances from common Haplotype Inference benchmarks both under the hypothesis of pure parsimony and entropy minimization. Results show that the approach achieves a good trade-off between solution quality and execution time and compares favorably with the state of the art.

1 Introduction

The role of genetic variation and inheritance in human diseases is extremely important, though still largely unknown [19]. To the aim of increasing this body of knowledge, the assessment of a full Haplotype Map of the human genome has become one of the current high priority tasks of human genomics [18]. A haplotype is one of the two non identical copies of a chromosome of a diploid organism, i.e., an organism that has two copies of each chromosome, one inherited from the father and one from the mother. The information haplotypes convey allows to perform association studies for the genetic variants involved in diseases and the individual responses to therapeutic agents. The most important variations are the *Single Nucleotide Polymorphisms* (SNPs), which occur when a nucleotide in the DNA sequence is replaced by another one. Technological limitations make it currently impractical to directly collect haplotypes by experimental procedures, but it is possible to collect *genotypes*, i.e., the conflation of a pair of haplotypes. Moreover, instruments can easily identify only whether the individual is *homozygous* (i.e., the alleles are the same) or *heterozygous* (i.e., the alleles are different)

T. Stützle (Ed.): LION 3, LNCS 5851, pp. 74–88, 2009.

at a given site. Therefore, haplotypes have to be inferred from genotypes in order to reconstruct the detailed information and trace the precise structure of DNA variations in a population. This process is called *Haplotype Inference* (also known as *haplotype phasing*) and the goal is to find a set of haplotype pairs (i.e., a *phasing*) so that all the given genotypes are *resolved*, that is, they can be obtained by combining a pair of haplotypes from the set.

The main methods to tackle the Haplotype Inference are either combinatorial or statistical. Both, however, being of non-experimental nature, need some genetic model that could provide criteria for evaluating the solution returned with respect to actual genetic plausibility. In the case of the combinatorial methods, which are the subject of the present work, the most often used criteria are *pure parsimony* and *entropy minimization*. The *pure parsimony* criterion [10] suggests to search for the smallest collection of distinct haplotypes that solves the Haplotype Inference problem. This criterion is consistent with current observations in natural populations for which the actual number of haplotypes is vastly smaller than the total number of possible haplotypes. Conversely, the rationale of the *entropy minimization* [12] criterion is to maximize phasing likelihood. The entropy is defined in terms of occurrences of a haplotype in the phasing.

Both criteria are widely adopted and the solutions found are considered as good and informative phasings. The adequacy of these model, indeed, has already been discussed elsewhere [9, 10]. Nevertheless, up to now, the techniques used to tackle the Haplotype Inference problem under either criterion are substantially different and developed *ad hoc* for the criterion chosen. In this paper, we present an approach that allows to easily accommodate different criteria in a single solver. The advantage of this approach is that the solver can be adapted to the criterion of choice and criteria can also be combined together without having to change the search strategy. The method we present is a local search metaheuristic[1] that tackles the Haplotype Inference problem as a constrained optimization problem with an objective function defined by the phasing evaluation criterion of choice (in this paper either *pure parsimony* or *entropy minimization*). The approach extends and improves upon a previous work in which a local search method was presented for the Haplotype Inference problem [5].

Current approaches for solving the solving the Haplotype Inference (HI) problem under the pure parsimony hypothesis (HI_{par}) include simple greedy heuristic [4] and exact methods such as Integer Linear Programming [3, 10, 11, 15], Semidefinite Programming [13, 14], SAT models [16, 17] and Pseudo-Boolean Optimization algorithms [8]. At present, complete approaches, i.e., the ones that guarantee to return an optimal solution, such as SAT-based ones, are very effective but they do not scale very well with respect to the instance size. Hence, the need for approximate algorithms, such as metaheuristics, that trade completeness for efficiency. Moreover, a motivation for studying and applying approximate algorithms is that the criteria used to evaluate the solutions provide an approximation of the actual solution quality, therefore a proof of optimality is not particularly important. To the best of our knowledge, besides our previous

---

[1] For an introduction to metaheuristics, see, e.g., [2].

work [5], the only attempt to employ metaheuristic techniques for HI_{par} is a recently proposed Genetic Algorithm [20]. However, also the cited paper does not report results on real size instances.

As far as we know, the state of the art for the problem under the entropy minimization criterion (HI_{ent}) is an iterative greedy algorithm [9] that can tackle also large instances (long genotypes) by dividing the set of genotypes into groups and solving the problem by considering overlapping windows. Moreover, it can also deal with pedigree information.

The remainder of this paper is structured as follows. We formally introduce the problem in Section 2. In Section 3 we describe the metaheuristic approach we developed that exploits problem structure. Experimental results are discussed in Section 4, where we compare our technique against the state of the art for HI_{par} and HI_{ent}. Finally, we discuss some possible improvements and future work in Section 5.

2   The Haplotype Inference Problem

In the Haplotype Inference problem we deal with *genotypes*, that is, strings of length m that correspond to a chromosome with m sites. Each value in the string belongs to the alphabet $\{0, 1, 2\}$. A position in the genotype is associated with a site of interest on the chromosome (e.g., a SNP) and it has value 0 (wild type) or 1 (mutant) if the corresponding chromosome site is a homozygous site (i.e., it has that state on both copies) or the value 2 if the chromosome site is heterozygous. A *haplotype* is a string of length m that corresponds to only one copy of the chromosome (in diploid organisms) and whose positions can assume the symbols 0 or 1.

2.1   Genotype Resolution

Given a chromosome, we are interested in finding an unordered[2] pair of haplotypes that can explain the chromosome according to the following definition:

Definition 1 (Genotype resolution). *Given a chromosome g, we say that the unordered pair $\langle h, k \rangle$ resolves (or covers) g, and we write $\langle h, k \rangle \triangleright g$ (or $g = h \oplus k$), if the following conditions hold (for $j = 1, \ldots, m$):*

$$g[j] = 0 \Rightarrow \qquad h[j] = 0 \wedge k[j] = 0 \tag{1a}$$

$$g[j] = 1 \Rightarrow \qquad h[j] = 1 \wedge k[j] = 1 \tag{1b}$$

$$g[j] = 2 \Rightarrow (h[j] = 0 \wedge k[j] = 1) \vee (h[j] = 1 \wedge k[j] = 0) \tag{1c}$$

If $\langle h, k \rangle \triangleright g$ we indicate the fact that the haplotype h (respectively, k) contributes in the resolution of the genotype g writing $h \trianglelefteq g$ (resp., $k \trianglelefteq g$). We also say that h is a resolvent of g). This notation can be extended to a set of haplotypes, writing $H = \{h_1, \ldots, h_l\} \trianglelefteq g$, with the meaning that $h_i \trianglelefteq g$ for all $i = 1, \ldots, l$. The operator $\oplus$ is defined accordingly.

---

[2] In the problem there is no distinction between the maternal and paternal haplotypes.

Conditions (1a) and (1b) require that both haplotypes must have the same value in all homozygous sites, while condition (1c) states that in heterozygous sites the haplotypes must have different values.

Observe that, according to the definition, for a single genotype string the haplotype values at a given site are predetermined in the case of homozygous sites, whereas there is a freedom to choose between two possibilities at heterozygous places. This means that for a genotype string with l heterozygous sites there are 2^{l-1} possible pairs of haplotypes that resolve it.

As an example, consider the genotype $g = (0212)$, then the possible pairs of haplotypes that resolve it are $\langle(0110), (0011)\rangle$ and $\langle(0010), (0111)\rangle$.

After these preliminaries we can state the *Haplotype Inference* problem as follows:

Definition 2 (Haplotype Inference problem). *Given a population of n individuals, each of them represented by a genotype string g_i of length m we are interested in finding a set ϕ of n pairs of (not necessarily distinct) haplotypes $\phi = \{\langle h_1, k_1\rangle, \ldots, \langle h_n, k_n\rangle\}$, so that $\langle h_i, k_i\rangle \triangleright g_i, i = 1, \ldots, n$, such that a given objective function $F(\{h_1, \ldots, h_n, k_1, \ldots k_n\})$ is optimized. We call H the set of haplotypes used in the construction of ϕ, i.e., $H = \{h_1, \ldots, h_n, k_1, \ldots, k_n\}$.*

From the mathematical point of view, there are many possibilities for building the set H, since there is an exponential number of possible haplotypes for each genotype. Therefore, a criterion should be added to the model for evaluating the quality of the solutions.

One natural model of the Haplotype Inference problem is the already mentioned *pure parsimony* approach that consists in searching for a solution that minimizes the total number of distinct haplotypes used or, in other words, $|H|$, the cardinality of the set H. A trivial upper bound for $|H|$ is $2n$ in the case of all genotypes resolved by a pair of distinct haplotypes. It has been shown that the Haplotype Inference problem under the pure parsimony hypothesis is APX-hard [15] and therefore NP-hard.

The second criterion we consider in this work is entropy minimization. The entropy of a phasing ϕ is defined upon the concept of *coverage* of a haplotype h in ϕ, informally defined as the number of genotypes resolved by h.

Definition 3 (Coverage). *Given a set of genotypes G, a phasing ϕ and a haplotype $h \in \phi$, the coverage of h in ϕ is:*

$$cvg(h, \phi) = |\{g \in G | \exists k \neq h : \langle h, k\rangle \triangleright g\}| \qquad (2)$$
$$+ 2|\{g \in G | \langle h, h\rangle \triangleright g\}|$$

The entropy $\mathcal{E}$ of a phasing ϕ is defined as follows:

Definition 4 (Entropy of a phasing)

$$\mathcal{E}(\phi) = \sum_{h : cvg(h, \phi) \neq 0} -\frac{cvg(h, \phi)}{2|G|} \log \frac{cvg(h, \phi)}{2|G|} \qquad (3)$$

Thus, the objective in this problem formulation is to find ϕ (hence a set of haplotypes H) such that $\mathcal{E}(\phi)$ is minimized. Also this variant of the Haplotype Inference problem was proven to be APX-hard [12].

It is important to stress, at this point, that finding a proven optimal solution is not particularly relevant, because the criteria defining the objective functions are an approximation of an (unknown) actual quality function. Therefore, approximate approaches that are able to return solutions of a good quality, even if not optimal, are of notable practical importance.

2.2   Compatibility and Complementarity

It is possible to define a graph that expresses the compatibility between genotypes, so as to avoid unnecessary checks in the determination of the resolvents.[3] In the graph $\mathcal{G} = (G, E)$, the set of vertexes coincides with the set of the genotypes; genotypes g_1, g_2 are connected by an edge if they are *compatible*, i.e., one or more common haplotypes can resolve both of them. The formal definition of this property is as follows.

Definition 5 (Genotypes compatibility). *Let g_1 and g_2 be two genotypes, g_1 and g_2 are compatible if, for all $j = 1, \ldots, m$, the following conditions hold:*

$$g_1[j] = 0 \Rightarrow g_2[j] \in \{0, 2\} \tag{4a}$$
$$g_1[j] = 1 \Rightarrow g_2[j] \in \{1, 2\} \tag{4b}$$
$$g_1[j] = 2 \Rightarrow g_2[j] \in \{0, 1, 2\} \tag{4c}$$

The same concept can be expressed also between a genotype and a haplotype as in the following definition.

Definition 6 (Compatibility between genotypes and haplotypes). *Let g be a genotype and h a haplotype, g and h are compatible if, for all $j = 1, \ldots, m$, the following conditions hold:*

$$g[j] = 0 \Rightarrow h[j] = 0 \tag{5a}$$
$$g[j] = 1 \Rightarrow h[j] = 1 \tag{5b}$$
$$g[j] = 2 \Rightarrow h[j] \in \{0, 1\} \tag{5c}$$

We denote this relation with $h \mapsto g$, and we write $h[j] \mapsto g[j]$ when the conditions hold for the single SNP j. Moreover with an abuse of notation we indicate with $h \mapsto \{g_1, g_2, \ldots\}$ the set of all the genotypes that are compatible with haplotype h.

Notice that the compatibility between a genotype g and a haplotype h does not necessarily require h to be a resolvent of g.

---

[3] In some cases, also a graph representing incompatibilities between genotypes can provide useful information.

It is worth to observe also that the set of compatible genotypes of a haplotype can contain only mutually compatible genotypes (i.e., they form a clique in the compatibility graph).

Another useful property is the following:

Proposition 1 (Haplotype complement). *Given a genotype g and a haplotype* $h \mapsto g$*, there exists a unique haplotype k such that* $h \oplus k = g$*. The haplotype k is called the* complement *of h with respect to g and is denoted with* $k = g \ominus h$*.*

3   Flexible Stochastic Local Search for Haplotype Inference

Local search is a search process that iteratively modifies the current candidate solution by applying move operators trying to follow trajectories in the search space leading to good solutions. Local search techniques belong to the family of metaheuristics [2] and are usually stochastic as they involve decisions taken according to a probabilistic distribution.

In this work, we present a stochastic local search designed for tackling the Haplotype Inference problem both under the pure parsimony and minimum entropy criteria. The solver is designed so as to separate the search strategy from the cost function that guides the search. The latter is clearly dependent on the optimization criterion chosen, whereas the search strategy could be described in a more general way.

The high level search strategy we chose is *tabu search* that exhaustively explores the neighborhood of the current solution, by trying all the possible moves, and chooses as new solution the best among the neighbors evaluated. The peculiarity of tabu search is that the neighborhood is restricted by forbidding recently performed moves.

The design process of metaheuristics involves the definition of the local search model and the choice of the search strategy. In the following we detail the design and implementation choices of our approach. For a discussion on alternative metaheuristic approaches for the Haplotype Inference problem we forward the interested reader to [5, 6].

The local search model is defined by specifying three entities, namely the *search space*, the *cost function* and the *neighborhood relation*.

In the approach we propose, the search space is composed of the pairs of haplotypes $\langle h, k \rangle$ that resolve genotype g, for all $g \in G$. Therefore in this representation all the genotypes are fully resolved at each state by construction. Thus, the search space is the collection of sets ϕ defined as in the problem statement.

The cost function is a measure of solution quality including components related both to the solution evaluation criteria of choice and to heuristic information that could guide search toward good solutions.

The component related to the evaluation criterion is an objective function defined as the cardinality $|H|$ of the set of haplotypes employed in the resolution or the entropy of the phasing, respectively in the case of pure parsimony or minimum entropy. In formulae:

$$f_1^{\mathrm{par}}(\phi) = |H| \qquad (6)$$

$$f_1^{\mathrm{ent}}(\phi) = \mathcal{E}(\phi) \qquad (7)$$

Moreover, we also include a heuristic measure related to the potential quality of the solution. In this respect, we counted the number of incompatible sites between each genotype/haplotype pair and the component of the cost function is expressed by the following formula:

$$f_2 = \sum_{h \in H} \sum_{g \in G} \sum_{j=1}^{m} 1 - \chi(h[j] \mapsto g[j]) \qquad (8)$$

In the formula, χ denotes the truth indicator function, whose value is 1 when the proposition in parentheses is true and 0 otherwise.

The cost function F is then the weighted sum of the two components:

$$F = \alpha_1 f_1^T + \alpha_2 f_2 \qquad (9)$$

in which $T \in \{\mathrm{par}, \mathrm{ent}\}$ and the weights α_1 and α_2 must be chosen for the problem at hand to reflect the trade-offs between the different components. In our experimentation we chose the values $\alpha_1 = \alpha_2 = 1$.

3.1   Neighborhood Relation and Search Strategy

We designed a stochastic local search technique, based on the *tabu search* meta-heuristic template. The strategy is defined in Algorithm 1. The algorithm starts with a set of randomly generated haplotypes of cardinality at most $2n$, where n is the number of genotypes. Then, a reduction procedure is called whose aim is to reduce the number of haplotypes by exploiting the structure of the compatibility graph. This procedure was first presented in our previous work [5] and tries to remove from H those haplotypes that are not necessary to resolve some genotype. The heuristic reduction procedure used in Algorithm 1 is based on the following proposition:[4]

Proposition 2 (Haplotype local reduction). *Given n genotypes $G = \{g_1, \ldots, g_n\}$ and the resolvent set $R = \{\langle h_1, k_1 \rangle, \ldots, \langle h_n, k_n \rangle\}$, so that $\langle h_i, k_i \rangle \triangleright g_i$. Suppose there exist two genotypes $g, g' \in G$ such that:*

$$g \triangleleft \begin{cases} h & \mapsto \{g, g', \ldots\} \\ k & \mapsto \{g, \ldots\} \end{cases}, \quad g' \triangleleft \begin{cases} h' & \mapsto \{g', \ldots\} \\ k' & \mapsto \{g', \ldots\} \end{cases} \qquad (10)$$

and $h \neq h'$, $h \neq k'$, $h' \trianglelefteq A$, $k' \trianglelefteq B$, where A and B are the sets of genotypes currently resolved, respectively, by h' and k'.

The replacement of $\langle h', k' \rangle$ with $\langle h, g' \ominus h \rangle$ in the resolution of g' is a correct resolution that employs a number of distinct haplotypes according to the following criteria:

---

[4] For proof and discussion see [5].

- *if $|A| = 1$ and $|B| = 1$, the new resolution uses* at most *one less distinct haplotype;*
- *if $|A| > 1$ and $|B| = 1$ (or symmetrically, $|A| = 1$ and $|B| > 1$), the new resolution uses* at most *the same number of distinct haplotypes;*
- *in the remaining case the new resolution uses* at most *one more distinct haplotype.*

After this preprocessing phase, the solver explores the search space by iteratively modifying pairs of resolving haplotypes trying to reduce F. Then, the iterative process is repeated until a termination criterion is met: in our implementation, we either allotted a maximum runtime or we allow for a maximum number of *idle iterations* (i.e., iterations from last improvement). Tabu search explores all the neighbors of the incumbent solution s that can be reached by applying moves that are not in the tabu list and chooses the best neighbor (lines 5–7). A move is tabu if it or its inverse have been applied in the last tl iterations. The tabu concept is employed to prevent the algorithm from getting stuck in local minima and avoid cycling between a set of states.

Algorithm 1. High level scheme of Tabu search for Haplotype Inference

1: $s \leftarrow$ GenerateRandomInitialSolution()
2: $s \leftarrow$ Reduce()
3: $s_b \leftarrow s$
4: **while** termination conditions not met **do**
5: $\quad \mathcal{N}_a(s) \leftarrow \{s' \in \mathcal{N}(s) \mid s'$ does not violate the tabu condition$\}$
6: $\quad s' \leftarrow \arg\min\{F(s'') \mid s'' \in \mathcal{N}_a(s)\}$
7: $\quad s \leftarrow s' \qquad$ {i.e., s' replaces s}
8: $\quad$ **if** $F(s) < F(s_b)$ **then**
9: $\qquad s_b \leftarrow s$
10: $\quad$ **end if**
11: **end while**
12: Return best solution found s_b

The main strength of the local search we designed is to be found in the move operator, that exploits the instance structure in order to consider the most promising neighbors of a solution. In fact, in a previous work [5], we defined the neighborhood on the basis of the unitary Hamming distance between haplotypes. Even though this has shown to be an effective choice, it did not exploit the structure of the instance. On the contrary, the move we present in this work explicitly takes into account the compatibility graph. Indeed, the neighborhood of a solution is defined by considering, for each genotype, one of its resolving haplotypes and trying to employ it for covering other genotypes. We call this neighborhood *CommonHaplotypes*.

For example, given the haplotype $h = (00110) \trianglelefteq g = (00112)$, it could be employed also to solve genotypes $g_1 = (22112)$ and $g_2 = (02212)$, which are compatible with g. The whole neighborhood is defined on all possible haplotype/genotype pairs, excluding the trivial ones (i.e., non-compatible or identical phasings).

The worst case complexity of the neighborhood exploration is $O(n^2)$, however, the average case is better, because the number of resolving haplotypes for a given pair of compatible haplotype/genotype pairs is usually less than n^2. This move considerably improves the solver performance with respect to our previous work. In the following, we will distinguish between the local search with Hamming neighborhood and the move focusing on common resolving haplotypes by referring to them as TS^H and TS^C, respectively.

4   Experimental Results

We developed our solver with EASYLOCAL++ [7], a framework for the development of local search algorithms. The algorithms have been implemented in C++ and compiled with *gcc 3.2.2* and run on a Intel Xeon CPU 2.80GHz machine with SUSE Linux 2.4.21-278-smp. According to the guidelines of [1], each algorithm was run on every instance one time for each combination of parameters. We allotted 300 seconds for each execution of the algorithms.

In this section, we present and discuss the results of the comparison of our new algorithm with our previous one and with the state of the art for HI_{par} and HI_{ent}.

The benchmark instances are composed of three sets of instances employed in [3]. The main characteristics of the instance sets are summarized in Table 1. The first two datasets, namely Harrower uniform and Harrower non-uniform, are composed of artificial instances created by Brown and Harrower using the ms software, which simulates neutral evolution and recombination. The Harrower hapmap dataset contains biological data extracted from the Hapmap project [18].

Table 1. A summary of the main characteristics of the benchmarks

Benchmark set	N. of instances	N. of genotypes	N. of sites
Harrower uniform	200	10÷100	30÷50
Harrower non-uniform	90	10÷100	30÷50
Harrower hapmap	24	5÷68	30÷75

The different algorithms were compared only on the basis of the chosen objective function, i.e., either in terms of the overall number of haplotypes employed (f_1^{par}) or in terms of the entropy value of their final solution (f_1^{ent}). The other cost components were disregarded in the comparison since they were used only for heuristic guidance.

4.1   Comparison against Local Search with Hamming Neighborhood

We first compare the quality of solutions returned by TS^H and TS^C within a CPU time-limit of 20 seconds. Results are plotted in Figure 1. From the plots

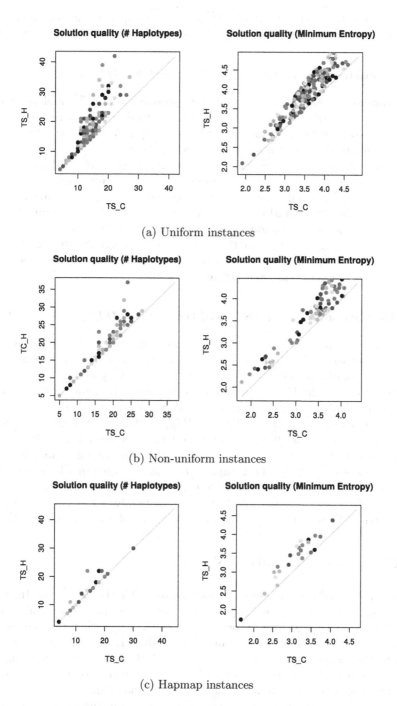

(a) Uniform instances

(b) Non-uniform instances

(c) Hapmap instances

Fig. 1. Comparison between TS^H and TS^C

it is clear that the performance of TS^C are always superior than those of TS^H, therefore, from now on, all the comparisons will be made with respect to TS^C.

4.2   Comparison with the State of the Art for HI_{par}

An estimation of the quality of solutions produced by TS^C, and its overall performance, in the case of HI_{par} can be obtained by comparing it against *rpoly* [8], a state-of-the-art exact solver for the HI_{par}. We run the solver on the same benchmark instances and on the same machine.

In Figure 2 the plots showing the comparison between TS^C and *rpoly* on the solved instances are drawn. We can observe that the solution quality achieved by TS^C is very high and it approximates the optimal solution quality returned by *rpoly* on some benchmarks. Nevertheless, it has to be observed that TS^C can return a good feasible solution to all the instances of the real-world Hapmap set and its running time is a fraction of the time required by the exact solver. Therefore, we expect that TS^C can scale very well on very large instances, with a high number of genotypes and heterozygous sites.

4.3   Comparison with the State of the Art for HI_{ent}

The local search approach discussed in this work has the advantage of enabling the developer to easily specify different objective functions or also a weighted combination of objectives. This characteristic can be very useful to explore different solutions to the Haplotype Inference, making it possible for biologists to compare different candidate solutions to the real problem. To this respect, we now analyze the performance of TS^C on HI_{ent} and contrast it with the greedy local search proposed in [9] that, to the best of our knowledge, is the state of the art for the Haplotype Inference problem under entropy minimization. The method proposed in [9] is a best improvement local search in which the neighborhood is composed of all the possible pairs resolving each genotype.

The results show that TS^C performs systematically better than the greedy local search of Gusev et al. [9], however at the price of a longer running time.

5   Discussion and Future Work

We have presented a metaheuristic approach to tackle the Haplotype Inference problem under two well-known hypotheses, namely pure parsimony and entropy minimization. The algorithm is designed in such a way that the criterion of choice can be changed and even combined with other criteria very easily, without changing the component implementing the search strategy. Our solver compares favorably against the state of the art for both the problem variants, as it achieves a good balance between solution quality and execution time.

This technique can be further improved by modifying the search strategy so as to take into account also pedigree information and possible unknown values in the data. We are working towards these goals.

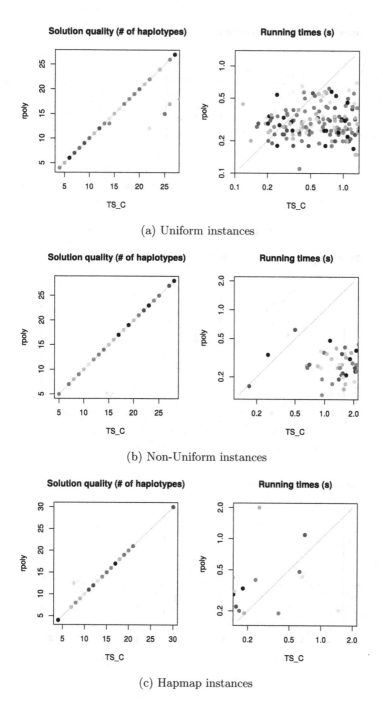

(a) Uniform instances

(b) Non-Uniform instances

(c) Hapmap instances

Fig. 2. Comparison between TS^C and rpoly for HI_{par}

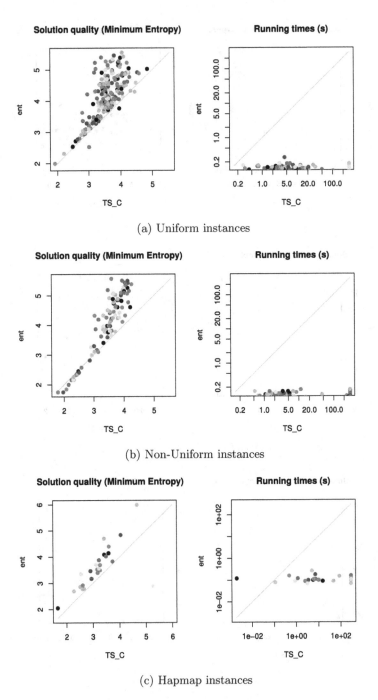

(a) Uniform instances

(b) Non-Uniform instances

(c) Hapmap instances

Fig. 3. Comparison between TS^C and ent for HI_{ent}

Acknowledgments. We thank Inês Lynce and Ana Sofia Graça for kindly providing us their instances and solvers. We also thank Ian M. Harrower for sending us his datasets.

References

1. Birattari, M.: On the estimation of the expected performance of a metaheuristic on a class of instances. how many instances, how many runs? Technical Report TR/IRIDIA/2004-01, IRIDIA, Univerisé Libre de Bruxelles (2004)
2. Blum, C., Roli, A.: Metaheuristics in combinatorial optimization: Overview and conceptual comparison. ACM Computing Surveys 35(3), 268–308 (2003)
3. Brown, D.G., Harrower, I.M.: Integer programming approaches to haplotype inference by pure parsimony. IEEE/ACM Transactions on Computational Biology and Bioinformatics 3(2), 141–154 (2006)
4. Clark, A.G.: Inference of haplotypes from PCR-amplified samples of diploid populations. Molecular Biology and Evolution 7, 111–122 (1990)
5. Di Gaspero, L., Roli, A.: Stochastic local search for large-scale instances of the haplotype inference problem by pure parsimony. Journal of Algorithms in Logic, Informatics and Cognition 63(1-3), 55–69 (2008)
6. Di Gaspero, L., Roli, A.: A preliminary analysis on metaheuristics methods applied to the haplotype inference problem. Technical Report DEIS-LIA-07-006, University of Bologna (Italy). LIA Series no. 84 (2007)
7. Di Gaspero, L., Schaerf, A.: EASYLOCAL++: An object-oriented framework for flexible design of local search algorithms. Software—Practice and Experience 33(8), 733–765 (2003)
8. Graça, A., Marques-Silva, J., Lynce, I., Oliveira, A.L.: Efficient haplotype inference with pseudo-boolean optimization. In: Anai, H., Horimoto, K., Kutsia, T. (eds.) AB 2007. LNCS, vol. 4545, pp. 125–139. Springer, Heidelberg (2007)
9. Gusev, A., Pasaniuc, B., Mandoiu, I.: Highly scalable genotype phasing by entropy minimization. IEEE/ACM Trans. on Computational Biology and Bioinformatics 5(2), 252–261 (2008)
10. Gusfield, D.: Haplotype inference by pure parsimony. In: Baeza-Yates, R., Chávez, E., Crochemore, M. (eds.) CPM 2003. LNCS, vol. 2676, pp. 144–155. Springer, Heidelberg (2003)
11. Halldórsson, B.V., Bafna, V., Edwards, N., Lippert, R., Yooseph, S., Istrail, S.: A survey of computational methods for determining haplotypes. In: Istrail, S., Waterman, M.S., Clark, A. (eds.) DIMACS/RECOMB Satellite Workshop 2002. LNCS (LNBI), vol. 2983, pp. 26–47. Springer, Heidelberg (2002)
12. Halperin, E., Karp, R.M.: The minimum-entropy set cover problem. Theoretical Computer Science 348(2-3), 240–250 (2005)
13. Huang, Y.-T., Chao, K.-M., Chen, T.: An approximation algorithm for haplotype inference by maximum parsimony. In: Haddad, H., Liebrock, L.M., Omicini, A., Wainwright, R.L. (eds.) Proceedings of the 2005 ACM Symposium on Applied Computing (SAC 2005), pp. 146–150. ACM Press, New York (2005)
14. Kalpakis, K., Namjoshi, P.: Haplotype phasing using semidefinite programming. In: BIBE, pp. 145–152. IEEE Computer Society, Los Alamitos (2005)
15. Lancia, G., Pinotti, M.C., Rizzi, R.: Haplotyping populations by pure parsimony: Complexity of exact and approximation algorithms. INFORMS Journal on Computing 16(4), 348–359 (2004)

16. Lynce, I., Marques-Silva, J.: Efficient haplotype inference with boolean satisfiability. In: Proceedings of the 21st National Conference on Artificial Intelligence and the Eighteenth Innovative Applications of Artificial Intelligence Conference. AAAI Press, Menlo Park (2006)
17. Lynce, I., Marques-Silva, J.: SAT in bioinformatics: Making the case with haplotype inference. In: Biere, A., Gomes, C.P. (eds.) SAT 2006. LNCS, vol. 4121, pp. 136–141. Springer, Heidelberg (2006)
18. The International HapMap Consortium. The international HapMap project. Nature 426, 789–796 (2003)
19. The International HapMap Consortium. A haplotype map of the human genome. Nature 437 (2005)
20. Wang, R.-S., Zhang, X.-S., Sheng, L.: Haplotype inference by pure parsimony via genetic algorithm. In: Zhang, X.-S., Liu, D.-G., Wu, L.-Y. (eds.) Operations Research and Its Applications: the Fifth International Symposium (ISORA 2005), Tibet, China, August 8-13. Lecture Notes in Operations Research, vol. 5, pp. 308–318. Beijing World Publishing Corporation, Beijing (2005)

A Knowledge Discovery Approach to Understanding Relationships between Scheduling Problem Structure and Heuristic Performance

Kate A. Smith-Miles[1], Ross J.W. James[2], John W. Giffin[2], and Yiqing Tu[1]

[1] School of Mathematical Sciences, Monash University, Melbourne, Australia
{kate.smith-miles,ytu}@sci.monash.edu.au
[2] Department of Management, University of Canterbury, Christchurch, New Zealand
{ross.james,john.giffin}@canterbury.ac.nz

Abstract. Using a knowledge discovery approach, we seek insights into the relationships between problem structure and the effectiveness of scheduling heuristics. A large collection of 75,000 instances of the single machine early/tardy scheduling problem is generated, characterized by six features, and used to explore the performance of two common scheduling heuristics. The best heuristic is selected using rules from a decision tree with accuracy exceeding 97%. A self-organizing map is used to visualize the feature space and generate insights into heuristic performance. This paper argues for such a knowledge discovery approach to be applied to other optimization problems, to contribute to automation of algorithm selection as well as insightful algorithm design.

Keywords: Scheduling, heuristics, algorithm selection, self-organizing map, performance prediction, knowledge discovery.

1 Introduction

It has long been appreciated that knowledge of a problem's structure and instance characteristics can assist in the selection of the most suitable algorithm or heuristic [1, 2]. The No Free Lunch theorem [3] warns us against expecting a single algorithm to perform well on all classes of problems, regardless of their structure and characteristics. Instead we are likely to achieve better results, on average, across many different classes of problem, if we tailor the selection of an algorithm to the characteristics of the problem instance. This approach has been well illustrated by the recent success of the algorithm portfolio approach on the 2007 SAT competition [4].

As early as 1976, Rice [1] proposed a framework for the algorithm selection problem. There are four essential components of the model:

- the problem space P represents the set of instances of a problem class;
- the feature space F contains measurable characteristics of the instances generated by a computational feature extraction process applied to P;

T. Stützle (Ed.): LION 3, LNCS 5851, pp. 89–103, 2009.

- the algorithm space A is the set of all considered algorithms for tackling the problem;
- the performance space Y represents the mapping of each algorithm to a set of performance metrics.

In addition, we need to find a mechanism for generating the mapping from feature space to algorithm space. The Algorithm Selection Problem can be formally stated as: For a given problem instance $x \in P$, with features $f(x) \in F$, find the selection mapping $S(f(x))$ into algorithm space A, such that the selected algorithm $\alpha \in A$ maximizes the performance mapping $y(\alpha(x)) \in Y$. The collection of data describing $\{P, A, Y, F\}$ is known as the meta-data.

There have been many studies in the broad area of algorithm performance prediction, which is strongly related to algorithm selection in the sense that supervised learning or regression models are used to predict the performance ranking of a set of algorithms, given a set of features of the instances. In the AI community, most of the relevant studies have focused on constraint satisfaction problems like SAT, QBF or QWH (P, in Rice's notation), using solvers like DPLL, CPLEX or heuristics (A), and building a regression model (S) to use the features of the problem structure (F) to predict the run-time performance of the algorithms (Y). Studies of this nature include Leyton-Brown and co-authors [5-7], and the earlier work of Horvitz et al. [8] that used a Bayesian approach to learn the mapping S. In recent years these studies have extended into the algorithm portfolio approach [4] and a focus on dynamic selection of algorithm components in real-time [9, 10].

In the machine learning community, research in the field of meta-learning has focused on classification problems (P), solved using typical machine learning classifiers such as decision trees, neural networks, or support vector machines (A), where supervised learning methods (S) have been used to learn the relationship between the statistical and information theoretic measures of the classification instance (F) and the classification accuracy (Y). The term meta-learning [11] is used since the aim is to learn about learning algorithm performance. Studies of this nature include [12-14] to name only three of the many papers published over the last 15 years.

In the operations research community, particularly in the area of constrained optimization, researchers appear to have made fewer developments, despite recent calls for developing greater insights into algorithm performance by studying search space or problem instance characteristics. According to Stützle and Fernandes [15], "currently there is still a strong lack of understanding of how exactly the relative performance of different meta-heuristics depends on instance characteristics".

Within the scheduling community, some researchers have been influenced by the directions set by the AI community when solving constraint satisfaction problems. The dynamic selection of scheduling algorithms based on simple low-level knowledge, such as the rate of improvement of an algorithm at the time of dynamic selection, has been applied successfully [16]. Other earlier approaches have focused on integrating multiple heuristics to boost scheduling performance in flexible manufacturing systems [17].

For many NP-hard optimization problems, such as scheduling, there is a great deal we can discover about problem structure which could be used to create a rich set of features. Landscape analysis (see [18-20]) is one framework for measuring the characteristics of problems and instances, and there have been many relevant developments in this direction, but the dependence of algorithm performance on these measures is yet to be completely determined [20].

Clearly, Rice's framework is applicable to a wide variety of problem domains. A recent survey paper [21] has discussed the developments in algorithm selection across a variety of disciplines, using Rice's notation as a unifying framework, through which ideas for cross-fertilization can be explored. Beyond the goal of performance prediction also lies the ideal of greater insight into algorithm performance, and very few studies have focused on methodologies for acquiring such insights. Instead the focus has been on selecting the best algorithm for a given instance, without consideration for what implications this has for algorithm design or insight into algorithm behaviour. This paper demonstrates that knowledge discovery processes can be applied to a rich set of meta-data to develop, not just performance predictions, but visual explorations of the meta-data and learned rules, with the goal of learning more about the dependencies of algorithm performance on problem structure and data characteristics.

In this paper we present a methodology encompassing both supervised and unsupervised knowledge discovery processes on a large collection of meta-data to explore the problem structure and its impact on algorithm suitability. The problem considered is the early/tardy scheduling problem, described in section 2. The methodology and meta-data is described in section 3, comprising 75,000 instances (P) across a set of 6 features (F). We compare the performance of two common heuristics (A), and measure which heuristic produces the lowest cost solution (Y). The mapping S is learned from the meta-data {P, A, Y, F} using knowledge derived from self-organizing maps, and compared to the knowledge generated and accuracy of the performance predictions using the supervised learning methods of neural networks and decision trees. Section 4 presents the results of this methodology, including decision tree rules and visualizations of the feature space, and conclusions are drawn in Section 5.

2   The Early/Tardy Machine Scheduling Problem

Research into the various types of E/T scheduling problems was motivated, in part, by the introduction of Just-in-Time production, which required delivery of goods to be made at the time required. Both early and late production are discouraged, as early production incurs holding costs, and late delivery means a loss of customer goodwill. A summary of the various E/T problems was presented in [22] which showed the NP-completeness of the problem.

2.1   Formulation

The E/T scheduling problem we consider is the single machine, distinct due date, early/tardy scheduling problem where each job has an earliness and tardiness penalty and due date. Once a job is dispatched on the machine, it runs to completion with no interruptions permitted. The objective is to minimise the total penalty produced by the schedule. The objective of this problem can be defined as follows:

$$\min \sum_{i=1}^{n} \left(\alpha_i \left| d_i - c_i \right|^+ + \beta_i \left| c_i - d_i \right|^+ \right) . \tag{1}$$

where n is the number of jobs to be scheduled, c_i is the completion time of job i, d_i is the due date of job i, α_i is the penalty per unit of time when job i is produced early, β_i is the penalty per unit of time when job i is produced tardily, and $|x|^+ = x$ if $x > 0$, or 0 otherwise. We also define p_i as the processing time of job i. The decision variable is the completion time c_i of job i, derived from the optimal starting sequence of jobs and their processing times.

The objective of this problem is to schedule the jobs as closely as possible to their due dates; however the difficulty in formulating a schedule occurs when it is not possible to schedule all jobs on their due dates, which also causes difficulties in managing the many tradeoffs between jobs competing for processing at a given time [23]. Two of the simplest and most commonly used dispatching heuristics for the E/T scheduling problem are the Earliest Due Date and Shortest Processing Time heuristics.

2.2   Earliest Due Date (EDD) Heuristic

The EDD heuristic orders the jobs based on the date the job is due to be delivered to the customer. The jobs with the earliest due date are scheduled first, while the jobs with the latest due date are scheduled last. After the sequence is determined, the completion times of each job are then calculated using the optimal idle time insertion algorithm of Fry, Armstrong and Blackstone [24]. For single machine problems the EDD is known to be the best rule to minimise the maximum lateness, and therefore tardiness, and also the lateness variance [25]. The EDD has the potential to produce optimal solutions to this problem, for example when there are few jobs and the due dates are widely spread so that all jobs may be scheduled on their due date without interfering with any other jobs. As there are no earliness or tardiness penalties, the objective value will be 0 and therefore optimal.

2.3   Shortest Processing Time (SPT) Heuristic

The SPT heuristic orders the jobs based on their processing time. The jobs with the smallest processing time are scheduled first, while the jobs with the largest processing time are scheduled last; this is the fastest way to get most of the jobs completed quickly. Once the SPT sequence has been determined, the job completion times are calculated using the optimal idle time insertion algorithm [24]. The SPT heuristic has been referred to as "the world champion" scheduling heuristic [26], as it produces schedules for single machine problems that are good at minimising the average time of jobs in a system, minimising the average number of jobs in the system and minimising the average job lateness [25]. When the tardiness penalties for the jobs are similar and the due dates are such that the majority of jobs are going to be late, SPT is likely to produce a very good schedule for the E/T scheduling problem, as it gets the jobs completed as quickly as possible. The "weighted" version of the SPT heuristic, where the order is determined by p_i/β_i, is used in part by many E/T heuristics, as this order can be proven to be optimal for parts of a given schedule.

2.4   Discussion

Due to the myopic nature of the EDD and SPT heuristics, neither heuristic is going to consistently produce high quality solutions to the general E/T scheduling problem. Both of these simple heuristics generate solutions very quickly however and therefore it is possible to carry out a large sample of problems in order to demonstrate whether or not the approach proposed here is useful for exploring the relative performance of two heuristics (or algorithms) and is able to predict the superiority of one heuristic over another for a given instance.

3   Methodology

In this section we describe the meta-data for the E/T scheduling problem in the form of {P, A, Y, F}. We also provide a description of the machine learning algorithms applied to the meta-data to produce rules and visualizations of the meta-data.

3.1   Meta-data for the E/T Scheduling Problem

The most critical part of the proposed methodology is identification of suitable features of the problem instances that reflect the structure of the problem and the characteristics of the instances that might explain algorithm performance. Generally there are two main approaches to characterizing the instances: the first is to identify problem dependent features based on domain knowledge of what makes a particular instance challenging or easy to solve; the second is a more general set of features derived from landscape analysis [27]. Related to the latter is the approach known in the meta-learning community as landmarking [28], whereby an instance is characterized by the performance of simple algorithms which serve as a proxy for more complicated (and computationally expensive) features. Often a dual approach makes sense, particularly if the feature set derived from problem dependent domain knowledge is not rich, and supplementation from landscape analysis can assist in the characterization of the instances. In the case of the generalised single machine E/T scheduling problem however, there is sufficient differentiation power in a small collection of problem dependent features that we can derive rules explaining the different performance of the two common heuristics. Extending this work to include a greater set of algorithms (A) may justify the need to explore landscape analysis tools to derive greater characterisation of the instances.

In this paper, each n-job instance of the generalised single machine E/T scheduling problem has been characterized by the following features.

1. Number of jobs to be scheduled in the instance, n
2. Mean Processing Time $\bar{p}$: The mean processing time of all jobs in an instance

3. Processing Time Range p_σ: The range (max – min) of the processing times of all jobs in the instance

4. Tardiness Factor τ: Defines where the average due date occurs relative to, and as a fraction of the total processing time of all jobs in the instance. A positive tardiness factor indicates the proportion of the schedule that is expected to be tardy, while a negative tardiness factor indicates the amount of idle time that is expected in the schedule as a proportion of the total processing time of all jobs in the sequence. Mathematically the tardiness factor was defined by Baker and Martin [29] as: $\tau = 1 - \dfrac{\sum d_i}{n \sum p_i}$

5. Due Date Range factor D_σ: Determines the spread of the due dates from the average due date for all jobs in the instance, normalized by the size of processing times. It is defined as $D_\sigma = \dfrac{(b-a)}{\sum p_i}$, where b is the maximum due date in the instance and a is the minimum due date in the instance, and is a fraction of the total processing time needed for the instance

6. Penalty Ratio ρ: The maximum over all jobs in the instance of the ratio of the tardy penalty to the early penalty.

Any instance of the problem, whether contained in the meta-data set or generated at a future time, can be characterized by this set of six features. It is not the only possible set of features but, as the results presented later in this paper demonstrate, it captures the essential variation in instances needed to accurately predict heuristic performance. Since we are comparing the performance of only two heuristics, we can create a single binary variable to indicate which heuristic performs best for a given problem instance. Let $Y_i=1$ if EDD is the best performing heuristic (lowest objective function) compared to SPT for problem instance i, and $Y_i=0$ otherwise (SPT is best). The meta-data then comprises the set of six-feature vectors and heuristic performance measure (Y), for a large number of instances, and the task is to learn the relationship between features and heuristic performance.

In order to provide a large and representative sample of instances for the meta-data, an instance generator was created to span a range of values for each feature. Problem instances were then generated for all combinations of parameter values. Note that these parameters are targets for the instances and the random generation process may create slight variation from these target values. The parameter settings used were:

- problem size (number of jobs, n): 20-100 with increments of 20 (5 levels)
- target processing time range p_σ: processing times randomly generated with a range ($p_{max} - p_{min}$) of 2-10 with increments of 2 (5 levels).
- target due date range factor D_σ as a proportion of total processing time: ranges from 0.2 to 1 in increments of 0.2 (5 levels)
- target tardiness factor τ as a proportion of total processing time: ranges from 0 (all jobs should complete on time) to 1 (all jobs should be late) in increments of 0.2 (6 levels)
- penalty ratio ρ: 1-10 with increments of 1 (10 levels)

From these parameters the following instance data can be generated:

- processing times p_i : calculated within the processing time range.
- processing time means $\bar{p}$: calculated from the randomly generated p_i
- due dates d_i : due dates randomly generated within the due date range and offset by the tardiness factor.

To calculate the actual p_σ, actual D_σ and actual τ we use the actual p_i, d_i of the problem rather than the target values. Ten instances using each parameter setting were then generated, giving a total of 5 (size levels) x 5 (processing time range levels) x 6 (tardiness factor levels) x 5 (due date range factor levels) x 10 (penalty ratio levels) x 10 (instances) = 75,000 instances.

A correlation analysis between the instance features and the Y values across all 75,000 instances reveals that the only instance features that appear to correlate (linearly) with heuristic performance are the tardiness factor (correlation = -0.59) and due date range factor (correlation = 0.44). None of the other instance features appear to have a linear relationship with algorithm performance. Clearly due date range factor and tardiness factor correlate somewhat with the heuristic performances, but it is not clear if these are non-linear relationships, and if either of these features with combinations of the others can be used to seek greater insights into the conditions under which one heuristic is expected to outperform the other.

Using Rice's notation, our meta-data can thus be described as:

- P = 75,000 E/T scheduling instances
- A = 2 heuristics (EDD and SPT)
- Y = binary decision variable indicating if EDD is best compared to SPT (based on objective function which minimizes weighted deviation from due dates)
- F = 6 instance features (problem size, processing time mean, processing time range, due date range factor, tardiness factor and penalty ratio).

Additional features could undoubtedly be derived either from problem dependent domain knowledge, or using problem independent approaches such as landscape analysis [28], landmarking [28], or hyper-heuristics [30]. For now though, we seek to learn the relationships that might exist in this meta-data.

3.2   Knowledge Discovery on the Meta-data

When exploring any data-set to discover knowledge, there are two broad approaches. The first is supervised learning (aka directed knowledge discovery) which uses training examples – sets of independent variables (inputs) and dependent variables (outputs) - to learn a predictive model which is then generalized for new examples to predict the dependent variable (output) based only on the independent variables (inputs). This approach is useful for building models to predict which algorithm or heuristic is likely to perform best given only the feature vector as inputs. The second broad approach to knowledge discovery is unsupervised learning (aka undirected knowledge discovery) which uses only the independent variables to find similarities and differences between the structure of the examples, from which we may then be able to infer relationships between these structures and the dependent variables. This second approach is useful for our goal of seeking greater insight into *why* certain heuristics might be better suited to certain instances and, rather than just building predictive models of heuristic performance.

In this section we briefly summarise the machine learning methods we have used for knowledge discovery on the meta-data.

Neural Networks
As a supervised learning method [31], neural networks can be used to learn to predict which heuristic is likely to return the smallest objective function value. A training dataset is randomly extracted (80% of the 75,000 instances) and used to build a non-linear model of the relationships between the input set (features F) and the output (metric Y). Once the model has been learned, its generalisation on an unseen test set (the remaining 20% of the instances) is evaluated and recorded as a percentage accuracy in predicting the superior heuristic. This process is repeated ten times for different random extractions of the training and test sets, to ensure that the results were not simply an artifact of the random number seed. This process is known as ten-fold cross validation, and the reported results show the average accuracy on the test set across these ten folds.

For our experimental results, the neural network implementation within the Weka machine learning platform [32] was used with 6 input nodes, 4 hidden nodes, and 2 output nodes utilising binary encoding. The transfer function for the hidden nodes was a sigmoidal function, and the neural network was trained with the backpropagation (BP) learning algorithm with learning rate = 0.3, momentum = 0.2. The BP algorithm stops when the number of epochs (complete presentation of all examples) reaches a maximum training time of 500 epochs or the error on the test set does not decrease after a threshold of 20 epochs.

Decision Tree
A decision tree [33] is also a supervised learning method, which uses the training data to successively partition the data, based on one feature at a time, into classes. The goal is to find features on which to split the data so that the class membership at lower leaves of the resulting tree is as "pure" as possible. In other words, we strive for leaves that are comprised almost entirely of one class only. The rules describing each class can then be read up the tree by noting the features and their splitting points. Ten-fold cross validation is also used in our experiments to ensure the generalisation of the rules.

The J4.8 decision tree algorithm, implemented in Weka [32], was used for our experimental results, with a minimum leaf size of 500 instances. The generated decision tree is pruned using subtree raising with confidence factor = 0.25.

Self-Organizing Maps
Self-Organizing Maps (SOMs) are the most well-known unsupervised neural network approach to clustering. Their advantage over traditional clustering techniques such as the k-means algorithm lies in the improved visualization capabilities resulting from the two-dimensional map of the clusters. Often patterns in a high dimensional input space have a very complicated structure, but this structure is made more transparent

and simple when they are clustered in a lower dimensional feature space. Kohonen [34] developed SOMs as a way of automatically detecting strong features in large data sets. SOMs find a mapping from the high dimensional input space to low dimensional feature space, so any clusters that form become visible in this reduced dimensionality. The architecture of the SOM is an multi-dimensional input vector connected via weights to a 2-dimensional array of neurons. When an input pattern is presented to the SOM, each neuron calculates how similar the input is to its weights. The neuron whose weights are most similar (minimal distance in input space) is declared the winner of the competition for the input pattern, and the weights of the winning neuron, and its neighbours, are strengthened to reflect the outcome. The final set of weights embeds the location of cluster centres, and is used to recognize to which cluster a new input vector is closest.

For our experiments we randomly split the 75000 instances into training data (50000 instances) and test data (25000 instances). We use the Viscovery SOMine software (www.eudaptics.com) to cluster the instances based only on the six features as inputs. A map of 2000 nodes is trained for 41 cycles, with the neighbourhood size diminishing linearly at each cycle. After the clustering of the training instances, the distribution of Y values is examined within each cluster, and knowledge about the relationships between instance structure and heuristic performance is inferred and evaluated on the test data.

4   Experimental Evaluation

4.1   Supervised Learning Results

Both the neural network and decision tree algorithms were able to learn the relationships in the meta-data, achieving greater than 97% accuracy (on ten-fold cross-validation test sets) in predicting which of the two heuristics would be superior based only on the six features (inputs). These approaches have an overall classification accuracy of 97.34% for the neural network and 97.13% for the decision tree. While the neural network can be expected to learn the relationships in the data more powerfully, due to its nonlinearity, its limitation is the lack of insight and explanation of those relationships. The decision tree's advantage is that it produces a clear set of rules, which can be explored to see if any insights can be gleaned. The decision tree rules are presented in the form of pseudo-code in Figure 1, with the fraction in brackets showing the number of instances that satisfied both the condition and the consequence (decision) in the numerator, divided by the total number of instances that satisfied the condition in the denominator. This proportion is equivalent to the accuracy of the individual rule.

The results allow us to state a few rules with exceptionally high accuracy:

1) If the majority of jobs are expected to be scheduled early (tardiness factor <= 0.5) then EDD is best in 99.8% of instances
2) If the majority of the jobs are expected to be scheduled late (tardiness factor > 0.7) then SPT is best in 99.5% of instances

3) If slightly more than half of the jobs are expected to be late (tardiness factor between 0.5 and 0.7) then as long as the tardiness penalty ratio is no more than 3 times larger than the earliness penalty ($\rho \le 3$), then EDD is best in 98.9% of the instances with a due date range factor greater than 0.2.

The first two rules are intuitive and can be justified from what we know about the heuristics - EDD is able to minimise lateness deviations when the majority of jobs can be scheduled before their due date, and SPT is able to minimise the time of jobs in the system and hence tardiness when the majority of jobs are going to be late [25]. The third rule reveals the kind of knowledge that can be discovered by adopting a machine learning approach to the meta-data. Of course other rules can also be explored from Figure 1, with less confidence due to the lower accuracy, but they may still provide the basis for gaining insight into the conditions under which different algorithms can be expected to perform well.

```
If (τ  <= 0.7) Then
   If (τ  <= 0.5) Then EDD best (44889/45000 = 99.8%)
   If (τ > 0.5) Then If (D_σ <= 0.2) Then If (ρ <= 3) Then EDD best (615/750 = 82.0%)
                                        Else SPT best (1483/1750 = 84.7%)
               Else If (ρ <= 3) Then EDD best (5190/5250 = 98.9%)
                    Else If (τ <= 0.6) Then EDD best (8320/8750 = 95.1%)
                         Else If ( p̄  <= 2) Then EDD best (556/700 = 79.4%)
                              Else If (n <= 60) Then SPT best (1150/1680 = 68.4%)
                                   Else EDD best (728/1120 = 65%)
Else SPT best (9950/10000 = 99.5%)
```

Fig. 1. Pseudo-code for the decision tree rule system, showing the accuracy of each rule

4.2 Unsupervised Learning Results

After training the SOM, the converged map shows 5 clusters, each of which contains similar instances defined by Euclidean distance in feature space. Essentially, the six-dimensional input vectors have been projected onto a two-dimensional plane, with topology-preserving properties. The clusters can be inspected to understand what the instances within each cluster have in common. The statistical properties of the 5 clusters can be seen in Table 1. The distribution of the input variables (features), and additional variables including the performance of the heuristics, can be visually explored using the maps shown in Figure 2. A k-nearest neighbour algorithm (with k=7) is used to distribute additional data instances (from the test set) or extra variables (Y values) across the map.

Looking first at the bottom row of Table 1, it is clear that clusters 1, 2 and 3 contain instances that are best solved using EDD (Y=1). These clusters are shown visually in the bottom half of the 2-d self-organizing map (see Figure 2a for cluster boundaries, and Figure 2b to see the distribution of Y across the clusters). These three clusters of instances account for 70.2% of the 75,000 instances (see Table 1). The remaining clusters 4 and 5 are best solved, on average, by SPT. The maps shown in Figure 2c – 2h enable us to develop a quick visual understanding of how the clusters differ from each other, and to see which features are prominent in defining instance structure.

Table 1. Cluster statistics for training data (test data in brackets) - mean values of input variables, and heuristic performance variable Y, as well as cluster size

	Cluster 1	**Cluster 2**	**Cluster 3**	**Cluster 4**	**Cluster 5**	**All Data**
instances	17117 (8483)	10454 (5236)	7428 (3832)	8100 (4000)	6901 (3449)	50000 (25000)
instances (%)	34.23 (33.93)	20.91 (20.94)	14.86 (15.33)	16.2 (16.0)	13.8 (13.8)	100 (100)
n	60.65 (61.03)	59.73 (59.73)	58.73 (58.96)	57.8 (57.7)	63.39 (61.56)	60.0 (59.97)
$\bar{p}$	2.77 (2.76)	5.24 (5.22)	5.08 (5.07)	5.12 (5.11)	2.70 (2.71)	4.0 (3.99)
p_σ	3.54 (3.52)	8.48 (8.45)	8.16 (8.13)	8.24 (8.21)	3.41 (3.41)	6.0 (5.99)
τ	0.31 (0.31)	0.36 (0.35)	0.21 (0.21)	0.72 (0.73)	0.72 (0.72)	0.43 (0.42)
$D\sigma$	0.70 (0.70)	0.88 (0.88)	0.38 (0.38)	0.40 (0.39)	0.40 (0.40)	0.6 (0.59)
ρ	5.89 (5.88)	4.93 (4.99)	5.37 (5.41)	5.24 (5.19)	5.87 (5.72)	5.5 (5.49)
Y	1.00 (0.99)	1.00 (1.00)	0.99 (0.99)	0.36 (0.36)	0.42 (0.41)	0.82 (0.82)

By inspecting the maps shown in Figure 2, and the cluster statistics in Table 1, we can draw some conclusions about whether the variables in each cluster are above or below average (compared to the entire dataset), and look for correlations with the heuristic performance metric Y. For instance, cluster 2 is characterized by instances with above average values of processing time mean and range, below average tardiness factor, and above average due date range factor. The EDD heuristic is always best under these conditions (Y=1). Instances in cluster 3 are almost identical, except that the due date range factor tends to be below average. Since cluster 3 instances are also best solved by the EDD heuristic, one could hypothesize that the due date range factor does not have much influence in predicting heuristic performance. An inspection of the maps, however, shows this is not the case.

The distribution of Y across the map (Figure 2b) shows a clear divide between the clusters containing instances best solved using EDD (bottom half) and the clusters containing instances best solved using SPT (top half). Inspecting the distribution of features across this divide leads to a simple observation that, if the tardiness factor τ is below average (around 0.5 represented by white to mid-grey in Figure 2f), then EDD will be best. But there are small islands of high Y values in clusters 4 and 5 that overlay nicely with the medium grey values of due date range factor. So we can observe another rule that EDD will also be best if the tardiness factor is above average and the due date range factor is above average. Also of interest, from these maps we can see that problem size and the penalty ratio do not influence the relative

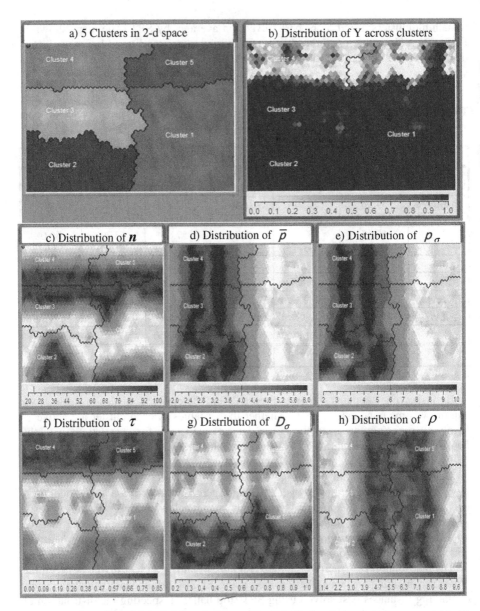

Fig. 2. Self-Organizing Map showing 5 clusters (fig. 2a), the heuristic performance variable Y (fig 2b), and the distribution of six features across the clusters (fig 2c – fig 2h). The grey scale shows a minimum value as white, and maximum value as black.

performance of these heuristics. As neither heuristic considers the penalty ratio (it is used within the optimal idle time insertion algorithm [24], common to both heuristics, but not used by the EDD or SPT heuristics themselves), its not being a factor in the clusters is not surprising.

Within Viscovery SOMine, specific regions of the map can be selected, and used as the basis of a classification. In other words, we can define regions and islands to be predictive of one heuristic excelling based on the training data (50,000 instances). We can then test the generalization of the predictive model using the remaining 25,000 instances as a test set, and applying the k-nearest neighbour algorithm to determine instances that belong to the selected region. We select the dark-grey to black regions of the Y map in Figure 2b, and declare that any test instances falling in the selected area are classified as Y=1, while any instances falling elsewhere in the map are classified as Y=0. The resulting accuracy on the test set is 95% in predicting which heuristic will perform better. The self-organizing map has proven to be useful for both visualization of feature space and predictive modeling of heuristic performance, although the accuracy is not quite as high as the supervised learning approaches.

5   Conclusions and Future Research

In this paper we have illustrated how the concepts of Rice's Algorithm Selection Problem can be extended within a knowledge discovery framework, and applied to the domain of optimization in order that we might gain to insights into optimization algorithm performance. This paper represents one of the first attempts to apply this approach to understand more about optimisation algorithm performance. While only two very simple heuristics have been used to illustrate the approach, we expect full generalization of the methodology to consider a broader range of complex heuristics and meta-heuristics. A large meta-data set comprising 75,000 instances of the E/T scheduling problem has been used to explore what can be learned about the relationships between the features of the problem instances and the performance of heuristics. Both supervised and unsupervised learning approaches have been presented, each with their own advantages and disadvantages made clear by the empirical results. The neural network obtained the highest accuracy for performance prediction, but its weakness is the lack of explanation or interpretability of the model. Our goal is not merely performance prediction, but to gain insights into the characteristics of instances that make solution by one heuristic superior than another. Decision trees are also a supervised learning method, and the rules produced demonstrate the potential to obtain both accurate performance predictions and some insights. Finally, the self-organizing map demonstrated its benefits for visualization of the meta-data and relationships therein.

One of the most important considerations for this approach to be successful for any arbitrary optimization problem is the choice of features used to characterize the instances. These features need to be carefully chosen in such a way that they can characterize instance and problem structure as well as differentiate algorithm performance.

There is little that will be learned via a knowledge discovery process if the features selected to characterize the instances do not have any differentiation power. The result will be supervised learning models of algorithm performance that predict average behaviour with accuracy measures no better than the default accuracies one could obtain from using a naïve model. Likewise, the resulting self-organizing map would show no discernible difference between the clusters when superimposing Y values (unlike in Figure 2b where we obtain a clear difference between the top and bottom halves of the map). Thus the success of any knowledge discovery process depends on the quality of the data, and in this case, the meta-data must use features that serve the purpose of differentiating algorithm performance. In this paper we have used a small set of problem-dependent features, related to the E/T Scheduling Problem, which would be of no use to any other optimization problem. For other optimization problem like graph colouring or the Travelling Salesman Problem, recent developments in phase transition analysis (e.g. [35]) could form the foundation of the development of useful features. Landscape analysis [20, 27] provides a more general (problem independent) set of features, as do ideas from landmarking [28] and hyper-heuristics [30]. It is natural to expect that the best results will be obtained from a combination of generic and problem dependent features, and this will be the focus of our future research. In addition, we plan to extend the approach to consider the performance of a wider variety of algorithms, especially meta-heuristics, where we will also be gathering meta-data related to the features of the meta-heuristics themselves (e.g. hill-climbing capability, tabu list, annealing mechanism, population-based search, etc.). This will help to close the loop to ensure that any insights derived from such an approach are able to provide inputs into the design of new hybrid algorithms that adapt the components of the meta-heuristic according to the instance features – an extension of the highly successful algorithm portfolio approach [4].

References

1. Rice, J.R.: The Algorithm Selection Problem. Adv. Comp. 15, 65–118 (1976)
2. Watson, J.P., Barbulescu, L., Howe, A.E., Whitley, L.D.: Algorithm Performance and Problem Structure for Flow-shop Scheduling. In: Proc. AAAI Conf. on Artificial Intelligence, pp. 688–694 (1999)
3. Wolpert, D.H., Macready, W.G.: No Free Lunch Theorems for Optimization. IEEE T. Evolut. Comput. 1, 67 (1997)
4. Xu, L., Hutter, F., Hoos, H., Leyton-Brown, K.: Satzilla-07: The Design and Analysis of An Algorithm Portfolio For SAT. In: Bessière, C. (ed.) CP 2007. LNCS, vol. 4741, pp. 712–727. Springer, Heidelberg (2007)
5. Leyton-Brown, K., Nudelman, E., Shoham, Y.: Learning the Empirical Hardness of Optimization Problems: The Case of Combinatorial Auctions. In: Van Hentenryck, P. (ed.) CP 2002. LNCS, vol. 2470, pp. 556–569. Springer, Heidelberg (2002)
6. Leyton-Brown, K., Nudelman, E., Andrew, G., McFadden, J., Shoham, Y.: A Portfolio Approach to Algorithm Selection. In: Proc. IJCAI, pp. 1542–1543 (2003)
7. Nudelman, E., Leyton-Brown, K., Hoos, H., Devkar, A., Shoham, Y.: Understanding Random SAT: Beyond the Clauses-To-Variables Ratio. In: Wallace, M. (ed.) CP 2004. LNCS, vol. 3258, pp. 438–452. Springer, Heidelberg (2004)
8. Horvitz, E., Ruan, Y., Gomes, C., Kautz, H., Selman, B., Chickering, M.: A Bayesian Approach to Tackling Hard Computational Problems. In: Proc. 17th Conf. on Uncertainty in Artificial Intelligence, pp. 235–244. Morgan Kaufmann, San Francisco (2001)
9. Samulowitz, H., Memisevic, R.: Learning to solve QBF. In: Proc. 22nd AAAI Conf. on Artificial Intelligence, pp. 255–260 (2007)
10. Streeter, M., Golovin, D., Smith, S.F.: Combining multiple heuristics online. In: Proc. 22nd AAAI Conf. on Artificial Intelligence, pp. 1197–1203 (2007)
11. Vilalta, R., Drissi, Y.: A Perspective View and Survey of Meta-Learning. Artif. Intell. Rev. 18, 77–95 (2002)
12. Michie, D., Spiegelhalter, D.J., Taylor, C.C. (eds.): Machine Learning, Neural and Statistical Classification. Ellis Horwood, New York (1994)

13. Brazdil, P., Soares, C., Costa, J.: Ranking Learning Algorithms: Using IBL and Meta-Learning on Accuracy and Time Results. Mach. Learn. 50, 251–277 (2003)
14. Ali, S., Smith, K.: On Learning Algorithm Selection for Classification. Appl. Soft Comp. 6, 119–138 (2006)
15. Stützle, T., Fernandes, S.: New Benchmark Instances for the QAP and the Experimental Analysis of Algorithms. In: Gottlieb, J., Raidl, G.R. (eds.) EvoCOP 2004. LNCS, vol. 3004, pp. 199–209. Springer, Heidelberg (2004)
16. Carchrae, T., Beck, J.C.: Applying Machine Learning to Low Knowledge Control of Optimization Algorithms. Comput. Intell. 21, 373–387 (2005)
17. Shaw, M.J., Park, S., Raman, N.: Intelligent Scheduling With Machine Learning Capabilities: The Induction of Scheduling Knowledge. IIE Trans. 24, 156–168 (1992)
18. Knowles, J.D., Corne, D.W.: Towards Landscape Analysis to Inform the Design of a Hybrid Local Search for the Multiobjective Quadratic Assignment Problem. In: Abraham, A., Ruiz-Del-Solar, J., Koppen, M. (eds.) Soft Computing Systems: Design, Management and Applications, pp. 271–279. IOS Press, Amsterdam (2002)
19. Merz, P.: Advanced Fitness Landscape Analysis and the Performance of Memetic Algorithms. Evol. Comp. 2, 303–325 (2004)
20. Watson, J., Beck, J.C., Howe, A.E., Whitley, L.D.: Problem Difficulty for Tabu Search in Job-Shop Scheduling. Artif. Intell. 143, 189–217 (2003)
21. Smith-Miles, K.A.: Cross-Disciplinary Perspectives on Meta-Learning For Algorithm Selection. ACM Computing Surveys (in press, 2009)
22. Baker, K.R., Scudder, G.D.: Sequencing With Earliness and Tardiness Penalties: A Review. Ops. Res. 38, 22–36 (1990)
23. James, R.J.W., Buchanan, J.T.: A Neighbourhood Scheme with a Compressed Solution Space for The Early/Tardy Scheduling Problem. Eur. J. Oper. Res. 102, 513–527 (1997)
24. Fry, T.D., Armstrong, R.D., Blackstone, J.H.: Minimizing Weighted Absolute Deviation in Single Machine Scheduling. IIE Transactions 19, 445–450 (1987)
25. Vollmann, T.E., Berry, W.L., Whybark, D.C., Jacobs, F.R.: Manufacturing Planning and Control for Supply Chain Management, 5th edn. McGraw Hill, New York (2005)
26. Krajewski, L.J., Ritzman, L.P.: Operations Management: Processes and Value Chains, 7th edn. Pearson Prentice Hall, New Jersey (2005)
27. Schiavinotto, T., Stützle, T.: A review of metrics on permutations for search landscape analysis. Comput. Oper. Res. 34, 3143–3153 (2007)
28. Pfahringer, B., Bensusan, H., Giraud-Carrier, C.G.: Meta-Learning by Landmarking Various Learning Algorithms. In: Proc. ICML, pp. 743–750 (2000)
29. Baker, K.B., Martin, J.B.: An Experimental Comparison of Solution Algorithms for the Single Machine Tardiness Problem. Nav. Res. Log. 21, 187–199 (1974)
30. Burke, E., Hart, E., Kendall, G., Newall, J., Ross, P., Schulenburg, S.: Hyper-heuristics: An Emerging Direction in Modern Search Technology. In: Glover, F., Kochenberger, G. (eds.) Handbook of Meta-heuristics, pp. 457–474. Kluwer, Norwell (2002)
31. Smith, K.A.: Neural Networks for Prediction and Classification. In: Wang, J. (ed.) Encyclopaedia of Data Warehousing and Mining, vol. 2, pp. 865–869. Information Science Publishing, Hershey (2006)
32. Witten, I.H., Frank, E.: Data Mining: Practical Machine Learning Tools and Techniques, 2nd edn. Morgan Kaufmann, San Francisco (2005)
33. Quinlan, J.R.: C4.5: Programs for Machine Learning. Morgan Kaufmann, San Francisco (1993)
34. Kohonen, T.: Self-Organized Formation of Topologically Correct Feature Maps. Biol. Cyber. 43, 59–69 (1982)
35. Achlioptas, D., Naor, A., Peres, Y.: Rigorous Location of Phase Transitions in Hard Optimization Problems. Nature 435, 759–764 (2005)

Fitness Landscape Analysis for the Resource Constrained Project Scheduling Problem

Jens Czogalla and Andreas Fink

Helmut-Schmidt-University, Hamburg, Germany
{czogalla,andreas.fink}@hsu-hamburg.de

Abstract. The fitness landscape of the resource constrained project scheduling problem is investigated by examining the search space position type distribution and the correlation between the quality of a solution and its distance to an optimal solution. The suitability of the landscape for search with evolutionary computation and local search methods is discussed.

1 Introduction

The problem considered in this paper is the non-preemptive single mode resource-constrained project scheduling problem (RCPSP). It consists in scheduling a set of activities with deterministic processing times, resource requirements, and precedence relations between activities. The aim is to find a schedule with minimum makespan (total project duration) observing both precedence relations and resource limits. The RCPSP is a classical problem in project scheduling. It is related to and subsumes many other scheduling problems (e.g., the job shop scheduling problem (JSP) as a special case [1]). The RCPSP is encountered in diverse contexts, including production, service industry, software development, and civil engineering.

The RCPSP is an NP-hard optimization problem [2]. Hence, the majority of state-of-the-art algorithms are based on metaheuristics. Kolisch and Hartmann [3] present a comprehensive experimental evaluation of heuristic approaches for the RCPSP where the best performing algorithms are population-based metaheuristics (evolutionary methods) such as genetic algorithms, discrete particle swarm optimization (DPSO), and scatter search. Nowadays most effective population-based methods use some kind of hybridization to improve individual solutions; see, e.g., [4]. Consequently, the best algorithms in [3] utilize a heuristic procedure to improve single schedules, in particular forward-backward improvement (FBI, see [5], [6]). However, general local search methods, which depend on an underlying neighborhood structure and move selection rules, such as tabu search and simulated annealing, are not competitive under the experimental settings in [3].

The concept of a fitness landscape and its statistical analysis has been shown to be useful for understanding the behavior of search algorithms and can help in predicting their performance [7]. The aim of this paper is to analyze the fitness

T. Stützle (Ed.): LION 3, LNCS 5851, pp. 104–118, 2009.
© Springer-Verlag Berlin Heidelberg 2009

landscape of the RCPSP and to contribute to the explanation of the empirical results presented by Kolisch and Hartmann [3].

2   The Resource Constrained Project Scheduling Problem

A project consists of a set $\mathcal{J}$ of N activities, $\mathcal{J} = \{1, \ldots, N\}$ and a set $\mathcal{R}$ of K renewable resources, $\mathcal{R} = \{1 \ldots, K\}$. In general the dummy start activity 1 and the dummy termination activity N are added to the project and act as source and sink of the project, respectively. The duration or processing time of activity $j \in \mathcal{J}$ is d_j with $d_1 = d_N = 0$. Each activity has to be processed without interruption. Precedence constraints force activity j not to be started before all its immediate predecessors in the set $P_j \subset \mathcal{J}$ have been finished. The structure of a project can be represented by an activity-on-node (AON) network $G = (\mathcal{V}, \mathcal{A})$, where $\mathcal{V}$ is the set of activities $\mathcal{J}$ and $\mathcal{A}$ is the set of precedence relationships [8]. While being processed, activity j requires $r_{j,k}$ units of resource $k \in \mathcal{R}$ in every time unit of its duration (with $r_{1,k} = r_{N,k} = 0, k = 1, \ldots, K$). For each renewable resource k there is a limited capacity of R_k at any point in time. The values d_j, R_k, and $r_{j,k}$ (duration of activities, availability of resources, and resource requirements of activities) are assumed to be nonnegative and deterministic.

A schedule can be presented as $S = (s_1, \ldots, s_N)$, where s_j denotes the start time of activity j with $s_1 = 0$. The objective is to determine the start time of each activity, so that the project makespan (total project duration) is minimized, and both the precedence and the resource constraints are satisfied. Effective search methods for the RCPSP are mostly based on a genotype search space which consists of precedence-feasible activity lists (permutations). The serial schedule generation scheme may be used to derive a schedule from the activity list [9].

As a generalization of the classical job shop scheduling problem the RCPSP belongs to the class of NP-hard optimization problems [2] and it is noted as $PS|prec|C_{max}$ according to the common classification and notation described in [1]. Hartmann and Kolisch [3,10] present basic components of heuristic approaches and evaluate the state-of-the-art of the design and application of metaheuristics for the RCPSP on a benchmark set of test instances.

3   Fitness Landscape Analysis

Given some search space, a fitness landscape is induced by a particular operator which defines a neighborhood structure [11]. More precisely, a fitness landscape is a labeled, directed graph [12, Chap. 2]. Since we employ an unary neighborhood operator (see Sec. 4) we can simplify Jones's model [12]: The vertices of the graph correspond to genotype solutions (i.e., permutations) of the RCPSP. A directed edge connecting vertex Π to Π' indicates that Π' is reachable from Π with one application of the neighborhood operator (i.e., Π is a neighbor of Π' under the considered neighborhood operator). The vertices of the graph are labeled with the corresponding fitness values, giving raise to the landscape image when thought of as heights [12, Chap. 2]. Based on this definition seven *position types* for points in the search space can be defined according to the topology of their local neighborhood [13, p. 211f]:

- strict local minima (SLMIN, all neighbors have larger fitness values)
- local minima (LMIN, no neighbor has a smaller fitness value)
- interior plateau (IPLAT, all neighbors have the same fitness values)
- ledge (LEDGE, different kinds of neighbors)
- slope (SLOPE, no neighbor has the same fitness value)
- local maxima (LMAX, no neighbor has a larger fitness value)
- strict local maxima (SLMAX, all neighbors have smaller fitness values)

Every search space position can be assigned to exactly one position type. For small instance sizes the exact *position type distribution* (PTD) can be determined by total enumeration. Sampling methods have to be applied for larger instance sizes. For a detailed discussion we refer to [13, p. 213].

Several global features of fitness landscapes influence the performance of heuristic optimization algorithms [7]. The *ruggedness* of a landscape is a measure for the correlation of fitness values of neighboring points in the search space and for the number of local optima. Observed in a time-series generated by a random walk it can be used to predict the performance of local search procedures incorporated in evolutionary methods [14]. A prerequisite for useful statistics based on random walks is the regularity of the search space, i.e., all elements of the landscape are visited by a random walk with equal probability. Precedence constraints cause the fitness landscape of the RCPSP to be non-regular rendering this statistical method meaningless. Directed stochastic search performed on such a landscape may be biased. As shown for the JSP in [14] a negative correlation between solution quality and the number of neighbors (minimization problem) can guide local search toward regions of the fitness landscape containing high-quality solutions. In [14] the correlation between solution quality and the number of neighbors was termed *drift*.

In [12] the *fitness distance correlation* (FDC) as a measure of problem difficulty for evolutionary methods is proposed. FDC requires the definition of a distance measure (see below). FDC is the correlation between the quality of a solution and its distance to an optimal solution. In case of two or more optimal solutions the distance to the closest optimal solution may be considered. The FDC states how closely fitness and distance to an optimal solution are related. If fitness increases when the distance to the optimum becomes smaller, search is expected to be relatively easy for selection-based algorithms, since the evolution of the population is guided to a global optimum via solutions with increasing fitness [7]. The FDC coefficient can be computed based on a sample of m solutions f and the corresponding distances d to the optimal solution. The FDC coefficient is defined as [13, p. 222f]

$$\varrho(f, d) = \frac{cov(f, d)}{\sigma(f)\sigma(d)} \tag{1}$$

where

$$cov(f, d) = \frac{1}{m-1} \sum_{i=1}^{m} (f_i - \overline{f})(d_i - \overline{d}) \qquad (2)$$

$$\sigma(f) = \sqrt{\frac{1}{m-1} \sum_{i=1}^{m} (f_i - \overline{f})^2}, \qquad \sigma(d) = \sqrt{\frac{1}{m-1} \sum_{i=1}^{m} (d_i - \overline{d})^2}. \qquad (3)$$

High ϱ values indicate that fitness and distance to the optimum are related and that search promises to be relatively easy for evolutionary methods [7]. In [15] a classification of problem difficulty based on FDC coefficients is suggested. A problem with $\varrho \geq 0.15$ (in [15] maximization problems were investigated) is called *straightforward* and *should* be suitable for evolutionary methods.

The FDC coefficient can be estimated using randomly generated samples [12, Chap. 5] or solutions that are improved by means of problem-specific procedures. Fitness distance scatter plots can support the evaluation of the FDC coefficient by avoiding misinterpretations.

Based on FDC analysis a *big valley* structure was reported for some combinatorial optimization problems, e.g., the traveling salesman problem [16], the graph bisection problem [16], the permutation flowshop sequencing problem [11], the no-wait flow-shop scheduling problem [17], the quadratic assignment problem [7], and the capacitated vehicle routing problem [18]. The big valley structure means that local optima tend to be relatively close to each other and to a global optimum. High FDC coefficients are an indicator for the presence of the big valley structure [19]. Further evidence for a big valley structure can be obtained from results on the correlation between the solution quality and the average distance between one element and any other element of a given set of locally optimal solutions [16]. The presence of a big valley structure should support the performance of recombination in population-based search. Usually the initial population of solutions is uniformly distributed over the complete search space. Recombination should focus the search in a region with good solutions. In a big valley structure recombination can potentially drive the search towards the optimal solution [19].

In the following we review distances derived from the interpretation of the permutation representation where either the adjacency relation among the elements of the permutation, or the relative order of the elements, or the absolute position of the elements may be of relevance.

The unidirectional version of the *adjacency distance* (R-type distance in [20], adjacency based distance in [21]) is defined as the number of times a pair of activities i, j is adjacent in both Π and Π' [20]:

$$d(\Pi, \Pi') = \sum_{i=1}^{n-1} y_i \qquad \text{with } y_i = \begin{cases} 1 & \text{for } \pi_i = \pi'_j \text{ and } \pi_{i+1} = \pi'_{j+1} \\ 0 & \text{otherwise} \end{cases} \qquad (4)$$

with $d_{adj,max} = n-1$. The *precedence distance* is the number of times n_{pre} some activity j is preceded by activity i in both Π and Π' [11]:

$$d_{pre}(\Pi, \Pi') = \frac{n(n-1)}{2} - n_{pre} \tag{5}$$

with $d_{pre,max} = (n(n-1))/2$. The *absolute position distance* (exact match distance in [22]) is the number of exact positional matches of activities [22]:

$$d_{abs}(\Pi, \Pi') = n - \sum_{k=1}^{n} y_i \quad \text{with } y_i = \begin{cases} 1 & \text{if } \pi_i = \pi'_i \\ 0 & \text{otherwise} \end{cases} \tag{6}$$

with $d_{abs,max} = n$. The *deviation distance* (position-based metric in [11]) requires the definition of the inverse permutation Σ of sequence Π, where the position of activity π_i is given by $\sigma_{\pi_i} = i$. The deviation distance is the amount of positional deviation and can be calculated as [11]

$$d_{dev}(\Pi, \Pi') = \sum_{j=1}^{n} |\sigma_j - \sigma'_j| \tag{7}$$

with $d_{dev,max} = n^2/2$ if n is even, or $d_{dev,max} = (n^2-1)/2$ if n is odd [22]. Given a set of solutions P with $p = |P|$, the *average distance between all solutions* is defined as [23]

$$d_{o,avg}(P) = \frac{\sum_{i=1}^{p} \sum_{j=i+1}^{p} d_o(\Pi_i, \Pi_j)}{p(p-1)/2} \tag{8}$$

with Π_i as the i-th solution of the set P.

In Section 4 we will present normalized distances $d_o = d_o(\Pi, \Pi')/d_{o,\max}$.

4   Computational Experiments

We use benchmark instances for the RCPSP according to [24] and [25]. The instances have been generated by a project generator (ProGen) using a full factorial design of the instance parameters *network complexity* (NC), *resource strength* (RS), and *resource factor* (RF) defining precedence relations, resource demands, and resource availabilities respectively. There are 480 instances with 30 and 60 activities and 600 instances with 120 activities. The instances and best known solutions are available from the project scheduling library PSPLIB[1].

A solution for the RCPSP is represented as an activity list which is assumed to be a precedence feasible permutation of the set of activities $\mathcal{J}$ (that is, the solution space consist of all precedence feasible permutations). In order to derive a schedule from the activity list a schedule generation scheme (SGS), either serial or parallel, is used as decoding procedure. In this work we use the serial SGS, which proceeds in N stages. At each stage the next unscheduled activity in the activity list, with every predecessor activity scheduled, is selected and

---

[1] Available at http://129.187.106.231/psplib/

scheduled at its earliest precedence and resource feasible start time. The algorithm terminates when all activities are scheduled. For a detailed description of the serial and parallel generation scheme see [9]. Note that in addition to this forward scheduling technique backward scheduling can be applied which might result in schedules that cannot be obtained by forward scheduling. In this paper we employ only the forward scheduling technique. The result of the serial SGS is an active schedule, i.e., no activity can be started earlier without delaying some other activity [10]. For a formal definition of active schedules see [26]. The set of active schedules will always contain an optimal solution, i.e., the serial SGS does not exclude optimal schedules a priori [10]. There is some redundancy in the search space since distinct activity lists may be related to the same schedule. Additionally different schedules might possess the same makespan. Consequently, multiple optimal solutions are to be expected.

In order to generate optimal solutions for the instance set J30 and near optimal solutions for the instance set J60[2] a DPSO (as proposed in [27]) was run at most 1,500 times and stopped as soon as 500[3] distinct optimal/near optimal solutions were found. We start with randomly generated initial populations. In order to limit a possible bias in the distribution of the solutions, due to the used DPSO, we employ the *lbest* topology (see [28]), which facilitates subpopulations in different regions of the search space. In order to examine the distribution of optimal/near optimal solutions we calculate the *average distance ratio*

$$r_{o,avg} = \frac{d_{o,avg}(O)}{d_{o,avg}(A)} \tag{9}$$

with O as the set of optimal/near optimal solutions and A as the set of all solutions (i.e., optimal/near optimal and random solutions). In case the set of optimal/near optimal solutions O contains only one solution[4] the average distance ratio was set to zero. A ratio close to 1.0 indicates that the optimal/near optimal solutions are not clustered at some point in the search space but rather are as widely spread as the random solutions. Figure 1 shows the histograms of $r_{pre,avg}$ for the instance sets J30 and J60. Furthermore we present some percentiles of the average distance ratio (precedence distance) in Table 1. The rational behind the choice of the precedence distance will become clear later. The results show a high average distance ratio (precedence distance) for the majority of the instances suggesting that the DPSO does not introduce a large bias[5].

---

[2] As near optimal solutions for the instance set J60 the best known heuristic solutions from PSPLIB (as of August 2008) were used. For 14 instances those solutions could not be generated.

[3] Those numbers were selected as a compromise between meaningful statistics and computational requirements after some preliminary tests.

[4] This is the case for one instance from the instance set J30 and for 11 instances from the instance set J60.

[5] Since precedence constraints reduce the search space a simple calculation of maximal distances between solutions does not promise meaningful results since the actual size of the search space in terms of the precedence distances is not clear.

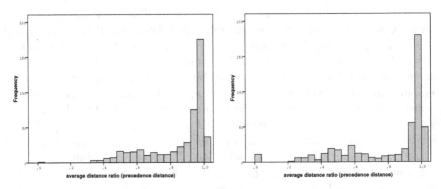

Fig. 1. Histogram of *average distance ratio* for the instance sets J30 (left, $N = 480$) and J60 (right, $N = 480$)

Table 1. Percentiles for *average distance ratio* (*precedence distance*)

instance set	P_{10}	P_{20}	P_{25}	P_{30}	P_{40}	P_{50}	P_{60}	P_{70}	P_{75}	P_{80}	P_{90}
J30	0.559	0.693	0.766	0.839	0.914	0.949	0.968	0.982	0.987	0.992	0.999
J60	0.439	0.562	0.605	0.706	0.917	0.951	0.980	0.990	0.993	0.996	1.000

Based on randomly generated solutions and their distances to the nearest optimal/near optimal solution the FDC coefficient ϱ (Eq. 1) for different distances was calculated. A *randomization test* [29] was used to determine the significance of the obtained FDC coefficients at a significance level of $\alpha = 0.05$ (as in [11]). Since most effective algorithms for the RCPSP use FBI it was applied to the random solutions and ϱ was calculated as well. For the instances with an RS value of 1.0 the application of the serial SGS to a precedence feasible permutation generated always schedules with optimal makespan. For an RS value of 1.0 the capacity of resources is defined as the amount of resources needed when the project is realized according to the earliest start time schedule. Consequently the optimal makespan equals the critical path lower bound obtained by computing the length of a critical path in the resource relaxation of the problem.

The histograms of the number of optimal solutions for the instance set J30 and near optimal solutions for the instance set J60 are shown in Figs. 2 and 3 without the instances with an RS value of 1.0. The vertical separation indicates the value 500. Table 2 shows the number of FDC coefficients that are larger than 0.15 and are significant at a significance level of $\alpha = 0.05$ for the distance measures reviewed in Sec. 3. For the instances with only optimal/near optimal solutions ϱ was set to 1.0. We will use the precedence distance for the remainder of this paper since the most relevant correlation coefficients were found using this distance measure.

In Figs. 4 and 5 the histograms of FDC coefficients (significant at $\alpha = 0.05$) for the instance sets J30 and J60 are depicted without the instances with an RS value of 1.0. The vertical separation marks the value of $\varrho = 0.15$. The results of the FDC analysis presented in Table 2 and Figs. 4 and 5 indicate that the fitness

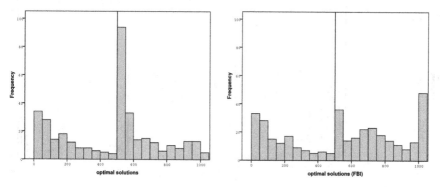

Fig. 2. Histogram of optimal solutions for instance set J30; random solutions (left, $N = 360$) and FBI (right, $N = 360$)

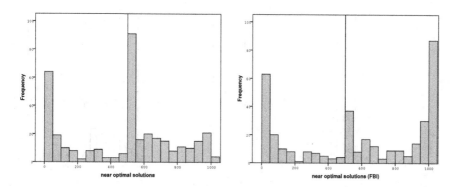

Fig. 3. Histogram of near optimal solutions for instance set J60; random solutions (left, $N = 360$) and FBI (right, $N = 360$)

Table 2. *Fitness distance correlation* coefficients for different distance functions. Shown are the number of instances for which correlations above 0.15 were found. The numbers in brackets show the number of instances for which a significant correlation at a significance level of $\alpha = 0.05$ was found.

	absolute dist.	adjacency dist.	precedence dist.	deviation dist.
J30 (random)	392 (444)	355 (425)	**413** (442)	405 (437)
J30 (FBI)	443 (464)	441 (465)	**453** (467)	443 (463)
J60 (random)	389 (454)	344 (445)	**422** (452)	416 (444)
J60 (FBI)	456 (463)	460 (467)	**467** (471)	462 (470)

landscape of the RCPSP is suitable for evolutionary methods according to [15] and explain the effectiveness of evolutionary methods as noted by Kolisch and Hartmann in [3]. Whereas the FDC coefficients indicate the presence of a big valley structure the average distance ratio contradicts the finding of a big valley structure. It seems that the decoding of precedence feasible permutations by

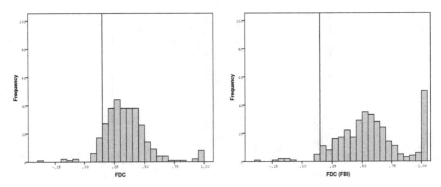

Fig. 4. Histogram of *fitness distance correlation* coefficients (significant at $\alpha = 0.05$) for instance set J30; random solutions (left, $N = 322$) and FBI (right, $N = 347$)

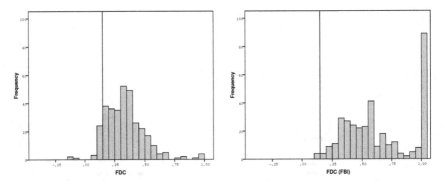

Fig. 5. Histogram of *fitness distance correlation* coefficients (significant at $\alpha = 0.05$) for instance set J60; random solutions (left, $N = 332$) and FBI (right, $N = 351$)

means of the serial SGS "creates" peaks (that is optimal/near optimal solutions) in different regions of the search space. Thus good solutions tend to be close to other good solutions but it seems that good solutions are not clustered in some region of the search space.

In order to examine the influence of instance characteristics on FDC coefficients the non-parametric *Kruskal-Wallis* one-way analysis of variance [30] was used for testing equality among groups. The results are presented in Table 3. In case of significant differences at a significance level of $\alpha = 0.05$ the results of *Mann-Whitney* tests were used to classify FDC coefficients into groups that are homogeneous in a sense that there is no significant difference among group members. The groups are indicated by small numbers. Note that we present the means of FDC coefficients rather than rank means for better readability. The results show that resource strength, that is the amount of available resources, has a significant effect on FDC. With more available resources (larger value for RS) the evolutionary search should become easier. Additionally FBI does support the evolutionary search.

Table 3. Statistical analysis of the influence of different instance characteristics on *fitness distance correlation*

	J30 (rand.)			J30 (FBI)			J60 (rand.)			J60 (FBI)		
	N	mean		N	mean		N	mean		N	mean	
NC												
1.50	147	0.542	-	152	0.716	-	147	0.553	-	156	0.746	-
1.80	151	0.507	-	151	0.698	-	149	0.501	-	158	0.733	-
2.10	144	0.525	-	150	0.699	-	156	0.489	-	157	0.707	-
RF												
0.25	104	0.540	-	107	0.764	-	107	0.476	1	115	0.722	-
0.50	109	0.506	-	115	0.678	-	116	0.537	3	119	0.761	-
0.75	115	0.539	-	116	0.689	-	113	0.513	2	119	0.737	-
1.00	114	0.515	-	115	0.692	-	116	0.527	2	118	0.694	-
RS												
0.20	110	0.309	1	110	0.497	1	118	0.297	1	119	0.430	1
0.50	111	0.367	1/2	109	0.584	2	113	0.367	2	114	0.621	2
0.70	101	0.367	2	114	0.709	3	101	0.356	2	118	0.858	3
1.00	120	1.000	3	120	1.000	4	120	1.000	3	120	1.000	4

In order to determine the position type distribution 1,000 random solutions were examined for each instance. As neighborhood structure the restricted shift move was used. Formally, let r_i denote the position of the ith activity in the current list, P_i the set of its immediate predecessors, S_i the set of its immediate successors, $L_i = \max\{r_j, \forall j \in P_i\}$, and $H_i = \min\{r_j, \forall j \in S_i\}$. The activity i can be shifted to positions between (and including) the position $L_i + 1$ and the position $H_i - 1$ [31]. In our experiments we found only solutions of the types LMIN, IPLAT, LEDGE, and LMAX. A high amount of position type LEDGE is favorable for local search procedures since improving moves are always possible from those positions (the same holds true for LMAX since we investigate a minimization problem). A high percentage of types LMIN and IPLAT impedes local search procedures since non-improving steps are necessary in order to continue the search process. For instances with an RS value of 1.0 we found only search space positions of the type IPLAT confirming the results obtained with the FDC analysis. In order to investigate the influence of the different instance characteristics on position type distribution we use the Kruskal-Wallis test for testing equality among groups. The results are presented in Tables 4, 5, and 6. The results indicate a rugged and challenging landscape for local search procedures due to a relatively large amount of position types LMIN and IPLAT. The landscapes of the instances from the instance set J120 appear not as rugged as those of the smaller instances. Note the different levels of RS for the instance set J120.

Table 4. Statistical analysis of the influence of different instance characteristics on *position type distribution* for instance set J30

	N	LEDGE		LEDGE+LMAX		LMIN		IPLAT	
NC									
1.50	160	0.573	-	0.644	-	0.093	-	0.260	-
1.80	160	0.563	-	0.632	-	0.103	-	0.263	-
2.10	160	0.581	-	0.655	-	0.086	-	0.265	-
RF									
0.25	120	0.354	1	0.521	1	0.195	1	0.284	2
0.50	120	0.593	2	0.650	2	0.093	2	0.257	1
0.75	120	0.665	3	0.698	3	0.046	3	0.256	1
1.00	120	0.677	3	0.705	3	0.041	3	0.254	1
RS									
0.20	120	0.909	4	0.944	4	0.053	2	0.003	1
0.50	120	0.762	3	0.859	3	0.127	3	0.015	2
0.70	120	0.618	2	0.772	2	0.195	4	0.033	3
1.00	120	0.000	1	0.000	1	0.000	1	1.000	4

Table 5. Statistical analysis of the influence of different instance characteristics on *position type distribution* for instance set J60

	N	LEDGE		LEDGE+LMAX		LMIN		IPLAT	
NC									
1.50	160	0.629	-	0.672	-	0.066	-	0.262	-
1.80	160	0.627	-	0.672	-	0.065	-	0.263	-
2.10	160	0.658	-	0.700	-	0.046	-	0.254	-
RF									
0.25	120	0.513	1	0.617	1	0.111	1	0.272	2
0.50	120	0.658	2	0.692	2	0.053	2	0.256	1
0.75	120	0.690	2/3	0.711	3	0.036	2/3	0.253	1
1.00	120	0.690	3	0.706	3	0.036	3	0.259	1
RS									
0.20	120	0.974	4	0.986	4	0.014	2	0.000	1
0.50	120	0.890	3	0.938	3	0.060	3	0.003	2
0.70	120	0.688	2	0.802	2	0.162	4	0.037	3
1.00	120	0.000	1	0.000	1	0.000	1	1.000	4

Figure 6 shows the correlation between the number of shift neighbors and solution quality for the instance sets J30, J60, and J120. Depicted are the cases

Table 6. Statistical analysis of the influence of different instance characteristics on *position type distribution* for instance set J120

	N	LEDGE		LEDGE+LMAX		LMIN		IPLAT	
NC									
1.50	200	0.973	-	0.984	-	0.016	-	0.001	-
1.80	200	0.972	-	0.982	-	0.016	-	0.002	-
2.10	200	0.981	-	0.989	-	0.011	-	0.000	-
RF									
0.25	150	0.923	1	0.953	1	0.043	1	0.004	2
0.50	150	0.991	2	0.994	2	0.006	2	0.000	1
0.75	150	0.993	3	0.994	3	0.005	3	0.001	1
1.00	150	0.995	3	0.997	3	0.003	3	0.000	1
RS									
0.10	120	0.998	5	0.999	5	0.001	1	0.000	1
0.20	120	0.994	4	0.996	4	0.004	2	0.000	1
0.30	120	0.984	3	0.990	3	0.010	3	0.000	1
0.40	120	0.971	2	0.983	2	0.017	4	0.000	2
0.50	120	0.932	1	0.956	1	0.039	5	0.005	2

where a significant correlation according to a randomization test at a significance level of $\alpha = 0.05$ was found. The instances with an RS value of 1.0 are again left out since a useful correlation could of course not be calculated with equal makespan for all solutions. The majority of the instances show a negative drift indicating a bias in local search towards regions of good solutions.

As side result of the PTD and drift analysis we obtained the average number of shift neighbors as presented in Table 7. Additionally we present the number of shift neighbors without precedence constraints $((n-1)^2)$ considering only non-dummy activities. The reduction of the size of the search space clearly accommodates local search procedures.

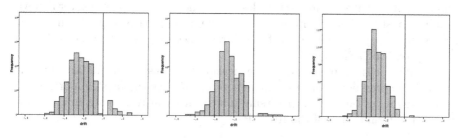

Fig. 6. Histogram of *drift* for instance sets J30 (left, $N = 336$), J60 (middle, $N = 328$), and J120 (right, $N = 589$)

Table 7. Average number of shift neighbors for different NC values

NC	J30	J60	J120
1,50	191.4	757.9	2,931.4
1,80	157.6	622.2	2,448.6
2,10	128.4	504.7	1,964.2
shift	841.0	3,481.0	14,161.0

5   Conclusions

In this paper we examined the fitness landscape of the RCPSP with the aim to investigate its suitability for evolutionary methods and local search procedures. The solution space consists of all precedence feasible permutations which are decoded by means of a serial SGS applied as forward scheduling technique.

The results from the fitness landscape analysis were obtained by creating random samples and additionally improving those solutions by the application of the forward-backward improvement heuristic. The results, firstly, indicate that evolutionary methods should be able to detect good solutions for the RCPSP and, secondly, are in line with the experimental evaluation of heuristic approaches for the RCPSP presented by Kolisch and Hartmann in [3]. The application of the FBI heuristic supports the evolutionary process and explain its success in [3].

The results of the analysis of the fitness landscape of the RCPSP regarding local search procedures is twofold. On the one hand a search space reduction due to precedence constraints could be observed. Furthermore a negative correlation between solution quality and numbers of shift neighbors for the majority of the instances could be ascertained. Both properties should support the performance of local search procedures. On the other hand the fitness landscape of the RCPSP is rugged featuring a large number of search space positions of types LMIN and IPLAT. Since non-improving steps at those positions have to be accepted sophisticated escape mechanisms as in tabu search have to be used; this leads to a large number of evaluated schedules. Consequently, local search procedures are not among the best performing approaches in [3] with an imposed termination criterium of 50,000 evaluated schedules. Multi-moves that perform several moves simultaneously in a single iteration might be an approach to guide the search process to more promising areas of the search space while avoiding the evaluation of not promising schedules.

Even if there is no superior distance measure (or measure for similarity) for the RCPSP the precedence distance might be regarded as the most important one (which might not be a big surprise since precedence constraints are an important part in the formulation of the RCPSP).

References

1. Brucker, P., Drexl, A., Möhring, R., Neumann, K., Pesch, E.: Resource-constrained project scheduling: Notation, classification, models, and methods. European Journal of Operational Research 112, 3–41 (1999)
2. Błażewicz, J., Lenstra, J.K., Rinnooy Kan, A.H.G.: Scheduling subject to resource constraints: Classification and complexity. Discrete Applied Mathematics 5, 11–24 (1983)
3. Kolisch, R., Hartmann, S.: Experimental investigation of heuristics for resource-constrained project scheduling: An update. European Journal of Operational Research 174, 23–37 (2006)
4. Ruiz, R., Maroto, C., Alcaraz, J.: Two new robust genetic algorithms for the flow-shop scheduling problem. Omega 34, 461–476 (2006)
5. Tormos, P., Lova, A.: A competitive heuristic solution technique for resource-constrained project scheduling. Annals of Operations Research 102, 65–81 (2001)
6. Valls, V., Ballestín, F., Quintanilla, S.: Justification and RCPSP: A technique that pays. European Journal of Operational Research 165, 375–386 (2005)
7. Merz, P., Freisleben, B.: Fitness landscape analysis and memetic algorithms for the quadratic assignment problem. IEEE Transactions on Evolutionary Computation 4, 337–352 (2000)
8. Valls, V., Ballestín, F.: A population-based approach to the resource-constrained project scheduling problem. Annals of Operations Research 131, 305–324 (2004)
9. Kolisch, R.: Serial and parallel resource-constrained project scheduling methods revisited: Theory and computation. European Journal of Operational Research 90, 320–333 (1996)
10. Hartmann, S., Kolisch, R.: Experimental evaluation of state-of-the-art heuristics for the resource-constrained project scheduling problem. European Journal of Operational Research 127, 394–407 (2000)
11. Reeves, C.R.: Landscapes, operators and heuristic search. Annals of Operations Research 86, 473–490 (1999)
12. Jones, T.: Evolutionary Algorithms, Fitness Landscapes and Search. PhD thesis, University of New Mexico, Albuquerque, New Mexico (1995)
13. Hoos, H.H., Stützle, T.: Stochastic Local Search: Foundations and Applications. Elsevier, Amsterdam (2005)
14. Bierwirth, C., Mattfeld, D.C., Watson, J.P.: Landscape regularity amd random walks for the job-shop scheduling problem. In: Gottlieb, J., Raidl, G.R. (eds.) EvoCOP 2004. LNCS, vol. 3004, pp. 21–30. Springer, Heidelberg (2004)
15. Jones, T., Forrest, S.: Fitness distance correlation as a measure of problem difficulty for genetic algorithms. In: Proceedings of the 6th International Conference on Genetic Algorithms, pp. 184–192 (1995)
16. Boese, K.D., Kahng, A.B., Muddu, S.: A new adaptive multi-start technique for combinatorial global optimizations. Operations Research Letters 16, 101–113 (1994)
17. Czogalla, J., Fink, A.: Fitness landscape analysis for the continuous flow-shop scheduling problem. In: Proceedings of the 3rd European Graduate Student Workshop on Evolutionary Computation, EvoPhD 2008, Naples, pp. 1–14 (2008)
18. Kubiak, M.: Distance measures and fitness-distance analysis for the capacitated vehicle routing problem. In: Doerner, K.F., Gendreau, M., Greistorfer, P., Gutjahr, W.J., Hartl, R.F., Reimann, M. (eds.) Metaheuristics: Progress in Complex Systems Optimization. Operations Research/Computer Science Interfaces Series, vol. 39, pp. 345–364. Springer, New York (2007)

19. Grahl, J., Radtke, A., Minner, S.: Fitness landscape analysis of dynamic multi-product lot-sizing problems with limited storage. In: Günther, H.O., Mattfeld, D.C., Suhl, L. (eds.) Managment logistischer Netzwerke. Entscheidungsunterstützung, Informationssysteme und OR-Tools, pp. 257–277. Physica-Verlag, Heidelberg (2007)

20. Martí, R., Laguna, M., Campos, V.: Scatter search vs. genetic algorithms. An experimental evaluation with permutation problems. In: Rego, C., Alidaee, B. (eds.) Metaheuristic Optimization Via Memory and Evolution. Tabu Search and Scatter Search. Operations Research/Computer Science Interfaces Series, pp. 263–282. Kluwer Academic Publishers, Boston (2005)

21. Schiavinotto, T., Stützle, T.: A review of metrics on permutations for search landscape analysis. Computers & Operations Research 34, 3143–3153 (2007)

22. Ronald, S.: More distance functions for order-based encodings. In: Proceedings of the 1998 IEEE International Conference on Evolutionary Computation, pp. 558–563 (1998)

23. Czogalla, J., Fink, A.: Design and analysis of evolutionary algorithms for the no-wait flow-shop scheduling problem. In: Geiger, M.J., Habenicht, W., Sevaux, M., Sörensen, K. (eds.) Metaheuristics in the Service Industry. Lecture Notes in Economics and Mathematical Systems. Springer, Berlin (to appear)

24. Kolisch, R., Sprecher, A., Drexl, A.: Characterization and generation of a general class of resource-constrained project scheduling problems. Management Science 41, 1693–1703 (1995)

25. Kolisch, R., Sprecher, A.: PSPLIB – A project scheduling problem library. European Journal of Operational Research 96, 205–216 (1996)

26. Sprecher, A., Kolisch, R., Drexl, A.: Semi-active, active, and non-delay schedules for the resource-constrained project scheduling problem. European Journal of Operational Research 80, 94–102 (1995)

27. Czogalla, J., Fink, A.: Particle swarm optimization for resource constrained project scheduling. Working Paper (2008)

28. Czogalla, J., Fink, A.: Particle swarm topologies for the resource constrained project scheduling problem. Accepted for publication in NICSO 2008 proceedings. Studies in Computational Intelligence. Springer, Heidelberg (2008) (to appear)

29. Edgington, E.S.: Randomization Tests, 2nd edn. STATISTICS: Textbooks and Monographs, vol. 77. Marcel Dekker, New York (1987)

30. Kruskal, W.H., Wallis, W.A.: Use of ranks on one-criterion variance analysis. Journal of the American Statistical Association 47, 583–621 (1952)

31. Boctor, F.F.: Resource-constrained project scheduling by simulated annealing. International Journal of Production Research 34, 2335–2351 (1996)

An ACO-Based Reactive Framework for Ant Colony Optimization: First Experiments on Constraint Satisfaction Problems

Madjid Khichane[1,2], Patrick Albert[1], and Christine Solnon[2,*]

[1] IBM France, Gentilly, France
{madjid.khichane,albertpa}@fr.ibm.com
[2] Université Lyon 1, LIRIS CNRS UMR5205, France
christine.solnon@liris.cnrs.fr

Abstract. We introduce two reactive frameworks for dynamically adapting some parameters of an Ant Colony Optimization (ACO) algorithm. Both reactive frameworks use ACO to adapt parameters: pheromone trails are associated with parameter values; these pheromone trails represent the learnt desirability of using parameter values and are used to dynamically set parameters in a probabilistic way. The two frameworks differ in the granularity of parameter learning. We experimentally evaluate these two frameworks on an ACO algorithm for solving constraint satisfaction problems.

1 Introduction

Ant Colony Optimization (ACO) has shown to be very effective to solve a wide range of combinatorial optimization problems [1]. However, when solving a problem (with ACO like with other metaheuristics), one usually has to find a compromise between two dual goals. On the one hand, one has to intensify the search around the most promising areas, that are usually close to the best solutions found so far. On the other hand, one has to diversify the search and favor exploration in order to discover new, and hopefully more successful, areas of the search space. The behavior of the algorithm with respect to this intensification/diversification duality (also called exploitation/exploration duality) can be influenced by modifying parameter values.

Setting parameters is a difficult problem which usually lets the user balance between two main tendencies. On the one hand, when choosing values which emphasize diversification, the quality of the final solution is often better, but the time needed to converge on this solution is also often higher. On the other hand, when choosing values which emphasize intensification, the algorithm often finds better solutions quicker, but it often converges on sub-optimal solutions. Hence, the best parameter values both depend on the instance to be solved and on the time allocated for solving the problem. Moreover, it may be better to

---

* This work has been partially financed by ILOG under the research collaboration contract ILOG/UCBL-LIRIS.

T. Stützle (Ed.): LION 3, LNCS 5851, pp. 119–133, 2009.

change parameter values during the solution process, depending on the search landscape around the current state, than to keep them fixed.

To improve the search process with respect to the intensification/diversification duality, Battiti et al [2] propose to exploit the past history of the search to automatically and dynamically adapt parameter values, thus giving rise to reactive approaches.

In this paper, we introduce a reactive framework for ACO to dynamically adapt some parameters during the search process. This dynamic adaptation is done with ACO: pheromone trails are associated with parameter values; these pheromone trails represent the learnt desirability of using parameter values and are used to dynamically set parameters in a probabilistic way. Our approach is experimentally evaluated on constraint satisfaction problems (CSPs) which basically involve finding an assignment of values to variables so that a given set of constraints is satisfied.

The paper is organized as follows. We first recall in section 2 some background on CSPs and ACO. We show in section 3 how to use ACO to dynamically adapt some parameters during the search process. In particular, we introduce two different reactive frameworks for ACO: a first framework where parameter values are fixed during the construction of a solution, and a second framework where parameters are tailored for each variable so that parameters are dynamically changed during the construction of a solution. We experimentally evaluate and compare these two reactive frameworks in section 4, and we conclude on some related work and further work in section 5.

2   Background

2.1   Constraint Satisfaction Problems (CSPs)

A *CSP* [3] is defined by a triple (X, D, C) such that X is a finite set of variables, D is a function that maps every variable $X_i \in X$ to its domain $D(X_i)$, that is, the finite set of values that can be assigned to X_i, and C is a set of constraints, that is, relations between some variables which restrict the set of values that can be assigned simultaneously to these variables.

An *assignment*, noted $\mathcal{A} = \{< X_1, v_1 >, \ldots, < X_k, v_k >\}$, is a set of variable/value couples such that all variables in $\mathcal{A}$ are different and every value belongs to the domain of its associated variable. This assignment corresponds to the simultaneous assignment of values $v_1, \ldots, v_k$ to variables $X_1, \ldots, X_k$, respectively. An assignment $\mathcal{A}$ is *partial* if some variables of X are not assigned in $\mathcal{A}$; it is *complete* if all variables are assigned.

The *cost* of an assignment $\mathcal{A}$, denoted by $cost(\mathcal{A})$, is defined by the number of constraints that are violated by $\mathcal{A}$. A *solution of a CSP* (X, D, C) is a complete assignment for all the variables in X, which satisfies all the constraints in C, that is, a complete assignment with zero cost.

Most real-life CSPs are over-constrained, so that no solution exists. Hence, the CSP framework has been generalized to maxCSPs [4]. In this case, the goal is no longer to find a consistent solution, but to find a complete assignment that maximizes the number of satisfied constraints. Hence, an *optimal solution of a maxCSP* is a complete assignment with minimal cost.

2.2 Ant Colony Optimization (ACO)

ACO is a metaheuristic [1] which has been successfully applied to a wide range of combinatorial optimization problems such as, *e.g.*, travelling salesman problems [5], quadratic assignment problems [6], or car sequencing problems [7]. The basic idea is to iteratively build solutions in a greedy randomized way. More precisely, at each cycle, each ant builds a solution, starting from an empty solution, by iteratively adding solution components until the solution is complete. At each iteration of this construction, the next solution component to be added is chosen with respect to a probability which depends on two factors:

- The pheromone factor reflects the past experience of the colony regarding the selection of this component. This pheromone factor is defined with respect to pheromone trails associated with solution components. These pheromone trails are reinforced when the corresponding solution components have been selected in good solutions; they are decreased by evaporation at the end of each cycle, thus allowing ants to progressively forget older experiments.
- The heuristic factor evaluates the interest of selecting this component with respect to the objective function.

These two factors are respectively weighted by two parameters α and β.
Besides α and β, an ACO algorithm is also parameterized by

- the number of ants, $nbAnts$, which determines the number of constructed solutions at each cycle;
- the evaporation rate, $\rho \in]0; 1[$, which is used at the end of each cycle to decrease all pheromone trails by multiplying them by $(1 - \rho)$;
- the lower and upper pheromone bounds, τ_{min} and τ_{max}, which are used to bound pheromone trails (when considering the MAX-MIN Ant System [6]).

The reactive framework proposed in this paper focuses on α and β which have a great influence on the solution construction process.

The weight of the pheromone factor, α, is a key parameter for balancing intensification and diversification. Indeed, the greater α, the stronger the search is intensified around solutions containing components with high pheromone trails, i.e., components that have been previously used to build good solutions. In particular, we have shown in [8] that the setting of α let us balance between two main tendencies. On the one hand, when limiting the influence of pheromone with a low pheromone factor weight, the quality of the final solution is better, but the time needed to converge on this value is also higher. On the other hand, when increasing the influence of pheromone with a higher pheromone factor weight, ants find better solutions during the first cycles, but after a few hundreds or so cycles, they are no longer able to find better solutions.

The weight of the heuristic factor, β, determines the greedyness of the search and its best setting also depends on the instance to be solved. Indeed, the relevancy of the heuristic factor usually varies from an instance to another. Moreover,

---

Algorithm 1. Ant Solver

Input: A CSP (X, D, C) and a set of parameters
$\{\alpha, \beta, \rho, \tau_{min}, \tau_{max}, nbAnts, maxCycles\}$
Output: A complete assignment for (X, D, C)
1 Initialize pheromone trails associated with (X, D, C) to τ_{max}
2 **repeat**
3   **foreach** k in $1..nbAnts$ **do**
4   Construct an assignment $\mathcal{A}_k$
5   Improve $\mathcal{A}_k$ by local search
6   Evaporate each pheromone trail by multiplying it by $(1 - \rho)$
7   Reinforce pheromone trails of $\mathcal{A}_{best} = arg\, min_{\mathcal{A}_k \in \{\mathcal{A}_1,...,\mathcal{A}_{nbAnts}\}}\, cost(\mathcal{A}_k)$
8 **until** $cost(\mathcal{A}_i) = 0$ *for some* $i \in \{1..nbAnts\}$ **or** *maxCycles reached* ;
9 **return** the constructed assignment with the minimum cost

---

for a given instance, the relevancy of the heuristic factor may vary during the solution construction process.

Note finally that not only the ratio between α and β matters, but also their absolute value. Let us consider for example the two following parameter settings: $p_1 = \{\alpha = 1, \beta = 2\}$ and $p_2 = \{\alpha = 2, \beta = 4\}$. In both settings, β is twice as high as α. However, p_2 emphasizes more strongly differences than p_1. Let us consider for example the case where ants have to choose between two components a and b which pheromone factors respectively are $\tau(a) = 1$ and $\tau(b) = 2$, and heuristic factors respectively are $\eta(a) = 2$ and $\eta(b) = 3$. When considering the p_1 setting, choice probabilities are

$$p(a) = \frac{1^1 \cdot 2^2}{1^1 \cdot 2^2 + 2^1 \cdot 3^2} = 0.18 \text{ and } p(b) = \frac{2^1 \cdot 3^2}{1^1 \cdot 2^2 + 2^1 \cdot 3^2} = 0.82$$

whereas when considering the p_2 setting, choice probabilities are

$$p(a) = \frac{1^2 \cdot 2^4}{1^2 \cdot 2^4 + 2^2 \cdot 3^4} = 0.05 \text{ and } p(b) = \frac{2^2 \cdot 3^4}{1^2 \cdot 2^4 + 2^2 \cdot 3^4} = 0.95$$

2.3   Solving Max-CSPs with ACO

The ACO algorithm considered in our comparative study is called Ant Solver (AS) and is described in algorithm 1. We briefly describe below the main features of this algorithm; more information can be found in [9,10].

Pheromone trails associated with a CSP (X, D, C) (line 1). We associate a pheromone trail with every variable/value couple $\langle X_i, v \rangle$ such that $X_i \in X$ and $v \in D(X_i)$. Intuitively, this pheromone trail represents the learned desirability of assigning value v to variable X_i. As proposed in [6], pheromone trails are bounded between τ_{min} and τ_{max}, and they are initialized at τ_{max}.

Construction of an assignment by an ant (line 4): At each cycle (lines 2-8), each ant constructs an assignment: starting from an empty assignment $\mathcal{A} = \emptyset$, it iteratively adds variable/value couples to $\mathcal{A}$ until $\mathcal{A}$ is complete. At each step, to select a variable/value couple, the ant first chooses a variable $X_j \in X$ that is not yet assigned in $\mathcal{A}$. This choice is performed with respect to the smallest-domain ordering heuristic, i.e., the ant selects a variable that has the smallest number of consistent values with respect to the partial assignment $\mathcal{A}$ under construction. Then, the ant chooses a value $v \in D(X_j)$ to be assigned to X_j with respect to the following probability:

$$p_{\mathcal{A}}(<X_j, v>) = \frac{[\tau(<X_j, v>)]^\alpha \cdot [\eta_{\mathcal{A}}(<X_j, v>)]^\beta}{\sum_{w \in D(X_j)} [\tau(<X_j, w>)]^\alpha \cdot [\eta_{\mathcal{A}}(<X_j, w>)]^\beta}$$

where

- $\tau(<X_j, v>)$ is the pheromone trail associated with $<X_j, v>$,
- $\eta_{\mathcal{A}}(<X_j, v>)$ is the heuristic factor and is inversely proportional to the number of new violated constraints when assigning value v to variable X_j, i.e., $\eta_{\mathcal{A}}(<X_j, v>) = 1/(1 + cost(\mathcal{A} \cup \{<X_j, v>\}) - cost(\mathcal{A}))$,
- α and β are the parameters that determine the relative weights of the factors.

Local improvement of assignments (line 5): Once a complete assignment has been constructed by an ant, it is improved by performing some local search, i.e., by iteratively changing some variable/value assignments. Different heuristics can be used to choose the variable to be repaired and the new value to be assigned to this variable. For all experiments reported below, we have used the min-conflict heuristics [11], i.e., we randomly select a variable involved in some violated constraint, and then we assign this variable with the value that minimizes the number of constraint violations. Such local improvements are iterated until reaching a locally optimal solution which cannot be improved by modifying one variable assignment.

Pheromone trails update (lines 6-7): Once every ant has constructed an assignment, and improved it by local search, the amount of pheromone laying on each variable/value couple is updated according to the ACO metaheuristic. First, all pheromone trails are uniformly decreased (line 6) in order to simulate some kind of evaporation that allows ants to progressively forget worse constructions. Then, pheromone is added on every variable/value couple belonging to the best assignment of the cycle, $\mathcal{A}_{best}$ (line 7) in order to further attract ants towards the corresponding area of the search space. The quantity of pheromone laid is inversely proportional to the number of constraint violations in $\mathcal{A}_{best}$, i.e., $1/cost(\mathcal{A}_{best})$.

3 Using ACO to Dynamically Adapt α and β

We now propose to use ACO to dynamically adapt the values of α and β. In particular, we propose and compare two different reactive frameworks. In the first framework, called AS($\mathcal{GPL}$) and described in 3.1, the setting of α and β is fixed during the construction of a solution and is adapted after each cycle, once every ant has constructed a solution. In the second framework, called AS($\mathcal{DPL}$) and described in 3.2, the setting of α and β is personalized for each variable so that it changes during the construction of a solution. These two frameworks are experimentally evaluated in 4.

3.1   Description of AS($\mathcal{GPL}$)

AS($\mathcal{GPL}$) (Ant Solver with Global Parameter Learning) basically follows algorithm 1 but integrates new features for dynamically adapting α and β. Hence, α and β are no longer given as input parameters of the algorithm, but their values are chosen with respect to the ACO metaheuristic at each cycle[1].

Parameters of AS($\mathcal{GPL}$). Besides the parameters of Ant Solver, i.e., the number of cycles *nbCycles*, the number of ants *nbAnts*, the evaporation rate ρ, and the lower and upper pheromone bounds τ_{min} and τ_{max}, AS($\mathcal{GPL}$) is parameterized by a set of new parameters that are used to set α and β, i.e.,

- two sets of values $\mathcal{I}_\alpha$ and $\mathcal{I}_\beta$ which respectively contain the set of values that may be considered for setting α and β;
- a lower and an upper pheromone bound, $\tau_{min_{\alpha\beta}}$ and $\tau_{max_{\alpha\beta}}$;
- an evaporation rate $\rho_{\alpha\beta}$.

Note that our reactive framework supposes that α and β take their values within two given discrete sets of values $\mathcal{I}_\alpha$ and $\mathcal{I}_\beta$ which must be known *a priori*. These two sets should contain good values, *i.e.*, those which allow Ant Solver to find the best results for every possible instance. As discussed in Section 4, we propose to choose the values of $\mathcal{I}_\alpha$ and $\mathcal{I}_\beta$ by running Ant Solver with different settings for α and β on a representative set of instances, and by keeping in $\mathcal{I}_\alpha$ and $\mathcal{I}_\beta$ the values that allowed Ant Solver to find the best results on these instances.

Pheromone structure. We associate a pheromone trail $\tau_\alpha(i)$ with every value $i \in \mathcal{I}_\alpha$ and a pheromone trail $\tau_\beta(j)$ with every value $j \in \mathcal{I}_\beta$. Intuitively, these pheromone trails represent the learnt desirability of setting α and β to i and j respectively. During the search process, these pheromone trails are bounded between the two bounds $\tau_{min_{\alpha\beta}}$ and $\tau_{max_{\alpha\beta}}$. At the beginning of the search process, they are initialized to $\tau_{max_{\alpha\beta}}$.

---

[1] We have experimentally compared two variants of this reactive framework: a first variant where the values are chosen at the beginning of each cycle (between lines 2 and 3) so that every ant uses the same values during the cycle, and a second variant where the values are chosen by ants before constructing an assignment (between lines 3 and 4). The two variants obtain results that are not significantly different. Hence, we only consider the first variant which is described in this section.

Choice of values for α and β. At each cycle (i.e., between lines 2 and 3 of algorithm 1), α (resp. β) is set by choosing a value $i \in \mathcal{I}_\alpha$ (resp. $i \in \mathcal{I}_\beta$) with respect to a probability $p_\alpha(i)$ (resp. $p_\beta(i)$) which is proportional to the amount of pheromone laying on i, i.e.,

$$p_\alpha(i) = \frac{\tau_\alpha(i)}{\sum_{j \in \mathcal{I}_\alpha} \tau_\alpha(j)} \quad (\text{resp. } p_\beta(i) = \frac{\tau_\beta(i)}{\sum_{j \in \mathcal{I}_\beta} \tau_\beta(j)})$$

Pheromone trails update. The pheromone trails associated with α and β are updated at each cycle, between lines 7 and 8 of algorithm 1. First, each pheromone trail $\tau_\alpha(i)$ (resp. $\tau_\beta(i)$) is evaporated by multiplying it by $(1 - \rho_{\alpha\beta})$. Then the pheromone trail associated with α (resp. β) is reinforced. The quantity of pheromone laid on $\tau_\alpha(\alpha)$ (resp. $\tau_\beta(\beta)$) is inversely proportional to the number of constraint violations in $\mathcal{A}_{best}$, the best assignment built during the cycle. Therefore, the values of α and β that have allowed ants to build better assignments will receive more pheromone.

3.2   Description of AS($\mathcal{DPL}$)

The reactive framework described in the previous section dynamically adapts α and β at every cycle, but it considers the same setting for all assignment constructions within a same cycle. We now describe another reactive framework called AS($\mathcal{DPL}$) (Ant Solver with Distributed Parameter Learning). The basic idea is to choose new values for α and β at each step of the construction of an assignment, i.e., each time an ant has to choose a value for a variable. The goal is to tailor the setting of α and β for each variable of the CSP.

Parameters of AS($\mathcal{DPL}$). The parameters of AS($\mathcal{DPL}$) are the same as the ones of AS($\mathcal{GPL}$).

Pheromone structure. We associate a pheromone trail $\tau_\alpha(X_k, i)$ with every variable $X_k \in X$ and every value $i \in \mathcal{I}_\alpha$ and a pheromone trail $\tau_\beta(X_k, j)$ with every variable $X_k \in X$ and every value $j \in \mathcal{I}_\beta$. Intuitively, these pheromone trails respectively represent the learnt desirability of setting α and β to i and j when choosing a value for variable X_k. During the search process, these pheromone trails are bounded between the two bounds $\tau_{min_{\alpha\beta}}$ and $\tau_{max_{\alpha\beta}}$. At the beginning of the search process, they are initialized to $\tau_{max_{\alpha\beta}}$.

Choice of values for α and β. At each step of the construction of an assignment, before choosing a value v for a variable X_k, α (resp. β) is set by choosing a value $i \in \mathcal{I}_\alpha$ (resp. $i \in \mathcal{I}_\beta$) with respect to a probability $p_\alpha(X_k, i)$ (resp. $p_\beta(X_k, i)$) which is proportional to the amount of pheromone laying on i for X_k, i.e.,

$$p_\alpha(X_k, i) = \frac{\tau_\alpha(X_k, i)}{\sum_{j \in \mathcal{I}_\alpha} \tau_\alpha(X_k, j)} \quad (\text{resp. } p_\beta(X_k, i) = \frac{\tau_\beta(X_k, i)}{\sum_{j \in \mathcal{I}_\beta} \tau_\beta(X_k, j)})$$

Pheromone trails update. The pheromone trails associated with α and β are updated at each cycle, between lines 7 and 8 of algorithm 1. First, each pheromone trail $\tau_\alpha(X_k, i)$ (resp. $\tau_\beta(X_k, i)$) is evaporated by multiplying it by $(1 - \rho_{\alpha\beta})$. Then, some pheromone is laid on the pheromone trails associated with the values of α and β that have been used to build the best assignment of the cycle ($\mathcal{A}_{best}$): for each variable $X_k \in X$, if α (resp. β) has been set to i for choosing the value to assign to X_k when constructing $\mathcal{A}_{best}$, then $\tau_\alpha(X_k, i)$ (resp. $\tau_\beta(X_k, i)$) is incremented by $1/cost(\mathcal{A}_{best})$.

4   Experimental Results

4.1   Test Suite

We illustrate our reactive framework on a benchmark of maxCSP which has been used for the CSP 2006 competition [12]. We have considered the 686 binary maxCSP instances defined in extension. Among these 686 instances, 641 are solved to optimality[2] both by the static version of Ant Solver and the two reactive versions, whereas CPU times are not significantly different. Hence, we concentrate our experimental study of section 4.3 on the 25 harder instances that are not always solved to optimality. Among these 25 hard instances, we have chosen 10 representative ones which features are described in Table 1. We shall give more experimental results, for all instances of the benchmark of the competition, in section 4.4, when comparing our reactive ACO framework with the best solvers of the competition.

4.2   Experimental Setup

We have tuned parameters for Ant Solver by running it on a representative subset of 100 instances (including the 25 hardest ones) among the 686 instances of the competition, with different parameter settings. We have selected the setting that allowed Ant Solver to find the best results on average, i.e., $\alpha = 2$, $\beta = 8$, $\rho = 0.01$, $\tau_{min} = 0.1$, $\tau_{max} = 10$, and $nbAnts = 15$. We have set the maximum number of cycles to 10000, but the number of cycles needed to converge to the best solution is often much smaller. In this section, AS(Static) refers to Ant Solver with this static parameter setting.

We also have tuned α and β for every instance separately (while keeping the other parameters to the same values). In this section, AS(Tuned) refers to Ant Solver with the best static parameter setting for the considered instance.

For the two reactive variants of Ant Solver (AS($\mathcal{GPL}$) and AS($\mathcal{DPL}$)), we have kept the same parameter setting for the "old" parameters, i.e., $\rho = 0.01$, $\tau_{min} = 0.1$, $\tau_{max} = 10$, and $nbAnts = 15$. For the new parameters, that have been introduced to dynamically adapt α and β, we have set $\mathcal{I}_\alpha$ and $\mathcal{I}_\beta$ to the

---

[2] For most of these instances, the optimal solution is known. However, for a few instances optimal solutions are not known. For these instances, we have considered the best known solution.

Table 1. For each instance, Name, X, D, C, and B respectively give the name, the number of variables, the size of the variable domains, the number of constraints, and the number of violated constraints in the best solution found during the 2006 competition

Nb	Name	X	D	C	B	Nb	Name	X	D	C	B
1	brock-400-1	401	2	20477	378	5	rand-2-40-16-250-350-30	40	16	250	1
2	brock-400-2	401	2	20414	378	6	rand-2-40-25-180-500-0	40	25	180	1
3	mann-a27	379	2	1080	252	7	rand-2-40-40-135-650-10	40	40	135	1
4	san-400-0.5-1	401	2	40300	392	8	rand-2-40-40-135-650-22	40	40	135	1

Table 2. Experimental comparison of best found solutions. Each line gives, for each variant of Ant Solver, the number of violated constraints in the best found solution (average on 50 runs and standard deviation). For AS(Tuned), we also give the values of α and β that have been considered.

	AS(Tuned)			AS(Static)	AS($\mathcal{GPL}$)	AS($\mathcal{DPL}$)
Nb	#const (sdv)	α	β	#const (sdv)	#const (sdv)	#const (sdv)
1	374.84 (0.7)	1	6	374.92 (0.39)	**374.** (1.01)	**374.** (1.01)
2	373.12 (0.26)	1	5	374.68 (1.09)	**371.32** (1.09)	371.48 (1.31)
3	253.88 (0.26)	1	6	254.62 (0.49)	**253.74** (0.44)	253.96 (0.28)
4	387.2 (0.11)	1	8	388.04 (1.77)	**387.** (0.)	**387.** (0.)
5	**1.** (0.)	2	8	**1.** (0.)	1.02 (0.14)	**1.** (0.)
6	1.02 (0.02)	2	6	1.02 (0.14)	1.04 (0.19)	**1.** (0.)
7	**1.** (0.)	1	6	1.12 (0.32)	1.66 (0.47)	1.48 (0.5)
8	**1.** (0.)	1	5	1.08 (0.27)	1.12 (0.32)	1.08 (0.27)

set of values that gave reasonably good results with Static Ant Solver, i.e., $\mathcal{I}_\alpha = \{0, 1, 2\}$ and $\mathcal{I}_\beta = \{0, 1, 2, 3, 4, 5, 6, 7, 8\}$. For the evaporation rate and the lower and upper pheromone bounds, we have used the same values as for static AS, i.e., $\rho_{\alpha\beta} = 0.01$, $\tau_{min_{\alpha\beta}} = 0.1$, $\tau_{max_{\alpha\beta}} = 10$.

4.3 Experimental Comparison of AS(Tuned), AS(Static), AS($\mathcal{GPL}$) and AS($\mathcal{DPL}$)

Table 2 gives the best setting for α and β that have been considered when running AS(Tuned). It shows us that this best setting is clearly different from one instance to another. We also noticed that, at the end of the search process of AS($\mathcal{GPL}$), pheromone trails used to set α and β have rather different values from one instance to another. This is more particularly true for β, thus showing that the relevancy of the heuristic factor depends on the considered instance.

Table 2 also compares the number of violated constraints in the best found solution after 10000 cycles, for the four variants of Ant Solver. As differences between the different variants are rather small on some instances, we have performed statistical significance tests. Table 3 gives the results of these statistical tests. It shows us that reactive variants are always at least as good as AS(Static), except for instances 5 and 7 which are better solved by AS(Static)

Table 3. Results of statistical significance tests: each line compares two variants X/Y and gives for every instance the result of the test for 50 runs, *i.e.*, = (resp. < and >) if X is not significantly different from Y (resp. worse and better than Y)

	1	2	3	4	5	6	7	8
AS($\mathcal{DPL}$)/AS($\mathcal{GPL}$)	=	=	=	=	>	>	=	>
AS($\mathcal{DPL}$)/AS(Static)	=	>	>	>	=	>	=	=
AS($\mathcal{DPL}$)/AS(Tuned)	=	>	=	=	>	<	<	
AS($\mathcal{GPL}$)/AS(Static)	=	>	>	>	<	=	<	=
AS($\mathcal{GPL}$)/AS(Tuned)	=	>	=	=	<	=	<	<
AS(Static)/AS(Tuned)	=	=	<	<	=	=	<	<

Table 4. Experimental comparison of the number of cycles (average and standard deviation on 50 runs) and the CPU time in seconds (average on 50 runs) spent to find the best solution

	AS(Tuned)		AS(Static)		AS($\mathcal{GPL}$)		AS($\mathcal{DPL}$)	
	cycles	time	cycles	time	cycles	time	cycles	time
Nb	avg (sdv)	avg	avg (sdv)	avg	avg (sdv)	avg	avg (sdv)	avg
1	44 (7)	4	30 (4)	2	2717 (516)	98	2501 (491)	93
2	2247 (477)	140	323 (194)	12	2668 (463)	96	3322 (415)	125
3	2193 (309)	542	1146 (335)	160	2714 (432)	399	2204 (295)	328
4	710 (213)	123	347 (174)	39	316 (38)	34	112 (14)	12
5	394 (32)	5	394 (32)	4	379 (44)	5	412 (37)	5
6	476 (19)	26	606 (23)	13	507 (49)	18	579 (23)	15
7	2436 (166)	160	1092 (152)	31	736 (126)	39	1557 (266)	52
8	1944 (120)	140	884 (65)	29	1302 (286)	66	1977 (252)	87

than AS($\mathcal{GPL}$). It also shows us that both reactive variants are able to reach the performances of AS(Tuned), and even outperform it, on many instances (all but 2 for AS($\mathcal{DPL}$) and all but 3 for AS($\mathcal{GPL}$)). Finally, it also shows us that AS($\mathcal{DPL}$) is not significantly different from AS($\mathcal{GPL}$) for 5 instances, and outperforms it on 3 instances.

Table 4 compares the number of cycles and the CPU time spent to find the best solution. We first note that the computational overhead due to the reactive framework is not significant so that the four Ant Solver variants spend comparable CPU times for performing one cycle on one given instance. We note also that the number of cycles needed to converge is different from one instance to another, but also from one variant of Ant Solver to another. In particular, AS(Static) often converges quicker than AS(Tuned).

In order to allow us to compare the four Ant Solver variants during the whole search process, and not only at the end of the 10000 cycles, Figure 1 plots the evolution of the percentage of runs that have found the optimal solution with respect to the number of cycles[3]. It shows that AS(Static) is able to solve

---

[3] As proof of optimality has not been done for all the considered instances, we consider the best known solution for the instances which optimal solution is not known.

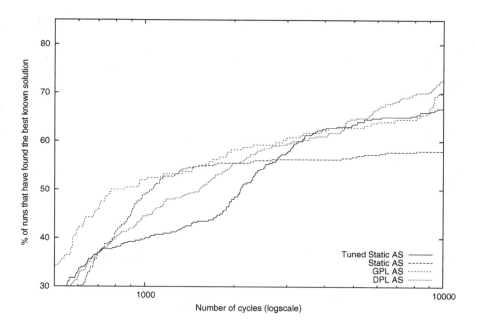

Fig. 1. Evolution of the percentage of runs that have found the optimal solution with respect to the number of cycles

to optimality more than half of the runs within the 2000 first cycles. However, after 2000 or so cycles, the percentage of runs solved to optimality by AS(Static) does not increase a lot. AS(Tuned) exhibits a rather different behavior: if it is able to solve to optimality less runs at the beginning of the search process, it has significantly outperformed AS(Static) at the end of the 10000 cycles. Let us consider for example instances 2, 3 and 8: Table 2 shows us that AS(Tuned) is able to find better solutions than AS(Static); however, Table 4 shows us that AS(Tuned) needs much more cycles than AS(Static) to find these solutions.

Figure 1 also shows us that AS($\mathcal{GPL}$) outperforms the three other variants during the 2000 first cycles, whereas after 2000 cycles AS($\mathcal{GPL}$), AS($\mathcal{DPL}$) and AS(Tuned) are rather close and all of them clearly outperform AS(Static). Finally, at the end of the search process AS($\mathcal{DPL}$) slightly outperforms AS($\mathcal{GPL}$) which itself slightly outperforms AS(Tuned).

Figure 2 plots the evolution of the percentage of runs that have found solutions that are close to optimality, *i.e.*, optimal solutions or solutions which violates one more constraint than the optimal solution. It shows that AS($\mathcal{GPL}$) more quickly finds nearly optimal solutions than AS($\mathcal{DPL}$), which itself is better than the static variants of Ant Solver. AS(Static) is better than AS(Tuned) at the beginning of the search process, but it is outperformed by AS(Tuned) after 1000 cycles or so.

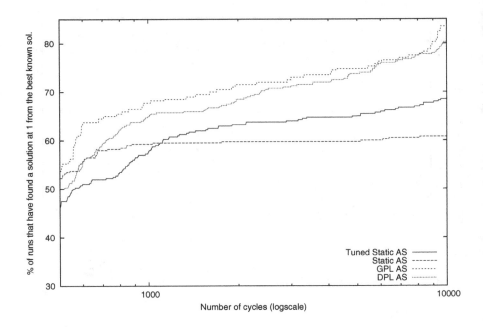

Fig. 2. Evolution of the percentage of runs that have found a solution that violates one more constraint than the optimal solution with respect to the number of cycles

4.4   Experimental Comparison of AS($\mathcal{DPL}$) with State-of-the-Art Solvers

We now compare AS($\mathcal{DPL}$) with the MaxCSP solvers of the 2006 competition. There was 9 solvers, among which 8 are based on complete branch and propagate approaches, and 1 is based on incomplete local search. For the competition, each solver has been given a time limit of 40 minutes on a 3GHz Intel Xeon (see [12] for more details). For each instance, we have compared AS($\mathcal{DPL}$) with the best result found during the competition (by any of the 9 solvers). We have also limited AS($\mathcal{DPL}$) to 40 minutes, but it has been run on a less powerful computer (a 1.7 GHz P4 Intel Dual Core). We do not report CPU times as they have been obtained on different computers. The goal here is to evaluate the quality of the solutions found by AS($\mathcal{DPL}$).

This comparison has been done on the 686 binary instances defined in extension. These instances have been grouped into 45 benchmarks. Among these 45 benchmarks, there was 31 benchmarks for which AS($\mathcal{DPL}$) and the best solver of the competition have found the same values for every instance of the benchmark. Hence, we only display results for the 14 benchmarks for which AS($\mathcal{DPL}$) and the best solver of the competition obtained different results (for at least one instance of the benchmark).

Table 5. Experimental comparison of AS($\mathcal{DPL}$) with the solvers of the 2006 competition. Each line gives the name of the benchmark, the number of instances in this benchmark (#I), the total number of constraints in these instances (#C), and the total number of violated constraints when considering, for each instance, its best known solution ($\sum$ best known). Then, we give the best results obtained during the competition: for each instance, we have considered the best result over the 9 solvers of the competition and we give the total number of constraints that are violated ($\sum cost$) followed by the number of instances for which the best known solution has been found (#I^{best}). Finally, we give the results obtained by AS($\mathcal{DPL}$): the total number of constraints that are violated ($\sum cost$) followed by the number of instances for which the best known solution has been found (#I^{best}).

				Competition		AS($\mathcal{DPL}$)	
Bench	#I	#C	$\sum$ best known	$\sum cost$	#I^{best}	$\sum cost$	#I^{best}
brock	4	56381	1111	1123	2	**1111**	**4**
hamming	4	14944	460	463	1	**460**	**4**
mann	2	1197	281	**281**	**2**	283	1
p-hat	3	312249	1472	1475	1	**1472**	**3**
san	3	48660	687	692	2	**687**	**3**
sanr	1	6232	182	183	0	**182**	**1**
dsjc	1	736	19	20	0	**19**	**1**
le	2	11428	2869	2925	1	**2869**	**2**
graphw	6	16993	416	420	4	**416**	**6**
scenw	27	29707	809	904	25	**809**	**27**
tightness0.5	15	2700	15	**15**	**15**	16	14
tightness0.65	15	2025	15	**15**	**15**	18	12
tightness0.8	15	1545	21	**22**	**13**	25	10
tightness0.9	15	1260	26	**30**	**11**	31	10

Table 5 gives results for these 14 benchmarks. It shows that AS($\mathcal{DPL}$) outperforms the best solvers of the competition for 9 benchmarks. More precisely, AS($\mathcal{DPL}$) has been able to improve the best solutions found by a solver of the competition for 19 instances. However, it has not found the best solution for 15 instances; among these 15 instances, 14 belong to the tightness* benchmarks which appear to be difficult ones for AS($\mathcal{DPL}$).

5 Conclusion

We have introduced two reactive frameworks for dynamically and automatically tuning the pheromone factor weight α and the heuristic factor weight β which have a strong influence on intensification/diversification of ACO searches. The goal is twofold: first, we aim at freeing the user from the unintuitive problem of tuning these parameters; second, we aim at improving the search process and reaching better performances on difficult instances.

First experimental results are very encouraging. Indeed, in most cases our reactive ACO reaches performances of a static variant, and even outperforms it on some instances.

Related work. There exists a lot of work on reactive approaches, that dynamically adapt parameters during the search process [2]. Many of these reactive approaches have been proposed for local search approaches, thus giving rise to reactive search. For example, Battiti and Protasi have proposed in [13] to use resampling information in order to dynamically adapt the length of the tabu list in a tabu search.

There also exists reactive approaches for ACO algorithms. In particular, Randall has proposed in [14] to dynamically adapt ACO parameters by using ACO and our approach borrows some features from this reactive ACO framework. However, parameters are learnt at a different level. Indeed, in [14] parameters are learnt at the ant level so that each ant evolves its own parameters and considers the same parameter setting during a solution construction. In our approach, paramaters are learnt at the colony level, so that every ant uses the same pheromone trails to set parameters. Moreover, we have compared two frameworks, a first one where the same parameters are used during a solution construction, and a second one where parameters are tailored for every variable, and we have shown that this second framework actually improves the search process on some instances, thus bringing to the show that, when solving constraint satisfaction problems, the relevancy of the heuristic and the pheromone factors depend on the variable to be assigned.

Further work. We plan to evaluate our reactive framework on other ACO algorithms in order to evaluate its genericity. In particular, it will be interesting to compare the two reactive frameworks on other problems: for some problems such as the Traveling Salesman Problem, it is most probable that tailoring parameters for every solution component is not interesting, whereas on other problems, such as the multidimensional knapsack or the car sequencing problems, we conjecture that this should improve the search process.

A limit of our reactive framework lies in the fact that the search space for the parameter values must be known in advance and discretized. As pointed out by a reviewer, it would be preferable to solve the meta-problem as what it is, *i.e.*, a continuous optimization problem. Hence, further work will address this issue.

Finally, we plan to integrate a reactive framework for dynamically adapting the other parameters, ρ, τ_{min}, and τ_{max} which have strong dependencies with α and β. This could be done, for example, by using intensification/diversification indicators, such as the similarity ratio or resampling information.

References

1. Dorigo, M., Stuetzle, T.: Ant Colony Optimization. MIT Press, Cambridge (2004)
2. Battiti, R., Brunato, M., Mascia, F.: Reactive Search and Intelligent Optimization. Operations research/Computer Science Interfaces. Springer, Heidelberg (2008)

3. Tsang, E.: Foundations of Constraint Satisfaction. Academic Press, London (1993)
4. Freuder, E., Wallace, R.: Partial constraint satisfaction. Artificial Intelligence 58, 21–70 (1992)
5. Dorigo, M., Gambardella, L.: Ant colony system: A cooperative learning approach to the traveling salesman problem. IEEE Transactions on Evolutionary Computation 1(1), 53–66 (1997)
6. Stützle, T., Hoos, H.: $\mathcal{MAX} - \mathcal{MIN}$ Ant System. Journal of Future Generation Computer Systems 16, 889–914 (2000)
7. Solnon, C.: Combining two pheromone structures for solving the car sequencing problem with Ant Colony Optimization. European Journal of Operational Research (EJOR) 191, 1043–1055 (2008)
8. Solnon, C., Fenet, S.: A study of aco capabilities for solving the maximum clique problem. Journal of Heuristics 12(3), 155–180 (2006)
9. Solnon, C.: Ants can solve constraint satisfaction problems. IEEE Transactions on Evolutionary Computation 6(4), 347–357 (2002)
10. Solnon, C., Bridge, D.: An ant colony optimization meta-heuristic for subset selection problems. In: System Engineering using Particle Swarm Optimization, Nova Science, pp. 7–29 (2006)
11. Minton, S., Johnston, M., Philips, A., Laird, P.: Minimizing conflicts: a heuristic repair method for constraint satistaction and scheduling problems. Artificial Intelligence 58, 161–205 (1992)
12. van Dongen, M., Lecoutre, C., Roussel, O.: Results of the second csp solver competition. In: Second International CSP Solver Competition, pp. 1–10 (2007)
13. Battiti, R., Protasi, M.: Reactive local search for the maximum clique problem. Algorithmica 29(4), 610–637 (2001)
14. Randall, M.: Near parameter free ant colony optimisation. In: Dorigo, M., Birattari, M., Blum, C., Gambardella, L.M., Mondada, F., Stützle, T. (eds.) ANTS 2004. LNCS, vol. 3172, pp. 374–381. Springer, Heidelberg (2004)

Selection of Heuristics for the Job-Shop Scheduling Problem Based on the Prediction of Gaps in Machines

Pedro Abreu[1], Carlos Soares[1,2], and Jorge M.S. Valente[1,2]

[1] LIAAD-INESC Porto LA, Porto, Portugal
[2] Faculdade de Economia, Universidade do Porto, Porto, Portugal
pedabreu@liaad.up.pt, {csoares,jvalente}@fep.up.pt

Abstract. We present a general methodology to model the behavior of heuristics for the Job-Shop Scheduling (JSS) that address the problem by solving conflicts between different operations on the same machine. Our models estimate the gaps between consecutive operations on a machine given measures that characteristics the JSS instance and those operations. These models can be used for a better understanding of the behavior of the heuristics as well as to estimate the performance of the methods. We tested it using two well know heuristics: Shortest Processing Time and Longest Processing Time, that were tested on a large number of random JSS instances. Our results show that it is possible to predict the value of the gaps between consecutive operations from on the job, on random instances. However, the prediction the relative performance of the two heuristics based on those estimates is not successful. Concerning the main goal of this work, we show that the models provide interesting information about the behavior of the heuristics.

1 Introduction

The complexity of optimization problems such as the Job-Shop Scheduling (JSS) makes it very difficult to understand the behavior of heuristic methods. For instance, little is known about the effect of different distributions of the duration of operations on the performance of any heuristic. This is true even for simple ones, such as the Shortest Processing Time [5]. An interesting research question is whether it is possible to create models that relate properties of JSS instances with the performance of different heuristics. The advantages of relating the properties of the instances to the performance of the algorithms are the possibility to:

1. develop automatic selection of optimization methods
2. obtain a better understanding of the conditions under which a certain method does not obtain good results
3. to support comparative studies

The first advantage represents the application of the model for practical purposes. The user is faced with a new instance of the JSS and must decide which

T. Stützle (Ed.): LION 3, LNCS 5851, pp. 134–147, 2009.

method to use. The characteristics of the instance are computed and the model is used to indicate which of the methods is expected to generate the best solution. The second enables researchers and practitioners to learn about the weak points of methods. The former can use this knowledge to improve the methods while the latter can use them as guidelines to help them choose which one to use. Finally, researchers usually compare the methods on benchmark instances. Perhaps, some new methods have excellent performance on many instances but it is discarded because poor performance on the set of benchmark instances used in the empirical comparison. Knowledge about the relation between the characteristics of the instances and the performance of the methods can be used to guide the selection of instances for the study.

In this paper, we address this problem using a Machine Learning approach to study the behaviour of two very simple and common heuristics to solve the JSS problem are considered here, namely the Shortest Processing Time (SPT) and the Longest Processing Time (LPT) methods [5].

In Section 2, we describe the background for this work and motivate it further. Our approach is described in Section 3 and the results obtained are presented in Section 4. We analyze the results and present our conclusions in Section 5.

2   The Job-Shop Scheduling Problem

The deterministic job-shop scheduling problem can be seen as the most general of the classical scheduling problems. Formally, this problem can be described as follows. A finite set J of n jobs $\{J_1, J_2, \ldots, J_n\}$ has to be processed on a finite set M of m machines $\{M_1, M_2, \ldots, M_m\}$. Each job J_i must be processed once on every machine M_j, so each job consists of a chain of m operations. Let O_{ij} represent the operation of job J_i on machine M_j, and let p_{ij} be the processing time required by operation O_{ij}.

The operations of each job J_i have to be scheduled in a predetermined given order, i.e. there are precedence constraints between the operations of each job J_i. Let $\prec$ be used to denote a precedence constraint, so that $O_{ik} \prec O_{il}$ means that job J_i has to be completely processed on machine M_k prior to being processed on machine M_l. Each job has its own flow pattern through the machines, so the precedence constraints between operations can be different for each job. Other additional constraints also have to be satisfied. Each machine can only process one job at a time (capacity constraints). Also, preemption is not allowed, so operations cannot be interrupted and must be fully processed once started. Let t_{ij} denote the starting time of operation O_{ij}. The objective is to determine starting times t_{ij} for all operations, in order to optimize some objective function, while satisfying the precedence, capacity and no-preemption constraints. The time when all operations of all jobs are complete is denoted as the makespan $C_{\max}$. In this paper, we consider as objective function the minimization of the makespan:

$$C^*_{\max} = \min (C_{\max})$$
$$= \min_{\text{feasibleschedules}} (\max (t_{ij} + p_{ij})),$$
$$\forall J_i \in J, M_j \in M.$$

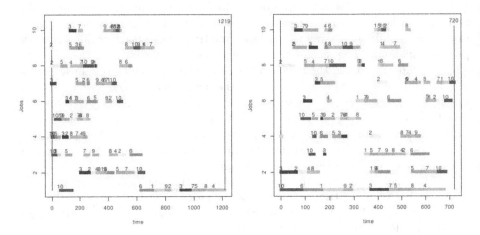

Fig. 1. Schedules generated for the same instance with LPT (left) and SPT (right)

The job-shop scheduling problem is NP-hard [3,7], and notoriously difficult to solve. Many papers have been published on the job-shop scheduling problem. A comprehensive survey of job shop scheduling techniques can be found in [5]. Given the complexity of the job-shop scheduling problem, the exact methods are limited to instances of small size. Metaheuristic algorithms have been successfully applied to instances of small to medium size. However, for large instances, dispatching rules are the only heuristic procedure that can provide a solution within reasonable computation times. Furthermore, dispatching rules are also often required for other heuristic procedures, e.g. metaheuristic algorithms frequently use dispatching rules to generate initial solutions.

The longest processing time (LPT) and shortest processing time (SPT) heuristics are two of the most well-known dispatching rules, and are widely used for the job-shop scheduling problem, as well as for a large number of other scheduling problems. In this paper, we consider these two rules, implemented with an active schedule generation algorithm [2,4]. This algorithm assigns operations to machines as soon as possible, taking into account the constraints described earlier. Following this strategy, *conflicts* will probably occur. This means that two or more operations will overlap on a given machine if they are scheduled as soon as possible. In that case, the algorithm uses a rule to choose the order in which the operations will be assigned to the machine. Many rules can be used for that purpose. In this work, we use the dispatching rules LPT and SPT for that purpose. The LPT rule schedules the operation with the longest processing time and SPT chooses the operation with the shortest processing time. As illustrated in Figure 1, the performance of these heuristics differs across different instances. SPT achieves the best result in this case as in 63.7% of the 1000 instances used in this work while LPT is the best in the 36.1%.

3 Learning the Behavior of Heuristics

Our goal is to use machine learning methods to induce models that relate the characteristics of JSS instances with the behavior of these dispatch rules. Given that the schedule generated by them is the result of a series of decisions, one for every conflict that occurs, the problem addressed here is the prediction of the gaps separating two consecutive operations of the same job generated by each heuristic.

Besides providing information about the behavior of the heuristics, these predictions can also be used to predict the makespan of the schedule generated by each heuristic, and thus, be able to predict which heuristic will generate the best schedule.. Our hypothesis is that by breaking this problem into its basic – and possibly simpler – sub-problems (i.e., predicting the gaps) rather than addressing it directly [1], better results can be obtained. Therefore, the makespan is the combination of the gaps between operations, as illustrated in Figure 1. Based on this assumption, our method is divided into two steps, which are discussed in the following sections:

1. Predict the gaps separating two consecutive operations of the same job, on the schedules generated by the heuristic;
2. Calculate the makespan, using the predicted gaps.

3.1 Prediction of the Individual Gaps

Our goal is to predict the gaps generated by the two heuristics considered, SPT and LPT, for each pair of consecutive operations, $O_{jm_{o-1}}$ and O_{jm_o}, defined as $Gap^{SPT}(O_{jm_o})$ and $Gap^{LPT}(O_{jm_o})$, respectively. The $Gap^{sched}_{jm_o} = t_{jm_o} - e_{jm_{o-1}}$ is the length of the period between two consecutive operations in job j in a given schedule $sched$, $e_{jm_o} = t_{jm_o} + p_{jm_o}$ is the end time of operation O_{jm}, m_o is the machine in precedence order $o \in \{2, \ldots, |M|\}$ and $Gap^{sched}_{jm_1} = t_{jm_1}$. The value of Gap is non-zero when the beginning of the second operation (O_{jm_o}) has been delayed by a conflict, which was resolved by the heuristic in favour of another operation. In Figure 1 we observe two schedules generated by SPT and LPT, respectively. It illustrates how the use of different heuristics can yield large differences in gaps.

We address this as a regression problem [6]. In regression, the goal is to obtain a model that relates a set of independent variables, or features, X_i with a target variable, Y, based on a set of examples for which both the values of $\{X_1, X_2, \ldots\}$ and Y are known. In our case, an example is a characterization of two consecutive operations of a job, and the target variable, Y, is the corresponding gap, $Gap^{sched}(O_{jm_o})$.

3.2 Features to Describe Conflicts

In order to obtain a reliable model, regression methods must be provided with features, $\{X_1, X_2, \ldots\}$ which are good predictors of the target variable, Y. In other words, to be able to make accurate predictions, features that provide information about the target variable are required. To predict the gap between consecutive operations of a job, we need to design features that represent properties of the JSS instance that affect its value. Interesting features characterize the operations (e.g., duration), the machines (e.g., total processing time required) and relations between operations, jobs and machines (e.g., total duration of operations of the same job as the current operation that precede it.

Some of the measures used here are based on the Infinite Capacity Schedule (ICS). The ICS is the schedule obtained by relaxing the capacity constraints (i.e., the constraints that specify that each machine can only process one job at a time). This schedule can be easily constructed by scheduling the operations as specified by the precedence constraints. Based on the ICS, we compute a measure of the distance between two operations ($O_{j_1 m}$ and $O_{j_2 m}$) that are processed in the same machine as follows: $dist(O_{j_1,m}, O_{j_2,m}) = \frac{|h_{j_1 m} - h_{j_2 m} + 1|}{\frac{p_{j_1 m} + p_{j_2 m}}{2} + 1}$, where $h_{jm} = t_{jm} + \frac{p_{jm}}{2}$ is the time unit when half of operation O_{jm} is processed. When the operations are consecutive, the value of the measure is 1 and when the two operations are centered on the same time unit, the value depends only on the duration of operation. This means that if two operations are near and have a long duration, the value of this measure is low, so there is a high possibility of having a conflict when trying to generate a feasible schedule.

Some additional notation is: $d(S)$ is the set containing the duration of operations in S, $d(S) = \{p_{ij} : \forall O_{ij} \in S\}$; ld_{ij} (hd_{ij}) is the subset of M_j containing operations with shorter (longer) duration than O_{ij}, $ld_{ij} = \{O_{rj} \in M_j : p_{rj} < p_{ij}\}$ ($hd_{ij} = \{O_{rj} \in M_j : p_{rj} > p_{ij}\}$); $dist_{jm_o}(S)$ is the ICS-based distance measure between O_{jm_o} and the other operations in set S, $dist_{jm_o}(S) = \{dist(O_{jm_o}, O_{j_1 m_o}) : \forall O_{j_1 m_o} \in S\}$; $dist_{ij}^{[s,r]}$ is the subset of M_j containing operations with distance of ICS to O_{ij} between s and r. Additionally, some of the measures are computed for several operations in a machine and aggregated using functions sum, average, minimum, maximum and variance, represented below as f, and aggregated using functions sum, average, variance, represented below as g, for simplicity. In the following list, we describe the groups of features using the format: name of the feature, schema of the short name used to refer to it, an informal explanation and one or more formulas describing its exact calculation. The schema for short names is based on the tags: $< aggregation function >$, $< distance >$ and $< position >$. For example, the short name for the group of features "Processing Time in machine" is $< aggregation function > DM$, which means that that the features are minDM, maxDM, avgDM, varDM and sumDM. The tags $< distance >$ and $< position >$ are used in "Duration with distance higher than r and less than s and Shorter/Longer duration" features group. For example, if we use the interval $[-3, -1.5]$ the tag $< distance >$ is "Near" and $< position >$ is "Neg" because the interval is between 1.5 and 3 ("Near") and negative ("Neg"). Therefore, the features used to characterize a given operation (O_{jm_o}) are:

- **Operation Duration (OPDuration).** The duration of the operation, o
- **Precedence Order (PO)** The pre-defined order of the operation in the job, o
- **Duration Rank in Machine (DRM).** The position of the operation in a ranking of the operations in the same machine in increasing order of duration, $|ld_{jm_o}| + 1$
- **Processing Time in Machine** ($< aggregation function > DM$). Aggregated duration of the operations to be processed on the same machine as O_{jm_o}, $f(d(M_{m_o}))$
- **Distance to Other Operations in Machine** ($< aggregation function > DistM$). Aggregated distance in the ICS between the operation and the other operations to be processed on the same machine, $f(dist_{jm_o}(M_{m_o} \setminus \{O_{jm_o}\}))$
- **Duration of Operations with Shorter/Longer Duration in Machine** ($< aggregation function > DLDM/DHDM$). Aggregated duration of operations to be processed on the same machine that have shorter/longer duration than operation O_{jm_o}, $f(d(ld_{jm_o}))/f(d(hd_{jm_o}))$
- **Distance to Other Operations in Machine with Shorter/Longer Duration** ($< aggregation function > DistLM/DistHM$) Aggregated distance in the ICS between the operation and the other operations with shorter/longer duration to be processed in the same machine, $f(dist_{jm_o}(ld_{jm_o}))/f(dist_{jm_o}(hd_{jm_o}))$
- **Duration with Distance between r and s and Shorter/Longer Duration** ($< aggregation function > < distance > < position > Lower / Higher Duration$)

  Aggregated duration of the operation with distance in the ICS to operation O_{jm_o} between $[r, s]$ and with duration shorter/longer than the operation O_{jm_o}, $g(S)$ where $S = dist_{jm_o}^{[r,s]} \cap ld_{jm_o}/hd_{jm_o}$. The values of $[r, s]$ are:$] - \infty, -3],] - 3, -1.5],] - 1.5, 0],]0, 1.5],]1.5, 3]$ and $]3, \infty[$

Some features were highly correlated with others, which may affect the performance of some regression models. Therefore, we have carried out a simple feature selection method. For sets of features with a correlation higher than 0.9 between themselves, we eliminated all but one. The features are: $MaxDLDM$, $MinDHDM$, $MaxDHDM$, $MeanDistHDM$, $MaxDistLDM$, $MinDistLDM$, $SumDM$, $SumDHDM$, $SumDistM$ and $SumVeryNearNegHigherDuration$. So, the total number of features used are 59.

3.3 Predicting the Makespan

The predictions of the gaps (Gap) for each schedule can be used to estimate the makespan of the schedule generated using the rule R for instance I as follows

$$MK_I^R = max_{j \in \{1,...,|J|\}} \left(\sum_{o=1}^{|M|} \left(p_{jm_o} + Gap^R(O_{jm_o}) \right) \right)$$

Note that these estimates can be used to select the best heuristic. SPT should be used if $MK_I^{SPT} < MK_I^{LPT}$, otherwise LPT should be used.

4   Experiments

We have tested two regression methods that are available in the R statistical package (www.r-project.org): regression trees, linear regression, support vector machines (SVM) and random forest. In order to analyse the results, we consider the empirical plots of the real against the predicted data, and also use the following analytical error measure - relative mean squared error ($RMSE$). This error is calculated as $RMSE = \frac{\sum_i (f_i - \hat{f}_i)^2}{\sum_i (f_i - \bar{f}_i)^2}$, where $\hat{f}_i$ is the prediction of the target feature for example i from the test dataset, f_i is the real value of the target feature of example i from the test dataset and the $\bar{f}_i$(baseline) is the average of the values of the target feature on the training dataset. We note that if the value of RMSE is greater than 1, it means that the learning algorithm did not learn a model capable of generalizing to new examples better than by using the mean value of the target on the training set.

We generated three types of datasets, namely Uniform, Gaussian and Beta datasets. These datasets differ in the method used to generate the JSS instances that are tested. In the Uniform and Gaussian datasets, the precedence order is generated using a uniformly random permutation, as described in Taillard [8]. Also, in the Uniform (Gaussian) dataset, the duration of each operation is generated using a uniform (Gaussian) distribution, such that these durations are not correlated with the machines or jobs, as described in [9]. The Beta dataset contains more diverse instances, with duration of operations either not correlated with machines or jobs, correlated with machines or correlated with jobs, and randomly generated parameters α and β. For the precedence order in the instances of the Beta dataset, instead of generating a new precedence order for each job, as in the Uniform and Gaussian datasets, a previously generated precedence order is sometimes repeated. The larger the number of repetitions, the closer the instance is to a flowshop instance, since in the flowshop problem the precedence constraints are identical for all jobs. Each instance has 10 jobs to be processed on 10 machines and the range of processing time values is between 1 and 99. Experiments were carried out separately for the data corresponding to each of the distributions. We have generated 1000 instances using each of the distributions, which corresponds to 100000 examples, each one representing one gap of consecutive operations. For each experiment the data corresponding to 500 instances were used for training and the remaining data were used for testing. Note that, given the large size of the test set, there is no need to employ re-sampling techniques, such as cross-validation, for estimating the error;

In the figures, we have on the x-axis the prediction based on the predictions obtained with the Random Forest model and on the y-axis, the real value.

4.1   Analysis of Errors

The results on the problem of predicting the gaps are generally positive, especially the ones obtained with the Random Forest algorithm (Table 1 and Figure 2). Better results are obtained in the prediction of the gaps generated

Table 1. Error results obtained with different models (Linear Regression and Random Forest) for each heuristic of the prediction of gaps (fourth column), end time jobs (fifth column) and makespan (sixth column) on different datsets (Uniform, Gaussian and Beta)

| | | | RMSE | | |
Dist.	Alg.	Heur.	Gap	End Time	Makespan
U	LR	SPT	0.58	0.45	1.33
U	LR	LPT	0.81	0.87	2.29
U	RF	SPT	0.52	0.42	1.31
U	RF	LPT	0.75	0.85	1.28
G	LR	SPT	0.63	0.55	1.50
G	LR	LPT	0.80	0.89	3.10
G	RF	SPT	0.56	0.53	1.55
G	RF	LPT	0.74	0.88	1.54
B	LR	SPT	0.63	0.27	0.22
B	LR	LPT	0.77	0.44	0.41
B	RF	SPT	0.55	0.27	0.23
B	RF	LPT	0.71	0.42	0.24

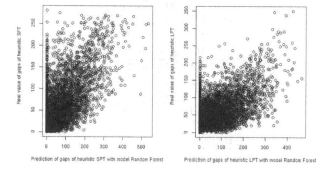

Fig. 2. True gap (y-axis) vs value predicted using Random Forest model (x-axis) for SPT(left) and LPT(right) of Beta dataset. Similar plots were obtained for the Uniform and Gaussian datasets.

with SPT than with LPT. This could either mean that the latter is a more difficult problem or that the features are more suitable for the former problem.

Some interesting observations can be made based on Figure 2 and further analysis of the results (not shown due to lack of space). First, there are a lot of examples (gaps) with a true value of 0. On the other hand, although the maximum value of the gaps is approximately 500, the predictions are generally below 300. We observe that the largest errors occur at the extremes of the range of real values. These observations indicate that the regression methods are not dealing with the distribution of target values appropriately.

As shown in Section 3.3, the predictions of the gaps can be used to predict the end time of jobs and the makespan of the schedule. In general, the errors

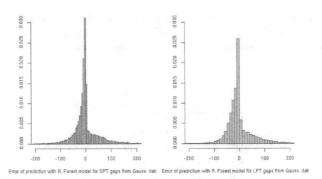

Fig. 3. Histograms of the errors obtained with the Random Forest model for SPT gaps (left) and LPT (right) on the Gaussian dataset

Table 2. Error of selecting the best heuristic based on the predictions of gaps obtained with different models (Linear Regression, Random Forest and Baseline), for all instances and for the instances for which there is a clear decision, i.e., the difference in the makespan predicted for the two heuristics is higher than 20% (column "Selected")

Dist.	Alg.	All	Selected
U	LR	0.38	0.25
U	RF	0.37	0.2
U	Bl	0.43	0.33
G	LR	0.41	0.29
G	RF	0.39	0.26
G	Bl	0.42	0.31
B	LR	0.44	0.38
B	RF	0.36	0.25
B	Bl	0.46	0.37

of predictions concerning LPT schedules are larger than the ones concerning SPT schedules. Additionally, we observe that the error of predicting the end time of jobs of the LPT schedules increases relative to the error of predicting the corresponding gaps. The opposite occurs for the SPT schedules, except for the instances generated with the beta distribution. Figure 3, which plots the distribution of the errors of predicting gaps for the two heuristics, provides a possible explanation for this. The distribution of errors on the SPT schedules is more symmetric than on the LPT schedules. So, when adding the gaps, the errors of the SPT heuristic may be cancelling themselves out because of the symmetry around 0. However, further analysis of results is necessary to confirm this. Comparing the errors in the predictions of the makespan and of the end time of jobs, there is an increase in the Uniform and Gaussian datasets, while on the Beta dataset they are both very similar (Table 1 and Figure 5).

These results indicate that it is possible to predict the gaps generated by these two heuristics, which is one of the main goals of this paper. However, we also

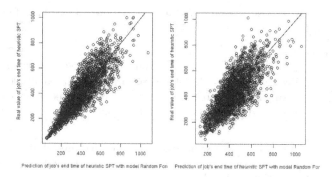

Fig. 4. True values of the end time of jobs (y-axis) vs the values predicted using Random Forest model (x-axis) for SPT (left) and LPT(right) of Beta dataset. Similar results were obtained on the Uniform and Gaussian datasets.

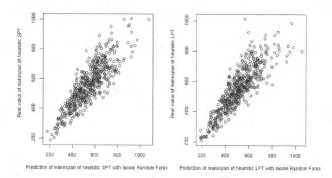

Fig. 5. True values of the makespan (y-axis) vs the value predicted using Random Forest model (x-axis) for SPT(left) and LPT(right) of Beta dataset. Similar results were obtained on the uniform and Gaussian datasets.

evaluated the accuracy of selecting the best heuristic based on the estimated makespans. The results in Table 2 show that this is also possible. However, the difference to the default accuracy (i.e., using a baseline method that always selects the heuristic that wins in the largest number of instances in the training set) is small. The table also shows that in instances with a larger difference of performance between the heuristics, the advantage of the prediction is more clear. These results indicate that we must improve the predictions of the gaps before being able to accurately predict the makespan and which method will obtain the best result. The Table 3 presents some information about the true (y) and the predicted values $(\hat{y})$. Based on these values, we observe that:

- more than half of the values are negative, when, in fact, gaps cannot be negative
- more than half of the true values are equal to zero while less than a quarter of the predicted values are near 0

Table 3. Descriptive statistics about the real value and predicted values of linear model for SPT and LPT

Heuristic Type		Min.	1Q	Median	3Q	Max
SPT	Real values	0.00	0.00	0.00	29.00	935.00
	Predicted values	-67.3800	0.4862	15.4000	56.7200	404.0000
LPT	Real values	0.00	0.00	0.00	53.00	870.00
	Predicted values	-85.88	12.45	33.88	63.77	302.40

Table 4. Values and statistical significance of the coefficients of the linear model on the Beta dataset. Each value represents the feature with name obtained by the corresponding: OPDuration, PO and DRM.

	Statistic			Coefficient		
Heuristic	OPDuration	PO	DRM	OPDuration	PO	DRM
SPT	*	*	***	0.154	-1.296	-3.189
LPT	***	**	***	-0.525	-1.800	-6.316

- the maximum value of the predictions is much lower than the maximum of true values

These results provide further evidence that, although we are able to predict the gaps, further improvements are necessary not only to the set of features used but also to the modeling processes.

4.2   Analysis of Models

The second main goal of this work is to obtain some information about the behavior of the heuristics by analyzing the models generated. For that purpose, we analyze the coefficients of the linear regression, also on the dataset containing instances of type Beta. Although linear regression does not generate the most accurate models in this problem, its models are easy to interpret. However, we cannot compare the coefficients directly since the scale is not the same for different features. On the other hand, we can test the significance of those coefficients. So, we test the hypothesis that a particular coefficient $\beta_j = 0$ using the t-test. Under the null hypothesis that $\beta_j = 0$, the t-value $= \frac{\hat{\beta}_j}{\hat{\sigma}}$ is distributed as t distribution with $N - p - 1$ degrees of freedom (t_{N-p-1}), where N is the number of sample and p the number of features. In Tables 4, 5, 6, 7, we present the coefficients and the level of significance, according to the t-test. The features with three stars have a probability (considering as true the null hypothesis) between 0 and 0.001, with two stars between 0.001 and 0.01, one star between 0.01 and 0.05 and those annotated with a dot, between 0.05 and 0.10. The tables show that:

- Particularly in the last set of features (Table 7), most of the features with significant coefficients for the SPT heuristic are relative to operations with less duration in the same machine, while for the LPT heuristic this is true

Table 5. Values and statistical significance of the coefficients of the linear model on the Beta dataset. Each value represents the feature with name obtained by the corresponding: $< aggregationfunction > DistLM/DistHM$ (Shorter/Longer) and $< aggregationfunction > DistM$ (Global).

Heuristic	Machine Part	Statistic					Coefficient				
		Min	Avg	Max	Sum	Var	Min	Avg	Max	Sum	Var
SPT	Global	***	***	-		***	0.202	22.713	-1.609	-	0.091
	Shorter	-		-	***		-	0.123	-	-2.213	0.009
	Longer	*	-	**	***	*	-0.535	-	0.746	-2.145	-0.075
LPT	Global	***	***	**	-		-1.851	17.57	-0.762	-	0.0003
	Shorter	-	***	-	**		-	0.073	-	-1.016	-0.0012
	Longer	***	-	***	**	***	2.498	-	-3.193	-1.074	0.217

Table 6. Values and statistical significance of the coefficients of the linear model on the Beta dataset. Each value represents the feature with name obtained by the corresponding: $< aggregationfunction > DLM/DHM$ (Shorter/Longer) and $< aggregationfunction > DM$ (Global).

Heuristic	Machine	Aggregation					Statistic				
		Min	Avg	Max	Sum	Var	Min	Avg	Max	Sum	Var
SPT	Global			.	-	***	-0.106	0.078	0.129	-	0.019
	Shorter	.	***	-	***		0.248	-0.596	-	0.568	0.006
	Longer	-	***	-	-	***	-	-0.300	-	-	0.021
LPT	Global	***	**	-		*	0.007	5.006	-0.256	-	0.0011
	Shorter			-	***		0.0004	-0.0186	-	-0.0425	0.0003
	Longer	-	***	-	-	.	-	-0.125	-	-	-0.001

for features describing operations with more duration in the same machine. This makes sense because, in both cases, these are the operations that can actually delay the execution of the current operation.

- Additionally, in many of the other features, the sign of the coefficient is different for the heuristics, which means that, although the feature is significant for both, the influence on the gap is in opposite direction. In other words, the same feature is associated with an increase of the value of the gap for one heuristic and a decrease for the other. This could also be expected due to the nature of the methods.

- A more detailed analysis of the coefficients provides information which, in some cases, is expected while in others, it is quite surprising. For instance, the largest coefficients are for the average distance in the ICS of the current operation to the other operations in the same machine (AvgDistM). These coefficients are positive in both cases (SPT and LPT) which means that the more isolated the operation is on the ICS, the larger is the gap. This is unintuitive, as it could be expected that the more isolated it is, the less is the probability that it will be in conflict.

Table 7. Values and statistical significance of the coefficients of the linear model on the Beta dataset. Each value represents the feature with name obtained by the corresponding: $< aggregation function >< distance >< position > Lower/Higher Duration$.

Heur	Mach	Position	Distance	Statistic			Coefficients		
				Avg	Sum	Var	Avg	Sum	Var
SPT	Lower	Pos	VeryNear	***	***	*	-0.286	0.090	0.006
			Near	***	***	**	-0.400	-0.208	-0.009
			Far	***	***	***	-0.280	-0.433	-0.014
		Neg	VeryNear	***	***	***	-0.366	0.295	0.014
			Near	***	***	.	-0.166	0.107	-0.006
			Far	***	***	***	0.273	-0.437	0.013
	Higher	Pos	VeryNear	**			0.053	-0.008	0.0001
			Near	**			-0.055	0.013	0.001
			Far	***	.	*	-0.098	0.020	0.006
		Neg	VeryNear	***	-		0.065	-	-0.006
			Near				-0.034	-0.001	0.001
			Far	***	**	*	-0.084	-0.033	-0.005
LPT	Lower	Pos	VeryNear	.			0.005	-0.0003	-0.0001
			Near	.			-0.006	-0.0008	-0.0002
			Far	***			-0.01	-0.001	0.000
		Neg	VeryNear	***			0.018	-0.003	0.004
			Near		**		0.005	-0.007	0.000
			Far	*	***		-0.007	-0.01	-0.001
	Higher	Pos	VeryNear	***		*	-0.025	0.0002	-0.001
			Near	***	***		-0.019	-0.016	0.000
			Far		***		-0.001	-0.03	-0.0001
		Neg	VeryNear	-	***		-0.001	-	0.003
			Near	***	***	*	-0.012	-0.009	-0.001
			Far	***	***	***	0.018	-0.037	0.001

5   Conclusion

In this work, we addressed the problem of learning models that are able to predict the behavior of different heuristics for the Job-Shop Scheduling problem. We propose a methodology that is novel and can be applied to any heuristic that solves conflicts individually. Our goal is to determine the properties of the instance that determine the performance of the heuristic. Our approach divides the problem into two sub-problems: 1) for every pair of consecutive operations, predict the size of the corresponding gap in the schedules generated by the heuristic and 2) predict the makespan of the schedule based on the predictions obtained by the models induced for sub-problem 1.

In our experiments, positive results were obtained in the two sub-problems. However, the results on the second problem are not entirely satisfactory. Our plan is to improve the set of features used in order to obtain more accurate predictions in the first sub-problem. We expect that better base-level predictions will enable better results on the prediction of the makespan of the schedules generated by the heuristics.

Also as future work, we plan to explore the generality of the method, by testing it with another common heuristic for the JSS problem, MWR, and on heuristic methods that generate non-delayed schedules. Finally, we will also extend this work to predict the behavior of meta-heuristics. In this case, these results are particularly interesting because, as these heuristics are more computationally complex, the problem of selecting beforehand which method to apply is quite relevant.

Acknowledgments. This work was partially supported by project *Hybrid optimization and simulation methods for the Job-Shop problem and its sub-problems*, funded by Fundação para a Ciência e Tecnologia (POCI/EGE/61823/2004).

References

1. Abreu, P., Soares, C.: Mapping charateristics of instances to evolutionary algorithm operators: an empirical study on the basic job-shop scheduling problem. In: Salido, M.A., Fdez-Olivares, J. (eds.) Proceedings of the CAEPIA 2007 Workshop on Planning, Scheduling and Constraint Satisfaction, November 2007, pp. 80–92 (2007)
2. Baker, K.R.: Introduction to Sequencing and Scheduling. Wiley, New York (1974)
3. Garey, M.R., Johnson, D.S.: Computers and Intractability: A Guide to the Theory of NP-Completeness. W. H. Freeman, San Francisco (1979)
4. Giffler, B., Thompson, G.L.: Algorithms for solving production scheduling problems. Operations Research 8(4), 487–503 (1960)
5. Jain, A.S., Meeran, S.: Deterministic job-shop scheduling: Past, present and future. European Journal of Operational Research 113, 390–434 (1999)
6. Kononenko, I., Kukar, M.: Introduction to Machine Learning and Data Mining: Introduction to Principles and Algorithms. Horwood Publishing (2007)
7. Lenstra, J.K., Rinnooy Kan, A.H.G.: Computational complexity of discrete pptimisation problems. Annals of Discrete Mathematics 4, 121–140 (1979)
8. Taillard, E.: Benchmarks for basic scheduling problems. European Journal of Operational Research 64, 278–285 (1993)
9. Watson, J.-P., Barbulescu, L., Howe, A.E., Whitley, D.: Algorithm performance and problem structure for flow-shop scheduling. In: AAAI/IAAI, pp. 688–695 (1999)

Position-Guided Tabu Search Algorithm for the Graph Coloring Problem

Daniel Cosmin Porumbel[1,2], Jin-Kao Hao[1], and Pascale Kuntz[2]

[1] LERIA, Université d'Angers, Angers, France
{porumbel,hao}@info.univ-angers.fr
[2] LINA, Polytech'Nantes, Nantes, France
pascale.kuntz@univ-nantes.fr

Abstract. A very undesirable behavior of any heuristic algorithm is to be stuck in some specific parts of the search space, in particular in the basins of attraction of the local optima. While there are many well-studied methods to help the search process escape a basin of attraction, it seems more difficult to prevent it from looping between a limited number of basins of attraction. We introduce a Position Guided Tabu Search (PGTS) heuristic that, besides avoiding local optima, also avoids re-visiting candidate solutions in previously visited regions. A learning process, based on a metric of the search space, guides the Tabu Search toward yet unexplored regions. The results of PGTS for the graph coloring problem are competitive. It significantly improves the results of the basic Tabu Search for almost all tested difficult instances from the DIMACS Challenge Benchmark and it matches most of the best results from the literature.

1 Introduction

It is well known that the performance of all heuristic algorithms is heavily influenced by the search space structure. Consequently, the design of an efficient algorithm needs to exploit, implicitly or explicitly, some features of the search space. For many heuristics, especially local searches, the difficulty is strongly influenced by the asperity of the local structures of local optima (e.g. isolated local optima, plateau structures, valley structures, etc.). A paradigmatic example of a difficult structure is the trap [1], i.e., a group of close local minima confined in a deep "well". If trapped into such a structure, even a local search with local optimum escape mechanisms can become locked looping between the local minima inside the well. Several global statistical indicators (i.e., convexity, ruggedness, smoothness, fitness distance correlation) have also been proposed to predict the performance of both local and evolutionary algorithms; we refer to [2, 3] for a summary of such measures and related issues.

Other research studies focus on the structural similarities between local optima (i.e., the "backbone" structures) or on their global arrangement (see [4] for a detailed summary of the research threads in search space analysis). Indeed, the different local optimum characteristics of the search space (the number of local

T. Stützle (Ed.): LION 3, LNCS 5851, pp. 148–162, 2009.

optima, their space distribution, the topology of their basins of attraction, etc.) may be very different from one problem to another and even from one instance to another. These specific properties have been investigated for several classical problems, such as: boolean satisfiability [5,6], the 0–1 knapsack problem [7], graph coloring [8, 9, 10], graph bi-partitioning [11], the quadratic assignment problem [12], job shop or flow shop scheduling [4,13], and arc crossing minimization in graph drawing [14]. All these studies conclude that the local optimum analysis has a great potential to give a positive impact on the performance.

However, the operational integration of specific search space information in a search process remains a difficult problem. To achieve this, a heuristic needs to learn how to make better local decisions using global information available at coarser granularity levels. Moreover, the search process has usually no information on the search space before actually starting the exploration. To overcome such difficulties, the integration of a learning phase in the optimization process ("learning while optimizing") seems very promising. This approach, using ideas of reactive search [15], aims at developing an algorithm capable of performing a self-oriented exploration.

In this paper, we focus on the graph coloring problem and we present such a reactive algorithm with two central processes: (i) a classical local search based on Tabu Search (TS) [16], (ii) a learning process that investigates the best configurations visited by the first process. This learning process has the role of effectively guiding TS toward yet unexplored regions. It integrates a positional orientation system based on a metric of the search space; for this, we use a distance function that indicates how many steps TS needs to perform to go from one configuration to another.

More precisely, the Position Guided Tabu Search (PGTS) algorithm employs an extended tabu list length whenever it detects that it is exploring the proximity of a previously visited configuration, i.e., so as to avoid re-exploring the region. This strategy does not strictly prevent the algorithm from revisiting such regions, but the probability of avoiding them is strongly increased by a reinforced diversification phase associated with the extended tabu list. Here, we propose for the graph coloring problem a strategy based on a tractable distance computation, with time complexity $O(|V|)$, where V is the vertex set of the graph. We show that PGTS significantly improves the performances of the basic TS algorithm on a well-known set of DIMACS instances, and that it competes well with the best algorithms from literature.

The rest of the paper is organized as follows. Section 2 briefly outlines the graph coloring problem and its traditional TS algorithm (Tabucol). The Position Guided Tabu Search for graph coloring and the distance definition are presented in section 3. Section 4 is devoted to experimental results. Section 5 presents related work and provides elements for the generalization of PGTS to other combinatorial optimization problems, followed by some conclusions in the last section.

2 The Graph Coloring Problem and Its Classical Tabu Search Algorithm

We briefly recall the basic notions and definitions related to the graph coloring problem and to the tabu search algorithm adapted to this problem.

2.1 Definitions

Let $G = (V, E)$ be a graph with V and E being respectively the vertex and edge set. Let k be a positive integer.

Definition 1. *(Graph coloring and k-coloring) The graph G is k-colorable if and only if there exists a conflict-free vertex coloring using k colors, i.e., a function $c : V \rightarrow \{1, 2, \cdots, k\}$ such that $\forall \{i,j\} \in E, c(i) \neq c(j)$. The graph coloring problem (COL) is to determine the smallest k (the chromatic number denoted by χ_G) such that G is k-colorable.*

Definition 2. *(Color (array) based representation) We denote any function $c : V \rightarrow \{1, 2, \cdots, k\}$ by $C = (c(1), c(2), \cdots, c(|V|))$. We say that C is a candidate solution (or configuration) for the k-coloring problem (G, k).*

Moreover, C is said to be a proper (conflict-free) or legal coloring if and only if $c(i) \neq c(j)$, $\forall \{i, j\} \in E$. Otherwise, C is an improper (conflicting) coloring. A legal coloring is also referred to as a *solution* of the k-coloring problem (G, k).

Definition 3. *(Partition representation) A k-coloring $C = (c(1), c(2), \cdots, c(|V|))$ is a partition $\{C^1, C^2, \ldots, C^k\}$ of V (i.e., a set of k disjoint subsets of V covering V) such that $\forall x \in V, x \in C^i \Leftrightarrow c(x) = i$.*

We say that C^i is the class color i induced by the coloring C, i.e., the set of vertices having color i in C. This partition based definition is particularly effective to avoid symmetry issues arising from the color based encoding. We will see (Section 3.2) that the distance between two colorings can be calculated as a set theoretic partition distance.

Definition 4. *(Conflict number and objective function) Given a configuration C, we call conflict (or conflicting edge) any edge having both ends of the same color in C. The set of conflicts is denoted by $CE(C)$ and the number of conflicts (i.e., $|CE(C)|$—also referred to as the conflict number of C) is the objective function $f_c(C)$. A conflicting vertex is a vertex $v \in V$, for which there exists an edge $\{v, u\}$ in $CE(C)$.*

In this paper, we deal with the k-coloring problem (k-COL), i.e., given a graph G and an integer k, the goal is to determine a legal k-coloring. From an optimization perspective, the objective is to find a k-coloring minimizing the conflict number $f_c(C)$.

Algorithm 1. Basic Tabu Search Algorithm for Graph Coloring

Input: G, k
Return value: $f_c(C_{best})$ (i.e., 0 if a legal coloring is found)
C: the current coloring; C_{best}: the best coloring ever found
Begin
1. Set C a random initial configuration
2. **While** a *stopping condition* is not met
    (a) Find the best non-tabu $C' \in N(C)$ (a neighbor C' is tabu
       if and only if the pair (i, i'), corresponding to the move
       $C \xrightarrow{C(i):=i'} C'$, is marked tabu)
    (b) Set $C = C'$ (i.e., perform move $C(i) := i'$)
    (c) Mark the pair (i, i') tabu for T_ℓ iterations
    (d) **If** $(f_c(C) < f_c(C_{best}))$
       $-$ $C_{best} = C$
End

2.2 Tabu Search for Graph Coloring

Following the general ideas of TS [16], Tabucol [17] is a classical algorithm for k-COL that moves from one configuration to another by modifying the color of a conflicting vertex. The main adaptation of the Tabu Search meta-heuristic to graph coloring consists in the fact that it does not mark as tabu a whole configuration, but only a color assignment. To check whether a specific neighbor is tabu or not, it is enough to test the tabu status of the color assignment that would generate the neighbor. The general skeleton of our Tabucol implementation is presented in Algorithm 1; the stopping condition is to find a legal coloring or to reach a maximum number of iterations (or a time limit). The most important details that need to be filled are the neighborhood relation N and the tabu list management.

Neighborhood N. Given a coloring problem $(G(V, E), k)$, the search space Ω consists of all possible colorings of G; thus $|\Omega| = |V|^k$. A simple neighborhood function $N : \Omega \to 2^\Omega - \{\emptyset\}$ can be defined as follows. For any configuration $C \in \Omega$, a neighbor C' is obtained by changing the color of a single *conflicting* vertex in C.

Tabu List Management. There are several versions of this basic algorithm in the literature, but their essential differences lie in the way they set the tabu tenure T_ℓ. In our case, it is dynamically adjusted by a function depending on the objective function (i.e., the conflict number $f_c(C) = |CE(C)|$—as in [18,19,20]), but also on the number m of the last consecutive moves that did not modify the objective function. More precisely, $T_\ell = \alpha * f_c(C) + random(A) + \lfloor \frac{m}{m_{max}} \rfloor$, where α is a parameter taking values from $[0, 1]$ and $random(A)$ indicates a function returning a random value in $\{1, 2, \ldots, A\}$. In our tests, as previously published in [20], we use $\alpha = 0.6$ and $A = 10$.

The last term constitutes a reactive component only introduced to change T_ℓ when the conflict number does not change for $m_{\max}$ moves; typically, this situation appears when the search process is completely blocked cycling on a plateau. Each series of consecutive $m_{\max}$ (usually $m_{\max} = 1000$) moves leaving the conflict number unchanged increments all subsequent values of T_ℓ—but only until the conflict number changes again; such a change resets m to 0.

3 Position Guided Tabu Search Algorithm

3.1 Generic Description

The main objective of the algorithm is to discourage the search process from visiting configurations in some space regions that are already considered as explored. Taking as a basis the classical Tabu Search for graph coloring (see Algorithm 1), the new algorithm PGTS (Position Guided Tabu Search) integrates a learning component (Step 4.(c) in Algorithm 2) that processes all visited configurations and records a series of search space regions $S(C)$ that cover the whole exploration path. Ideally, these recorded regions contain all colorings that are structurally related to the visited ones.

A statistical analysis of the search space, briefly described in Section 3.3, has led us to define $S(C)$ as the closed sphere centered at C of radius R:

Definition 5. *(Sphere) Given a distance function* $d : \Omega \times \Omega \longrightarrow I\!N$, *a configuration* $C \in \Omega$ *and a radius* $R \in I\!N$, *the* R-sphere $S(C)$ *centered at* C *is the set of configurations* $C' \in \Omega$ *such that* $d(C, C') \leq R$.

Here, the distance $d(C, C')$ (see also Section 3.2) can be interpreted as the shortest path of TS steps between C and C'. More formally, $d(C, C')$ is the minimal number n for which there exist $C_0, C_1, \ldots C_n \in \Omega$ such that: $C_0 = C, C_n = C'$ and $C_{i+1} \in N(C_i)$ for all $i \in [0 \ldots n - 1]$.

PGTS starts iterating as the basic TS does, but, with the learning component (Step 4.(c), Algorithm 2), it also records the center of the currently explored sphere. While the current configuration C stays in the sphere of the last recorded center C_p, we consider the search process "pivots" around point C_p. PGTS performs exactly the same computations as TS except checking the distance $d(C, C_p)$ that is performed each iteration (in Step 4.(c)).

As soon as the search leaves the current sphere, the learning component activates a global positioning orientation system. It first compares C to the list of all previously recorded configurations (procedure Already–Visited in Algorithm 2) to check whether it has already visited its proximity or not. If C is not in the sphere of a previously recorded configuration, it goes on only by changing the pivot; i.e., it replaces C_p with C and records it. Otherwise, this means the search is re-entering the sphere of a previously recorded configuration and that should be avoided. This is a situation that triggers a signal to make more substantial configuration changes: a diversification phase is needed. For this purpose, the chosen mechanism is to extend the classical tabu tenure T_ℓ with a T_c factor.

Algorithm 2. Position Guided Tabu Search

```
PROCEDURE ALREADY-VISITED
Input: current configuration C
Return value: TRUE or FALSE
1. Forall recorded configurations Crec:
     - If d(C, Crec) ≤ R
       • Return TRUE
2. Return FALSE
ALGORITHM POSITION-GUIDED TABU SEARCH
Input: the search space Ω
Return value: the best configuration Cbest ever visited
C: the current configuration
1. Choose randomly an initial configuration C ∈ Ω
2. Cp = C (the pivot, i.e., the last recorded configuration)
3. Tc = 0  (the value by which PGTS extends the tabu tenure Tℓ)
4. While a stopping condition is not met
   (a) Choose the best non-tabu neighbor C' in N(C)
   (b) C = C'
   (c) If d(C, Cp) > R (the Learning Component)
        - Cp = C
        - If ALREADY-VISITED(Cp)
          • Then Increment Tc
        - Else
          • Tc = 0
          • Record Cp
   (d) Mark C as tabu for Tℓ + Tc iterations
   (e) If (fc(C) < fc(Cbest))
        - Cbest = C
   (f) If (fc(C) < fc(Cp))
        - Replace Cp with C in the archive
        - Cp = C (i.e., "recentering" the current sphere)
5. Return Cbest
```

Using longer tabu lists makes configuration changes more diverse because the algorithm never repeats moves performed during the last $T_\ell + T_c$ iterations. As such, by varying the tabu tenure, we control the balance between diversification and intensification—a greater T_c value implies a stronger diversification of the search process. A suitable control of T_c guarantees that PGTS is permanently discovering new regions and that it can never be blocked looping only through already visited regions.

The performance of this algorithm depends on three factors: a fast procedure to compute the distance (Section 3.2), a suitable choice of the spherical radius R (Section 3.3), and a strategy to quickly check the archive (Section 3.4).

3.2   Distance Definition and Calculation Complexity

The definition of the sphere $S(C)$ in the search space Ω is based on the following distance: the minimal number of neighborhood operations that need to be applied on a coloring so that it becomes equal with the other. This distance reflects the structural similarity between two colorings, the smaller the distance the more similar the colorings are. The equality is defined on the partition definition of a coloring (Definition 3): two colorings C_a and C_b are equal if and only if they designate the same classes of colors, i.e., if there exists a *color relabeling* σ (a bijection $\{1, 2, \ldots, k\} \xrightarrow{\sigma} \{1, 2, \ldots, k\}$) such that $C_a^i = C_b^{\sigma(i)}$, with $1 \leq i \leq k$.

The coloring distance can thus be expressed as a set-theoretic *partition distance*: the minimal number of vertices that need to be transferred from one class to another in the first partition so that the resulting partition is equal to the second. This distance function was defined since the 60ies and it can be calculated with well-studied methods—for example, see [21] for a general set-theoretic approach or [22] for the graph coloring application. Most studies consider a $O(|V| + k^3)$ algorithm based on the Hungarian method. However, we recently proved that, under certain conditions [23], this distance can be computed in $O(|V|)$ time with a enhanced method. Indeed, a fast distance computation is crucial to the PGTS algorithm as it calculates at least one distance per iteration and the time complexity of an iteration is $O(|V| + k \times f_c(C))$ (mainly due to operation 2.(a) in Algorithm 1).

The $O(|V|)$ distance calculation method is a Las Vegas algorithm (i.e., an algorithm that either reports the correct result or informs about the failure) that could calculate more than 90% of the required distances in practice: only less than 10% of cases require using the Hungarian algorithm (of complexity between $O(|V|+k^2)$ and $O(|V|+k^3)$ in the worst case). Basically, the distance is calculated with the formula $d(C_a, C_b) = |V| - s(C_a, C_b)$, where s is the complementary function of similarity, i.e., the maximum number of elements of C_a that do not need to be transfered to other C_a classes in order to transform C_a into C_b. Our algorithm goes through each element $x \in V$ and increments a matching counter between color class $C_a(x)$ of C_a and $C_b(x)$ of C_b. Denoting by $C_b^{\sigma(i)}$ the best match (with the highest counter) of class C_a^i, the similarity is at most $\sum_{1 \leq i \leq k} |C_a^i \cap C_b^{\sigma(i)}|$. The computation time can be very often reduced as PGTS does not actually require the precise value of the distance; it only has to check whether it is greater than R or not. If the aforementioned sum is less than $|V| - R$, the distance is greater than R.

Let us note that, as the distance values are in $[0, |V|)$, we often report the distance value in terms of percentages of $|V|$.

3.3   Choice of the Spherical Radius

In the exploration process of the regions, the parameter R controls the size of the visited spheres and, indirectly, the number of recorded spheres. The extreme value $R = 0$ forces the algorithm to record all visited configurations and that compromises the solving speed (via the `Forall` loop of the Already-Visited procedure). For the other extreme value $R = |V|$, the whole search space is contained in a unique sphere (because the distance is always less than $|V|$) and the algorithm is equivalent to the basic TS.

The effective choice of R has been determined from an analysis of a classical TS scenario: start the exploration process from an initial local minimum C_0, and denote by C_0, C_1, $C_2, \ldots, C_n$ the best colorings it visits, i.e., the visited configurations satisfying $f_c(C_i) \leq f_c(C)$, with $0 \leq i \leq n$ (note that most of the $C's$ can be local optima, too). We recorded all these configurations up to $n = 40000$ and we studied the possible values of the distances between each C_i and C_j with $1 \leq i, j \leq n$.

The histogram of the statistical distribution of these distance values directly showed a bimodal distribution with many occurrences of very small values (around $0.05|V|$) and of some much larger values—see an example on the right figure. There exist some distant clusters of close points; the high distances correspond to inter-cluster distances and the small ones to intra-cluster distances. If we denote a "cluster diameter" by c_d, we observed that c_d varies from $0.07|V|$ to $0.1|V|$ depending on the graph; moreover we noticed that:

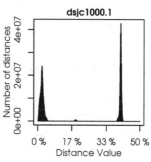

- there are numerous pairs (i, j) such that $d(C_i, C_j) < c_d$;
- there are numerous pairs (i, j) such that $d(C_i, C_j) > 2c_d$;
- there are very few (less than 1%) pairs (i, j) such that $c_d < d(C_i, C_j) < 2c_d$.

It is important to note that any two visited local minima situated at a distance of more than $0.1|V|$ are not in the same cluster because, ideally, they have different backbones. We assume that this observation holds on all sequences of colorings visited by TS; the value of R is set to $0.1|V|$ on all subsequent runs. Hence, as soon as PGTS leaves the sphere $S(C)$ of a visited configuration C, it avoids to re-visit later other configurations from the same cluster.

3.4   Archive Exploration

The exploration of the archive is a tricky stage for the computation time because of the numerous distance computations. Our objective is to keep the execution time of the learning component in the same order of magnitude as the exploring component. Due to the small bound of the distance computation time, computing one distance per iteration (i.e., in Step 4.(c)) is fast. The critical stage appears when PGTS needs to check the distance from the current coloring to *all* colorings from the archive (the `Forall` loop of Step 1, procedure Already-Visited in Algorithm 2). If the archive size exceeds a certain limit, the learning component execution time can become too long.

However, the processing of the archive may be tractable if we focus the learning component only the high quality configurations.

Definition 6. *(High-quality configuration) We say that configuration $C \in \Omega$ is high-quality if and only if $f_c(C) \leq B_f$, where B_f is a fitness boundary. Otherwise, we say that C is low-quality.*

The fitness boundary B_f is automatically set by PGTS so that the total number of iterations stays in the same order of magnitude as the number of distance computations. This proved to be a good "thumb rule" for obtaining an effective algorithm. To be specific, B_f directly controls the learning overhead because the whole learning component (Step 4.(c)) is now executed only if $f(C) < B_f$. In practice, B_f varies from 5 conflicts to 20 conflicts; for some problems we can even set $B_f = \infty$ and still obtain an acceptable speed. However, the algorithm automatically lowers and raises B_f according to the balance between the number of computed distances and the number of iterations.

4 Numerical Results

In this section, we show experimentally that the learning component helps the TS algorithm to obtain several colorings never found before by any other TS algorithm [17, 18, 19, 20, 24]. In fact, PGTS competes favorably with all existing local search algorithms.

4.1 Benchmark Graphs

We carry out the comparison only on the most difficult instances from the DIMACS Challenge Benchmark [25]: (i) *dsjc*1000.1, *dsjc*1000.5, *dsjc*1000.9, *dsjc*500.5 and *dsjc*250.5—classical random graphs [26] with unknown chromatic numbers (the first number is $|V|$ and the second denotes the density); (ii) *le*450.25c and *le*450.25d—the most difficult "Leighton graphs" [27] with $\chi = 25$ (they have at least one clique of size χ); (iii) *flat*300.28 and *flat*1000.76— the most difficult "flat" graphs [28] with χ denoted by the last number (generated by partitioning the vertex set in χ classes, and by distributing the edges only between vertices of different classes); (iv) *R*1000.1—a geometric graph constructed by picking points uniformly at random in a square and by setting an edge between all pairs of vertices situated within a certain distance.

For each graph, we present the results using a number of colors k such that the instance (G, k) is very difficult for the basic TS; most of the unselected graphs are less challenging.

4.2 Experimental Procedure

Note that PGTS is equivalent to TS in the beginning of the exploration, while the archive is almost empty. The learning process intervenes in the exploration process only after several millions of iterations, as soon as the exploration process starts to run into already explored spheres. Therefore, if the basic TS is able to solve the problem quite quickly without any guidance, PGTS does not solve it more rapidly; the objective of PGTS is visible in the long run.

In Table 1, we perform comparative tests of TS and PGTS by launching 10 independent executions with time limit of 50 hours each. Within this time limit, PGTS re-initializes its search with a random k-coloring each time it reaches 40 million iterations. All these restarts share the same archive of spheres for PGTS.

Table 1. Comparison of PGTS and basic TS for a time limit of 50 hours. Columns 1 and 2 denote the instance, the success rate (Columns 3 and 5 respectively) is the number of successful execution series out of 10; the time column presents the average number of hours needed to solve the problem (if the success rate is not 0).

Instance		PGTS		Basic TS	
Graph	K	Success rate	Time [h]	Success rate	Time [h]
dsjc250.5	28	10/10	< 1	10/10	< 1
dsjc500.5	48	2/10	35	0/10	–
dsjc1000.1	20	2/10	9	0/10	–
dsjc1000.5	87	5/10	28	0/10	–
dsjc1000.9	224	8/10	24	2/10	44
flat300_28_0	29	7/10	8	0/10	–
le450_25c	25	4/10	11	3/10	7
le450_25d	25	2/10	19	2/10	12
flat1000_76_0	86	3/10	33	0/10	–
r1000.1c	98	10/10	< 1	10/10	< 1

To guarantee that the comparison is unbiased, we impose the same running time limit of 50 hours for both algorithms.[1]

Generally speaking, a PGTS iteration is more computationally-expensive than a TS iteration, and, consequently, TS can perform many more iterations for the same CPU time. However, the learning process accounts for an important performance gain: in many cases in which the basic TS fails (or has a very low success rate in finding a solution, see Table 1, Column 5), PGTS (Column 3) solves the problem.

Comparison with the Best Algorithms. Table 2 reports the best results obtained by PGTS on our graph set, along with a comparison with the basic TS and with the state-of-the-art algorithms. Note that many presented k-colorings were never reported before by other local search algorithm. Among all local searches that we are aware of, only VSS and PartialCol (columns 5 and 6)—two very recent algorithms using an evolved neighborhood function and a enhanced representation, respectively, compete effectively with PGTS.

Let us mention that in the literature on graph coloring, it is a common practice to run a local search algorithm for hours in order to (try to) solve large problems. For example, the most recent coloring algorithms [24, 29] use running times of 10 hours for the largest instances. Another important point is that PGTS can continually explore new regions if it is given more computation time. Consequently, it is able to find better solutions by using the additional computational resources. Notice that this is a desirable characteristic which is not given by many existing algorithms. Very often, running them beyond some time (or iteration) threshold

---

[1] We used a 2.7GHz processor using the C++ programming language compiled with the $-O2$ optimization option under Linux. The source code is the same for both algorithms, the difference is only made by the learning component that is enabled for PGTS and disabled for TS.

Table 2. Comparison of the minimum number of colors for which a solution is found by: (i) the basic TS (Column 3), (ii) the new PGTS algorithm (Column 4) and (iii) state-of-the-art algorithms (Columns 5-10). Column 2 denotes the chromatic number (? if unknown) and the best k for which a legal coloring was ever reported in the literature. The colorings we report are publicly available on the Internet: www.info.univ-angers.fr/pub/porumbel/graphs/pgts/

Graph	χ, k^*	TS	PGTS	VSS [29] 2008	PCol [24] 2008	ACol [30] 2008	MOR [31] 1993	GH [20] 1999	MMT [32] 2008
*dsjc*250.5	?, 28	28	*28*	-	-	28	28	28	28
*dsjc*500.5	?, 48	49	*48*	48	48	48	49	48	48
*dsjc*1000.1	?, 20	21	*20*	20	20	20	21	20	20
*dsjc*1000.5	?, 83	88	87	88	88	84	88	83	83
*dsjc*1000.9	?, 224	224	*224*	224	225	224	226	224	226
*le*450.25c	25, 25	25	*25*	26	25	26	25	26	25
*le*450.25d	25, 25	25	*25*	26	25	26	25	26	25
*flat*300.28	28, 32	30	29	29	28	31	31	31	31
*flat*1000.76	76, 82	87	86	87	87	84	89	83	82
*r*1000.1c	?, 98	98	*98*	–	98	-	98	–	98

will not lead to better results simply because either the algorithms are trapped in deep local optima or because they re-explore again and again the same search space areas.

5 Discussion

Here, we discuss the properties of our new approach comparing to previous ones, and propose a generalization of PGTS to other combinatorial optimization problems.

5.1 Related Work

PGTS shares some basic ideas and objectives with the feature-based Guided Local Search [33] but our solving strategy is very different. We do not use explicit penalties and we do not need to identify specific solution features to penalize the evaluation function. In fact, we implicitly use a form of penalization (by encouraging the investigation unvisited regions) but at a higher level. Our method of avoiding certain regions is very targeted, in contrast with the penalty approach that might apply the same penalty (triggered by a situation in a particular region) to some very different and distant configurations.

A drawback of PGTS, when compared to the underlying basic TS, is that it might not sufficiently explore some spherical regions that are avoided after a first visit. This point could be completed by an algorithm that investigates only the interior of the spheres of the best recorded local minima. However, the new algorithm is still competitive even with the best known algorithms (see columns 5-10 in Table 2) from the literature.

Compared to other local search algorithms for graph coloring, one can see that PGTS resorts to a more global view of the exploration. Most previous local search algorithms focused on local level improvements, i.e., they use more powerful neighborhood relations, alternative solution encodings, specific evaluation functions, etc. Generally speaking, there are numerous such problem-specific techniques able to increase the performance of a combinatorial optimization heuristic. However, we showed that, by focusing on global-level learning techniques, one can more easily overcome the limitations to which the local-level improvement potential is inevitably exposed.

5.2   Toward a Generalization for Other Combinatorial Problems

A careful examination of the code of Position Guided Tabu Search (see Algorithm 2) shows that it contains no particular references to the coloring problem. The only required components are: a search space, a neighborhood function, an objective function and a search space metric. The performance of PGTS depends on three factors: a fast procedure to compute the distance, a suitable choice of the spherical radius R, and a strategy to quickly search the archive.

Hence, as long as there exists a distance measure whose computation time does not significantly outweighs the computation time of a TS iteration, PGTS can be applied effectively to any combinatorial problem. This search space distance should express the minimal required number of neighborhood moves to arrive from a configuration to the other. Ideally, one should be able to group in a R-sphere of a local optimum only "equivalent" local optima—i.e., configurations sharing a common "backbone" substructure.

One can find several examples of easy–to–compute distances that can also be defined in this manner by using some specific neighborhoods:

- the Hamming distance for problems with array representation using the 1-Flip neighborhood (i.e., constraint satisfaction problems with a neighborhood operator that consists in changing the value of a single variable of the current configuration),
- the Kendall tau distance [34] for problems with permutation-based representation using a neighborhood defined by adjacent transpositions (i.e., the travelling salesman problem considering a neighborhood in which a move inverses two adjacent cities—the adjacent pairwise interchange neighborhood),
- the edit distance for problems with an array representation and with the neighborhood defined using edit operations.

Concerning the archive processing time, it can be substantially reduced in at least three ways: (i) by focusing on high-quality configurations, (ii) by increasing the value of the radius R and (iii) by transforming the archive into a queue that removes the oldest element at each insert operation. In the later case, the algorithm becomes a Double Tabu Search with two lists: (1) the traditional list of the last visited configurations that are forbidden, (2) the tabu list of spheres, used to avoid revisiting spheres visited in the recent past. The significance of the expression "recent past" would depend on the size of the queue which should be tailored according to the learning component overhead.

6   Conclusions

We have presented a new local search algorithm that uses a learning process to guide the exploration process toward unvisited search space regions. It is possible to integrate this learning process in a classical tabu search with an acceptable overhead for all combinatorial optimization problems, provided that the distance computation is not too expensive. Moreover, the new algorithm does not necessarily introduce too many auxiliary user-provided parameters because the B_f value required in archive processing can be automatically set. The R value could be determined by calculating the distances between the local minima discovered during a classical search of the search space; for the graph coloring problem, we found that these local optima are typically grouped in clusters that can be confined in R-spheres with $R = 0.1|V|$. For other problems, R might be determined by finding the maximum distance between two configurations sharing an important backbone substructure.

This algorithm enabled us to improve the results of the basic TS for all graphs for which there is at least a different algorithm that ever reported better colorings than TS. Even compared to the best known algorithms from the literature (few of them local searches), PGTS proved to be very effective. Except very few graphs, it always finds the best known coloring. Moreover, in combination with another intensification algorithm, we found for the very first time a solution with 223 colors for the well-studied $dsjc1000.9$ graph.

Acknowledgments. This work is partially supported by the CPER project "Pôle Informatique Régional" (2000-2006) and the Régional Project MILES (2007-2009). The authors are grateful to the reviewers of the paper for their useful comments and questions.

References

1. Du, D., Pardalos, P.: Handbook of Combinatorial Optimization. Springer, Heidelberg (2007)
2. Kallel, L., Naudts, B., Reeves, C.: Properties of fitness functions and search landscapes. Theoretical Aspects of Evolutionary Computing, 175–206 (2001)
3. Merz, P.: Advanced fitness landscape analysis and the performance of memetic algorithms. Evolutionary Computation 12(3), 303–325 (2004)
4. Streeter, M., Smith, S.: How the landscape of random job shop scheduling instances depends on the ratio of jobs to machines. Journal of Artificial Intelligence Research 26, 247–287 (2006)
5. Zhang, W.: Configuration landscape analysis and backbone guided local search. Part i satisfiability and maximum satisfiability. Artificial Intelligence 158(1), 1–26 (2004)

6. Gerber, M., Hansen, P., Hertz, A.: Local optima topology for the 3-SAT problem. Cahiers du GERAD G-98-68 (1998)
7. Ryan, J.: The depth and width of local minima in discrete solution spaces. Discrete Applied Mathematics 56(1), 75–82 (1995)
8. Hertz, A., Jaumard, B., de Aragão, M.: Local optima topology for the k-coloring problem. Discrete Applied Mathematics 49(1-3), 257–280 (1994)
9. Hamiez, J., Hao, J.: An analysis of solution properties of the graph coloring problem. In: Metaheuristics computer decision-making, pp. 325–345. Kluwer, Dordrecht (2004)
10. Culberson, J., Gent, I.: Frozen development in graph coloring. Theoretical Computer Science 265, 227–264 (2001)
11. Merz, P., Freisleben, B.: Fitness landscapes, memetic algorithms, and greedy operators for graph bipartitioning. Evolutionary Computation 8(1), 61–91 (2000)
12. Merz, P., Freisleben, B.: Fitness landscape analysis and memetic algorithms for the quadratic assignment problem. IEEE Transactions on Evolutionary Computation 4(4), 337–352 (2000)
13. Reeves, C., Yamada, T.: Genetic algorithms, path relinking, and the flowshop sequencing problem. Evolutionary Computation 6(1), 45–60 (1998)
14. Kuntz, P., Pinaud, B., Lehn, R.: Elements for the description of fitness landscapes associated with local operators for layered drawings of directed graphs. In: Metaheuristics Computer Decision-Making, pp. 405–420. Kluwer, Dordrecht (2004)
15. Battiti, R., Brunato, R., Mascia, F.: Reactive Search and Intelligent Optimization. Springer, Heidelberg (2008)
16. Glover, F., Laguna, M.: Tabu Search. Springer, Heidelberg (1997)
17. Hertz, A., Werra, D.: Using tabu search techniques for graph coloring. Computing 39(4), 345–351 (1987)
18. Dorne, R., Hao, J.: Tabu search for graph coloring, t-colorings and set t-colorings. In: Voss, S., et al. (eds.) Meta-Heuristics Advances and Trends in Local Search Paradigms for Optimization, pp. 77–92. Kluwer, Dordrecht (1998)
19. Dorne, R., Hao, J.: A new genetic local search algorithm for graph coloring. In: Eiben, A.E., Bäck, T., Schoenauer, M., Schwefel, H.-P. (eds.) PPSN 1998. LNCS, vol. 1498, pp. 745–754. Springer, Heidelberg (1998)
20. Galinier, P., Hao, J.: Hybrid evolutionary algorithms for graph coloring. Journal of Combinatorial Optimization 3(4), 379–397 (1999)
21. Gusfield, D.: Partition-distance a problem and class of perfect graphs arising in clustering. Information Processing Letters 82(3), 159–164 (2002)
22. Glass, C., Pruegel-Bennett, A.: A polynomially searchable exponential neighbourhood for graph colouring. Journal of the Operational Research Society 56(3), 324–330 (2005)
23. Porumbel, C., Hao, J., Kuntz, P.: An efficient algorithm for computing the partition distance. Submitted, available on request (2008)
24. Blöchliger, I., Zufferey, N.: A graph coloring heuristic using partial solutions and a reactive tabu scheme. Computers and Operations Research 35(3), 960–975 (2008)
25. Johnson, D., Trick, M.: Cliques, Coloring, and Satisfiability Second DIMACS Implementation Challenge. DIMACS series in Discrete Mathematics and Theoretical Computer Science, vol. 26. American Mathematical Society (1996)
26. Johnson, D., Aragon, C., McGeoch, L., Schevon, C.: Optimization by simulated annealing an experimental evaluation; part ii, graph coloring and number partitioning. Operations Research 39(3), 378–406 (1991)
27. Leighton, F.: A graph coloring algorithm for large scheduling problems. Journal of Research of the National Bureau of Standards 84(6), 489–503 (1979)

28. Culberson, J., Luo, F.: Exploring the k-colorable landscape with iterated greedy. In: [25], pp. 345–284
29. Hertz, A., Plumettaz, A., Zufferey, N.: Variable space search for graph coloring. Discrete Applied Mathematics 156(13), 2551–2560 (2008)
30. Galinier, P., Hertz, A., Zufferey, N.: An adaptive memory algorithm for the k-coloring problem. Discrete Applied Mathematics 156(2), 267–279 (2008)
31. Morgenstern, C.: Distributed coloration neighborhood search. In: [25], pp. 335–358
32. Malaguti, E., Monaci, M., Toth, P.: A metaheuristic approach for the vertex coloring problem. INFORMS Journal on Computing 20(2), 302 (2008)
33. Voudouris, C., Tsang, E.: Guided local search. In: Glover, F., et al. (eds.) Handbook of metaheuristics, pp. 185–218. Kluwer Academic Publishers, Dordrecht (2003)
34. Kendall, M.: A new measure of rank correlation. Biometrika 30(1/2), 81–93 (1938)

Corridor Selection and Fine Tuning for the Corridor Method

Marco Caserta and Stefan Voß

Institute of Information Systems, University of Hamburg, Hamburg, Germany
{marco.caserta,stefan.voss}@uni-hamburg.de

Abstract. In this paper we present a novel hybrid algorithm, in which ideas from the genetic algorithm and the GRASP metaheuristic are cooperatively used and intertwined to dynamically adjust a key parameter of the corridor method, *i.e.*, the corridor width, during the search process. In addition, a fine-tuning technique for the corridor method is then presented. The response surface methodology is employed in order to determine a good set of parameter values given a specific problem input size. The effectiveness of both the algorithm and the validation of the fine tuning technique are illustrated on a specific problem selected from the domain of container terminal logistics, known as the blocks relocation problem, where one wants to retrieve a set of blocks from a bay in a specified order, while minimizing the overall number of movements and relocations. Computational results on 160 benchmark instances attest the quality of the algorithm and validate the fine tuning process.

1 The Corridor Method: An Introduction

The Corridor Method (CM) has been presented by [1] as a hybrid metaheuristic, linking together mathematical programming techniques with heuristic schemes. As illustrated in [2], the basic idea of the CM relies on the use of an exact method over restricted portions of the solution space of a given problem. Given an optimization problem P, the basic ingredients of the method are a very large feasible space $\mathcal{X}$, and an exact method M that could easily solve problem P if the feasible space were not too large. Since, in order to be of interest, problem P is assumed to be hard, the direct application of M to solve P usually becomes unpractical when dealing with real world as well as large scale instances.

The CM defines method-based neighborhoods, in which a neighborhood is build taking into account the method M used to explore it. Given a current feasible solution $\mathbf{x} \in \mathcal{X}$, the CM builds a neighborhood of $\mathbf{x}$, say $\mathcal{N}(\mathbf{x})$, which can effectively be explored by employing M. Ideally, $\mathcal{N}(\mathbf{x})$ should be exponentially large and built in such a way that it could be explored in (pseudo) polynomial time using M. In this sense the CM closely relates to very large scale neighborhood search as well as the so-called dynasearch; see, e.g., [3,4].

Typically, the corridor around an incumbent solution is defined by imposing exogenous constraints on the original problem. The effect of such constraints is to identify a limited portion of the search space. The selection of which portions

T. Stützle (Ed.): LION 3, LNCS 5851, pp. 163–175, 2009.

of the search space should be discarded can be driven by a number of factors, *in primis*, as the power of the method itself deployed to explore the resulting neighborhood. However, one could also envision the design of a stochastic mechanism that, after dividing the search space in portions, or limited regions, selects which of these subspaces should be included in the current corridor. Factors such as, *e.g.*, a greedy score, the cardinality of each subregion, etc., could be used to bias the stochastic selection mechanism that drives the definition of the corridor around the incumbent solution. The stochastic mechanism could be designed such that, on the one hand, the corridor selection is non-deterministic, so that at every step different corridors around the same incumbent solution could be created and, on the other hand, such selection is still influenced by a merit score, accounting for the attractiveness of each portion of the search space. Following such cooperative greedy stochastic corridor construction process, a balance between diversification and intensification is achieved, since, even though more promising regions of the search space have higher probabilities of being selected, not necessarily the best subregions will always be chosen.

In the next sections we will illustrate how this cooperative idea can be employed to design an effective algorithm for a well-known problem arising, e.g., at container ports. First, we illustrate the problem; then the algorithm is described, and subsequently numerical results are presented. More details about the problem as well as the steps of the algorithm are provided in [2].

2   The Blocks Relocation Problem

Relocation is one of the most important factors contributing to the productivity of operations at storage yards or warehouses [5]. A common practice aimed at effectively using limited storage space is to stack blocks along the vertical direction, whether they be maritime containers, pallets, boxes, or steel plates [6]. Given a heap of blocks, relocation occurs every time a block in a lower tier must be retrieved before blocks placed above it. Since blocks in a stack can only be retrieved following a LIFO (Last In First Out) discipline, in order to retrieve the low-tier block, relocation of all blocks on top of it will be necessary.

Let us consider a bay with m stacks and n blocks. In line with the available literature [6,7], we introduce the following assumptions:

A1: pickup precedences among blocks are known in advance. We indicate the pickup precedence with a number, where blocks with lower numbers have a higher precedence than blocks with higher numbers;

A2: when retrieving a target block, we are allowed to relocate only blocks found above the target block in the same stack using a LIFO policy;

A3: relocation is allowed only to other stacks within the same bay;

A4: relocated blocks can be put only on top of other stacks, *i.e.*, no rearrangement of blocks within a stack is allowed.

In the literature, the problem of arranging containers to maximize efficiency is extensively discussed; see, e.g., [8] for a recent survey on quantitative approaches to container terminal logistics. One approach to the problem is the retrieval problem considering relocations called blocks relocation problem (BRP). If the current configuration of the bay is taken as fixed, one might be interested in finding a sequence of moves to be executed, while retrieving blocks according to a given sequence, in order to minimize the overall number of relocation moves. While in the shuffling problems containers are rearranged but not removed, in this version of the problem, at each step, a container is removed from the bay, hence reducing the number of containers in the bay until all containers have been picked up from the bay. Exact as well as approximate algorithms have been proposed to minimize the number of relocations while retrieving blocks (see, e.g., [6,7,9,10,11,12]).

In this paper, we present an enhanced algorithm for the BRP. The algorithm is made up of (i) a corridor definition phase, during which exogenous constraints are imposed to construct a corridor around the incumbent solution; (ii) a neighborhood design and exploration phase, where the corridor is used to define the boundaries of the neighborhood to be explored; (iii) a move evaluation and selection phase, where a greedy rule is used to evaluate the fitness of the solutions in the neighborhood and to select a restricted pool of elite solutions and, finally, (iv) a trajectory fathoming phase, in which a logical test is employed to determine whether the current trajectory can be pruned without loosing any improving solution.

The major contribution of the paper is two-fold: on the one hand, we propose an effective way of enhancing the performance of a CM based algorithm by hybridizing such algorithm with ideas from genetic algorithms (GA) and GRASP; on the other hand, it is the first time that a statistically sound attempt to fine tune a CM inspired algorithm is carried out, hence contributing to the realm of metaheuristic calibration and fine tuning.

Let us first describe how the CM can be applied to the BRP. The basic idea of the CM is related to the imposition of exogenous constraints upon the original problem, such that the search space is reduced. Given an incumbent bay configuration $\mathcal{T}$, the goal of the BRP is to retrieve the block with highest priority in the bay. However, due to the LIFO policy assumption, whenever such target block is not in the uppermost tier of a stack, i.e., whenever other blocks are currently placed upon the target block, relocation operations will first be required in order to finally retrieve the target block. Given the uppermost block currently located in the same stack of the target, a decision regarding where to relocate such block must be made. Obviously, such decision affects the future retrieval process, since the relocated block might be placed on top of other blocks and, therefore, it might imply the need of further relocations in the next steps.

The size of the search space describing all the possible relocations of a set of blocks grows exponentially with respect to the number of blocks moved at each step. A simple way to limit the size of the search space describing the possible configuration that can be reached starting from the initial bay configuration is to impose "constraints," or limitations, upon the use of stacks. Let us suppose we are given a bay with m stacks. This implies that, when a relocation of a block

l is required to retrieve a target block k, block l can be placed on top of any stack different from the one where it is currently located, *i.e.*, $m - 1$ possible relocations arise. In turn, all of these $m - 1$ possible scenarios give raise to $m - 1$ new configurations, with an exponential growth in the number of configurations. However, let us assume that, whenever a relocation of a block is required in order to retrieve a target block, we are allowed to move such block on only $\delta < m$ of the available $m - 1$ stacks, hence creating δ different configurations. Figuratively, we could say that such exogenous constraint builds a horizontal corridor around the incumbent configuration, determining which configurations can be reached starting from the incumbent one. The value of parameter δ can be used to control the width of the corridor and, therefore, the growth of the search space.

In a similar fashion, a second parameter λ can be used to introduce a bound on the maximum height of a stack, *i.e.*, a vertical corridor. Consequently, a stack can be used to relocate blocks only if the maximum height has not yet been reached.

3   The Cooperative Algorithm

Given an incumbent bay configuration, let us indicate with $\mathcal{T}_i$ the ordered list of blocks in stack i, with $i = 1, \ldots, m$, where the first and the last elements of the list represent the blocks at the top and at the bottom of the stack, respectively. Consequently, we represent the incumbent bay configuration $\mathcal{T}$ as a sequence of stacks, *i.e.*, $\mathcal{T} = < \mathcal{T}_1, \ldots, \mathcal{T}_m >$.

Given an incumbent bay configuration $\mathcal{T}$, with a total of N blocks, let us indicate with $k \in [1, N]$ the target block, *i.e.*, the block with highest priority in $\mathcal{T}$. Index t, with $t \in [1, m]$, is used to indicate the stack in which block k is found. In addition, let us indicate with $\mathcal{L}$ the list of blocks above the target block in stack t, and with l the uppermost block in list $\mathcal{L}$, *i.e.*, the current block to be relocated.

Let us now define the concept of *forced relocations*. Given a current bay configuration, as in Figure 1, the number of forced relocations is given by the number of blocks in each stack currently on top of a block with higher priority. Such blocks will necessarily be relocated, in order to retrieve the block with higher priority located below. For example, in Figure 1, the number of forced relocations is equal to 4, as indicated by the shaded blocks. It is worth noting that the number of forced relocations in a bay constitutes a valid lower bound of the minimum number of relocations required to complete the retrieval operation.

The proposed algorithm terminates when only the block with the lowest priority is left in the bay and is made up of four phases:

Corridor definition: Given the incumbent configuration and the current block to be relocated, a *corridor* around the incumbent solution is defined. The corridor width is influenced by the value of parameter δ, which indicates the number of stacks available for relocation. As illustrated above, in order to reduce the size of the search space, we only allow relocations towards a subset of the available stacks, *i.e.*, only towards those stacks that are included in

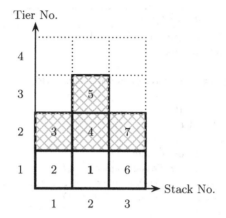

Fig. 1. An example of forced relocations within a bay. Blocks 3, 4, 5, and 7 will necessarily be relocated to retrieve blocks with higher priority.

the current corridor. A stochastic mechanism is used in order to identify the subset of stacks to be included in the current corridor;

Neighborhood design and exploration: Once the corridor around the incumbent solution is created, we define a neighborhood $\mathcal{N}(\mathcal{T}, k, t, l)$, which is made up by the set of solutions that can be reached via the application of an admissible relocation move. Subsequently, we thoroughly explore the neighborhood by evaluating the solutions contained in $\mathcal{N}(\mathcal{T}, k, t, l)$ with respect to a *greedy score* $g : \mathcal{N}(\mathcal{T}, k, t, l) \rightarrow \mathbb{R}$. In the spirit of the GRASP metaheuristic, a pool of *elite solutions* Ω is formed, *i.e.*, a restricted set of solutions representing the topmost quantile of the neighborhood population is identified;

Move evaluation and selection: A "roulette-type" probabilistic scheme is used to randomly select one solution from the elite set Ω and the corresponding move required to reach the new configuration is finally executed. Such mechanism allows for different solutions to be selected at each iteration, while still preserving a measure of attractiveness of each selection proportional to the greedy score;

Trajectory fathoming: given the best upper bound, we apply a simple logical test to detect whether the current trajectory is dominated by a previously found feasible solution. In such case, the current trajectory can be fathomed and the algorithm is restarted. Otherwise, if the logical test fails, the next iteration of the algorithm is performed.

The next sections describe each phase of the algorithm in detail. The overall algorithm is presented in Algorithm `Hybrid_brp()`. In the following, the four-phase algorithm is repeated until a complete trajectory is build, *i.e.*, until all blocks of the bay are retrieved in the right order. In turn, the trajectory construction scheme is iteratively repeated until a predefined `stopping_criterion()` is reached.

Algorithm 1. Hybrid_brp()

Require: initial configuration T^0, corridor width δ
Ensure: set of relocations
1: $z^* \leftarrow \infty$
2: **while** stopping_criterion() is not reached **do**
3:      $v \leftarrow 0$
4:      **for** $k = 1, \ldots, N$ **do**
5:          identify t and $\mathcal{L}$
6:          **while** $\mathcal{L} \neq \emptyset$ **do**
7:              $l \leftarrow$ uppermost element in $\mathcal{L}$
8:              $\Delta \leftarrow$ Corridor_Selection(T, k, t, l, δ)
9:              $\mathcal{N}(T, k, t, l) \leftarrow$ Neighborhood_Definition(T, Δ)
10:             $x \leftarrow$ Move_Selection($\mathcal{N}(T, k, t, l)$)
11:             $v \leftarrow v + 1$
12:             $T^v \leftarrow m(T^{v-1}, x)$
13:             $\mathcal{L} \leftarrow \mathcal{L} \setminus \{l\}$
14:             apply logical test Trajectory_Fathoming(T^v, z^*)
15:         **end while**
16:         retrieve target block k from T^v
17:     **end for**
18:     **if** $v < z^*$ **then**
19:         $z^* \leftarrow v$ and save best trajectory
20:     **end if**
21: **end while**

3.1 Corridor Definition

Let us assume an incumbent bay configuration T is given. With respect to such configuration, we identify the target block k, *i.e.*, the current block to be retrieved, the stack upon which block k is placed, stack t, the set of blocks on top of k, $\mathcal{L}$, and the uppermost block in list $\mathcal{L}$, *i.e.*, block l. Due to the LIFO policy assumption, in order to retrieve block k, it is mandatory to relocate each block in $\mathcal{L}$ from stack t to any other stack $\{1, \ldots, m\} \setminus \{t\}$. Iteratively, the uppermost block in list $\mathcal{L}$ will be picked from the list and relocated, until the list of blocking items is empty.

Given the incumbent configuration (T, k, t, l), we define a corridor around such configuration by selecting a subset of the stacks as admissible stacks. We identify three types of stacks with respect to the current configuration: *(i)* empty stacks, towards which it is always possible to relocate block l without generating any new forced relocation. Let us indicate with S_o the set of such stacks in the current configuration; *(ii)* no-deadlock stacks, *i.e.*, stacks for which the block with highest priority has a lower priority than block l. In other words, if we indicate with $\min(T_i)$ (in the following, $\min(i)$) the element with highest priority in stack i, then no-deadlock stacks are those stacks for which $\min(i) > l$. We indicate with S_1 the set of no-deadlock stacks; *(iii)* deadlock stacks, *i.e.*, stacks for which the block with highest priority has a priority higher than block l, $\min(i) < l$. In this case, relocating block l onto such a stack will generate a new relocation in the future, since we are going to retrieve one of the blocks in the stack prior to the retrieval of block l. Let S_2 indicate the set of deadlock stacks. For each type of stack, we compute the following score:

$$\sigma(i) = \begin{cases} \dfrac{1}{|S_0|}, & \text{if } i \in S_0 \\[3mm] \dfrac{\displaystyle\sum_{j \in S_1} \min(j)}{\min(i)}, & \text{if } i \in S_1 \\[3mm] \dfrac{\min(i)}{\displaystyle\sum_{j \in S_2} \min(j)}, & \text{if } i \in S_2 \end{cases} \tag{1}$$

The rationale behind such scores is that we try to capture the measure of attractiveness of each stack, based upon (i) whether a new deadlock is created after relocation of element l, and (ii) the impact of relocating block l onto the stack with respect to future relocations. Finally, after normalizing the scores, a roulette-wheel mechanism as in GA is used to select which stacks belong to the corridor.

3.2 Neighborhood Design and Exploration

Given a set of admissible stacks Δ, we define a neighborhood of the current configuration (T, k, t, l) by including in such a neighborhood only those configurations T' that can be created relocating block l onto a stack $i \in \Delta$.

Let us define an admissible move as a transformation function that operates upon a configuration T, such that $m : T \to T'$, where T' is the configuration obtained after relocating block l from stack t to any other stack, which is, $T' = m(T, x)$, where x indicates the stack onto which block l is relocated. A move $T' = m(T, x)$ is admissible if $x \in \Delta$.

Thus, a formal definition of the neighborhood is:

$$\mathcal{N}(T, k, t, l) = \{T' = m(T, x) : x \in \Delta\} \tag{2}$$

3.3 Move Evaluation and Selection

In order to evaluate each configuration, we define a greedy score. We employ a simple rule given by the number of forced relocations (deadlocks) within the current configuration, where a forced relocation is imposed every time a block with higher priority is found below a block with lower priority. Let us associate with each configuration T such greedy score $g : T \to \mathbb{R}$, where $g(T)$ is equal to the number of forced relocations within configuration T. We define the set of elite solutions in the current neighborhood Ω as the best 50% quantile of the overall set of solutions in $\mathcal{N}(T, k, t, l)$. Finally, we employ a roulette-wheel stochastic mechanism to select one solution from Ω, in a fashion similar to what is presented in Algorithm `Corridor_Selection()`.

This step of the algorithm somehow resembles parts of what is known as a semi-greedy heuristic [13] or a GRASP procedure [14]: at each iteration, the choice of the next element to be added is determined by ordering a set of candidate elements according to a greedy score. Such greedy score measures the myopic benefit of selecting a specific move. Such mechanism is adaptive in the sense that the score value associated with each candidate solution is changed to take into account the effect of the previous decisions. On the other hand, the roulette-wheel probabilistic mechanism allows to select different moves at each iteration of the algorithm, while still preserving a measure of attractiveness of each selection proportional to the value of the greedy score.

3.4 Trajectory Fathoming

Finally, after each step of the algorithm, we apply a logical test aimed at detecting whether the current bay configuration is dominated by a previously visited solution and, hence, could be abandoned. Given a new configuration T', if the total number of moves required to reach such configuration, $z(T')$, plus the number of forced relocations in T', $g(T')$ is greater than or equal to a given upper bound (the best solution found so far **ub**), the steps followed to reach T' will never be part of the optimal decision sequence. Consequently, the current trajectory path can be dropped.

4   Experimental Plan and Computational Results

In this section we present computational results on randomly generated instances. All tests presented in this section have been carried out on a Pentium IV Linux Workstation with 512Mb of RAM. The algorithm has been coded in C++ and compiled with the GNU C++ compiler using the -O option.

We designed an experiment that resembles that of [6] and of [7]. We focus our attention on tests on large scale instances, for which the optimal solution is unknown. The random generation process takes as input two parameters, the number of stacks m and the number of tiers h, and randomly generates a rectangular bay configuration of size $n = h \times m$, where n indicates the total number of blocks in the bay. For each combination of m and h we generated 40 different instances.[1]

The experiment plan is twofold: On the one hand, we want to measure the solution quality of the algorithm by comparing it with two benchmark algorithms from the literature, namely those proposed in [6] and [7]. On the other hand, we make the first attempt to fine tune and calibrate a CM inspired algorithm. We perform a statistical analysis based upon the "Response Surface Methodology" to draw conclusions about the relation between the algorithmic parameter δ (corridor width) and the objective function value. The key question addressed is: given the input size, in terms of bay width m and bay height h, is it possible

---

[1] The code and all the instances used during the experiment can be obtained from the authors upon request.

to determine what the value of parameter δ should be in order to obtain the best possible feasible solution?

This section is organized as follows: first, we illustrate how the response-surface methodology has been used in order to find "good" values of δ for any given instance size. Next, we validate the statistical model by collecting results of the heuristic scheme using the values suggested by the response surface, and comparing them with the solutions provided by the benchmark algorithms.

4.1 Response Surface Methodology

In this section we present how we adapted the response surface methodology [15] to fine tune the proposed algorithm. A more detailed description of such a procedure is provided in [16].

We first selected a set of factor levels, i.e., $\delta \in \{1, \ldots, m-1\}$. Next, we collected the objective function value of the algorithm on each run with varying bay size, in the intervals $m \in [6, 10]$ and $h \in [6, 10]$. We then tested the model:

$$Y_e = \beta_0 + \beta_1 \delta + \beta_2 \delta^2$$

Once the coefficients of the regression models were computed, an ANOVA analysis was used to check the significance level of the estimates. Next, the steepest ascent method was used to reset the center of the experiment in the direction indicated by the gradient of Y_e. Finally, we minimized the polynomial Y_e by setting the first order derivative to zero and checking that the point is a local minimum.

The minimum points, i.e., the values of δ that for any given input size minimized the objective function value, were thus used to generate the final regression model that describes what the value of δ should be. The model tested was:

$$\delta = \beta_0 + \beta_1 h + \beta_2 m + \beta_3 h^2 + \beta_4 m^2$$

Figure 2 and Table 1 present the fitting surface as well as the regression statistics for the model. As Table 1 shows, the coefficients of the polynomial are significant up to the second order and the model itself seems to be statistically significant, since the model has a p-value less than $2.2e - 16$. From the figure as well as the table, it is possible to observe that, while both instance parameters h and m have a bearing on the δ value, the value of m has a higher impact on the corridor width.

It is worth remembering, though, that the validity of the proposed model is limited to the interval studied, i.e., the model can be used to forecast the δ value for instances with $m \in [6, 10]$ and $h \in [6, 10]$.

4.2 Computational Results

In this section we finally present computational results that prove the effectiveness of the proposed algorithm. The value of δ is determined using the polynomial of Table 1 (we fixed $\lambda \leq h + 2$ throughout the experiment).

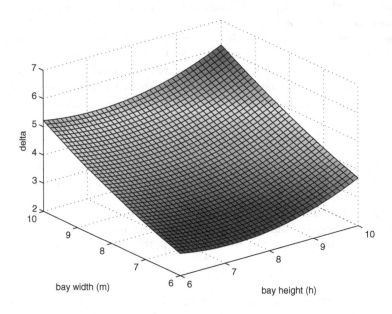

Fig. 2. Response surface: $\delta = f(h, m)$

Table 1. Regression coefficients and p-values

	Estimate	Std. Error	t value	Pr($> |t|$)
β_0	8.24093	3.45633	2.384	0.007636
β_1	-1.92831	0.63195	-3.051	0.002449
β_2	-0.02607	0.63195	-0.041	0.007114
β_3	0.13465	0.03939	3.418	0.000704
β_4	0.03962	0.03939	1.006	0.005203

In Table 2 we compare the results of the proposed scheme with those obtained running the code of [6] and of [7] on the same set of instances. It is worth noting that all values reported in the table are *average* values, computed over 40 different instances of the same class. This helps in offsetting instance specific biases in the reported results. In addition, we fixed a maximum computational time for the proposed algorithm of 60 seconds, after which the algorithm was stopped and the best solution found returned.

In Table 2, the first two columns define the instance size, in terms of number of tiers h and number of stacks m. Columns three and four report the results, in terms of number of relocations and computational time, required by the heuristic of [6]. Similarly, columns five and six report results of [7] on the same instance, both in terms of relocation moves and computational time, while columns seven and eight summarize the results of the proposed algorithm. Finally, the last two columns provide a measure of the corridor, in terms of width (δ) and height (λ).

Table 2. Computational results on large scale instances. Each class is made up of 40 instances of the same size. Reported results are average values over 40 runs.

Bay Size		KH		CM		H-BRP		Corridor	
h	m	No.	Time	No.	Time	No.	Time	δ^+	λ
6	6	37.3	0.1	32.4	7.94	30.85	0.01	3	$h+2$
6	10	75.1	0.1	49.5	15.72	46.17	3.21	5	$h+2$
10	6	141.6	0.1	102.0	30.13	76.55	6.33	4	$h+2$
10	10	178.6	0.1	128.3	65.42	105.5	18.37	6	$h+2$

$^+$: number of used stacks (not necessarily adjacent)

We solved each instance with different combinations of $\delta = \{1, 2, \ldots, \lfloor m/2 \rfloor\}$ and $\lambda = \{h + 1, h + 2\}$. We report the values used to obtain the best solution in the shortest computational time.

In addition, in order to further reduce the stochastic effects of the algorithm, we run the algorithm with the same set of parameters δ and λ five times on the same instance. In the table we report the average values over all runs of a given instance class.

As can be observed from Table 2, the proposed algorithm is competitive both in terms of solution quality and computational time, especially when it comes to dealing with larger instances. The merits of our proposed algorithm stem from the cooperative way of selecting the appropriate stacks to make up the corridor within the CM. In that sense it resembles an "educated" choice of a corridor based on a greedy rule rather than a somewhat arbitrary one which just assumes a figurative distance.

We now present statistical analysis aimed at asserting:

(i) whether there is significant difference in results among the algorithms; and
(ii) how the three algorithms rank in terms of solution quality.

In order to evaluate whether an algorithm outperforms the others used as benchmark, we select the Friedman Test. The Friedman Test ranks algorithms according to the objective function value obtained on each individual instance. In comparing each individual run, the best performing algorithm on that specific instance gets rank 1, the second best rank 2 and the third one rank 3. The null hypothesis of the test is that all algorithms are equivalent and, therefore, their ranking should be randomly distributed. In other words, if none of the algorithms were significantly better, we would expect similar average ranking for the three algorithms. Let r_{ij} be the rank of the jth algorithm on the ith instance. The average rank of an algorithm is thus defined as:

$$R_j = \frac{\sum_{i=1}^{N} r_{ij}}{N}$$

where N is the total number of benchmark instances, i.e., $N = 40 \times 4 = 160$.

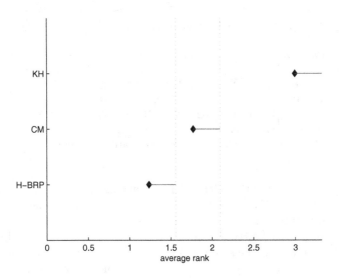

Fig. 3. Statistical analysis of results: Ranking with a 99% level of confidence

The Friedman Test produced a p-value of less than 0.0001, *i.e.*, it can be concluded that there is a significant difference among the three algorithms.Therefore, the null hypothesis can be rejected and we proceed, as indicated in [17], with a post-hoc test. The Nemenyi test can be used when all algorithms are compared to each other. The performance of two algorithms is significantly different if their corresponding average ranks differ by at least a critical distance (see [17] for more details).

Figure 3 provides a ranking of the three algorithms with a 99% level of confidence. In the figure, we can observe the average rank of each algorithm along with a line indicating such critical distance. Given an algorithm A with its critical distance value CD_A, we can assert that all the algorithms whose average rank falls beyond the critical distance value CD_A are outperformed by algorithm A. In other word, it is possible to claim that algorithm A is significantly better than those algorithms. As clearly shown in the picture, algorithm H–BRP outperforms the other two algorithms used as benchmark, since its critical value is less than the average rank of algorithms CM and KH.

5   Conclusions

In this paper we have illustrated the design of a novel CM based algorithm, in which ideas of the CM paradigm are hybridized with features of the GA as well as the GRASP. The algorithm has been employed to solve a challenging container terminal problem and its effectiveness has been proved on a set of large scale instances of such a problem. Finally, the performance of the proposed algorithm has been enhanced by the use of a thorough fine tuning phase.

A statistically sound technique, the response surface methodology, has been employed to determine "good" values of the parameter δ, *i.e.*, the corridor width around an incumbent solution, for any given problem input size. The statistical model derived by the fine tuning phase has finally been validated on a set of 160 randomly generated instances of the blocks relocation problem.

References

1. Sniedovich, M., Voß, S.: The corridor method: a dynamic programming inspired metaheuristic. Control and Cybernetics 35(3), 551–578 (2006)
2. Caserta, M., Voß, S.: A cooperative strategy for guiding the corridor method. In: Kacprzyk, J. (ed.) Studies in Computational Intelligence. Springer, Heidelberg (2009)
3. Ergun, O., Orlin, J.: A dynamic programming methodology in very large scale neighborhood search applied to the traveling salesman problem. Discrete Optimization 3, 78–85 (2006)
4. Potts, C., van de Velde, S.: Dynasearch - iterative local improvement by dynamic programming. Technical report, University of Twente (1995)
5. Yang, J.H., Kim, K.H.: A grouped storage method for minimizing relocations in block stacking systems. Journal of Intelligent Manufacturing 17, 453–463 (2006)
6. Kim, K.H., Hong, G.P.: A heuristic rule for relocating blocks. Computers & Operations Research 33, 940–954 (2006)
7. Caserta, M., Voß, S., Sniedovich, M.: An algorithm for the blocks relocation problem. Working Paper, Institute of Information Systems, University of Hamburg (2008)
8. Stahlbock, R., Voß, S.: Operations research at container terminals: a literature update. OR Spectrum 30, 1–52 (2008)
9. Watanabe, I.: Characteristics and analysis method of efficiencies of container terminal: an approach to the optimal loading/unloading method. Container Age 3, 36–47 (1991)
10. Castilho, B., Daganzo, C.: Handling strategies for import containers at marine terminals. Transportation Research B 27(2), 151–166 (1993)
11. Kim, K.H.: Evaluation of the number of rehandles in container yards. Computers & Industrial Engineering 32(4), 701–711 (1997)
12. Kim, K.H., Park, Y.M., Ryu, K.R.: Deriving decision rules to locate export containers in container yards. European Journal of Operational Research 124, 89–101 (2000)
13. Hart, J., Shogan, A.: Semi-greedy heuristics: an empirical study. Operations Research Letters 6, 107–114 (1987)
14. Festa, P., Resende, M.: An annotated bibliography of GRASP. Technical report, AT&T Labs Research (2004)
15. Box, G., Wilson, K.: On the experimental attainment of optimum conditions. Journal of the Royal Statistical Society Series B - 13, 1–45 (1951)
16. Caserta, M., Quiñonez, E.: A cross entropy-Lagrangean hybrid algorithm for the multi-item capacitated lot-sizing problem with setup times. Computers & Operations Research 36(2), 530–548 (2009)
17. Demšar, J.: Statistical comparisons of classifiers over multiple data sets. Journal of Machine Learning Research 7, 1–30 (2006)

Dynamic Multi-Armed Bandits and Extreme Value-Based Rewards for Adaptive Operator Selection in Evolutionary Algorithms

Álvaro Fialho[1], Luis Da Costa[2], Marc Schoenauer[1,2], and Michèle Sebag[1,2]

[1] Microsoft Research – INRIA Joint Centre, Orsay, France
[2] TAO team, INRIA Saclay – Île-de-France & LRI (UMR CNRS 8623), Orsay, France
`FirstName.LastName@inria.fr`

Abstract. The performance of many efficient algorithms critically depends on the tuning of their parameters, which on turn depends on the problem at hand. For example, the performance of Evolutionary Algorithms critically depends on the judicious setting of the operator rates. The Adaptive Operator Selection (AOS) heuristic that is proposed here rewards each operator based on the extreme value of the fitness improvement lately incurred by this operator, and uses a Multi-Armed Bandit (MAB) selection process based on those rewards to choose which operator to apply next. This Extreme-based Multi-Armed Bandit approach is experimentally validated against the Average-based MAB method, and is shown to outperform previously published methods, whether using a classical Average-based rewarding technique or the same Extreme-based mechanism. The validation test suite includes the easy One-Max problem and a family of hard problems known as "Long k-paths".

1 Introduction

Evolutionary Algorithms (EAs), remotely inspired from the Darwinian "survival of the fittest" principle, have been demonstrated to be efficient in tackling ill-posed optimization problems. Given a search space X, an objective function defined on X, referred to as *fitness*, and a set of elements in X, termed *population* of *individuals*, EAs iteratively proceed by (i) selecting some individuals, favoring those with better fitness; (ii) perturbing these individuals through some *variation operators*, thus generating *offspring*; (iii) evaluating the offspring fitness; (iv) replacing some individuals by some offspring, again favoring fitter offspring.

EAs have demonstrated their ability to address a wide range of optimization problems beyond the reach of standard methods, e.g. involving structured and mixed search spaces; irregular, noisy, rugged or highly constrained fitness functions. Their performance actually relies on tuning quite a few parameters (such as the population size, types of variation operators and respective application rates, types of selection mechanisms and other intrinsic parameters) depending on the problem at hand. This wealth of tunable parameters is the main reason

T. Stützle (Ed.): LION 3, LNCS 5851, pp. 176–190, 2009.

why EAs are still far away from being part of the standard optimization tool-boxes. Although knowledgeable users can benefit from this flexibility and take the most out of the evolutionary approach, the naive user will generally fail to appropriately tune an EA in a reasonable amount of time. Tuning the parameters to efficiently solve the problem at hand corresponds to an optimization problem *per se*, as noted in the very early days of the field [1]. Therefore, a mandatory step for EAs to "cross the chasm" and make it out of the research labs is to offer some automatic parameter setting capabilities. Accordingly, *Parameter Setting in EAs* was and still is one of the most active research topics in Evolutionary Computation [2] (section 2).

This paper specifically focuses on the control of the variation operators. Different operators play different roles in the search process, the importance of which depends on the current state: for instance, crossover operators ensure the *exploration* of wide regions of the search space in the early stages of evolution; meanwhile, mutation operators both ensure the *exploitation* and local search around the current best individuals, at any stage, and prevent the loss of diversity in the last stages of evolution. The so-called *Exploration vs Exploitation* trade-off thus relies on the mutation and crossover rates; in practice, these are most often defined by the user once for all, depending on his experience and intuition, although the need for exploration and exploitation clearly varies as the search goes on.

Adaptive Operator Selection (AOS) is meant to adaptively update the operator rates online, depending on e.g. the fitness improvement brought by the offspring. Since Davis' seminal work [3] (section 2), AOS proceeds by combining two main ingredients, illustrated in Fig. 1: The *Credit Assignment* mechanism associates a reward to each operator, reflecting the operator impact on the progress of the search (e.g. fitness improvement); the *Operator Selection* mechanism actually chooses one operator depending on the past associated rewards of all operators.

This paper investigates the combination of an *Operator Selection* and *Credit Assignment* heuristics, first described in [4] and [5] respectively. The proposed *Operator Selection* rule ([4], section 3.1), performs the dynamic operator selection by combining a Multi-Armed Bandit algorithm [6] with a statistical test for change point detection, the Page-Hinkley test [7], assuming an unbiased Credit Assignment mechanism. The proposed *Credit Assignment* ([5], section 3.2) considers the extreme values of the fitness improvement due to an operator, claiming that rare but highly beneficial "jumps" matter as much or more than frequent but small improvements.

A proof of principle of the AOS combining the above Extreme-Value-based *Credit Assignment* and the Dynamic Multi-Armed Bandit *Operator Selection* rule, referred to as *Ex-DMAB*, is presented in this paper; this proof of principle considers the easy One-Max problem and a family of hard problems, the Long k-paths [8]. Not only are the Long k-path landscapes more difficult than the One-Max ones (the former involves a single, exponentially long path leading to the global optimum, together with many shortcuts, while the latter involves many paths with linear length leading to the global optimum); overall, they

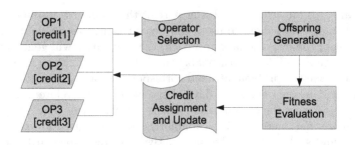

Fig. 1. General scheme of the Adaptive Operator Selection framework

can be considered to be deceptive in terms of operator selection (more on this in section 4.3). Section 4 reports on the experimental results of the approach, showing significant improvements with respect to baseline approaches. The paper concludes with some perspectives for further research.

2   Related Work

This section briefly describes and discusses the state of the art in Evolutionary Parameter Setting, focusing on *Adaptive Operator Selection*.

2.1   Parameter Setting in EAs

After [2,9,10], Parameter Setting in EAs includes two main categories of heuristics, respectively referred to as *Parameter Tuning* and *Parameter Control*:

- In *Parameter Tuning* (also known as *off-line* or *external* tuning), parameters are *tuned* before the run. This category mostly includes statistical methods derived from *Design Of Experiments* (see *e.g.*, [11,12,13,14]). Although more efficient than standard ANOVA methods, Parameter Tuning relies on extensive, computationally expensive, experiments. Furthermore, there is strong empirical evidence that the optimal parameter values actually vary between the beginning and the end of evolution; choosing the parameter values once for all thus results in a sub-optimal setting.
- In *Parameter Control* (also known as *on-line* or *internal* control), parameters are *controlled* during the run. This category can be further divided into three types of approaches:
    1. In *Deterministic Control*, parameter values are predefined functions of time, which clearly raises the question of how to define such functions (defining these *a priori* is but another Parameter Setting issue).
    2. In *Self-Adaptive Control*, parameters are encoded in the genotype, and therefore tuned and optimized "for free" by evolution itself. The main weakness of self-adaptive control is to aggregate the solution and the parameter spaces, thus increasing the overall complexity of the optimization problem.

3. In *Adaptive* (or *Feedback-Based*) *Control*, parameter values are predefined functions of the whole history of the run. Adaptive control has met significant successes in the last decade, specifically in the continuous optimization framework (see [15] and references therein).

Focusing on *Adaptive Operator Selection* (AOS), the history of the run is used to adjust the operator rates through two modules: a *Credit Assignment* module computes the operator reward based on its impact on the search progress; an *Operator Selection* module exploits these rewards and selects the operator to be applied next.

2.2   Credit Assignment

Several *Credit Assignment* mechanisms have been proposed in the literature, starting back in the late 80s with the seminal work of Davis [3]. In most approaches, the operator credit, aka *reward*, reflects the fitness of the offspring built by the operator. More specifically, the reward measures the fitness improvement over some reference fitness: that of the offspring parents [16,17,18], of the current best [3] or median [19] individual. Offspring which do not improve on the reference fitness are simply not taken into account.

In some cases however, fitness improvement is but one element relevant to the progress of evolution. Typically in multi-modal search landscapes, population diversity is equally important; it must mandatorily be preserved in order to avoid premature convergence to local optima. Based on this remark, the so-called *Compass* credit assignment [20] measures the operator ability to produce more fit individuals while preserving the population diversity.

While in all above approaches the operator reward is based on the current fitness improvement, or on the fitness improvement averaged over the last n offspring, another approach is proposed in [21]. This latter approach uses a statistical measure aimed at outlier detection, and the authors report significant improvement comparatively to other *Credit Assignment* on a set of continuous benchmark problems.

A last issue concerns the offspring contribution to the operator rewards. Most authors only reward the operator used to produce the current offspring [16,17,18]; other authors consider that it is only fair to reward the operators used to produce the offspring ancestors, e.g. using a bucket brigade algorithm [3,19].

2.3   Operator Selection Rules

The simplest and most widely used *Operator Selection* is *Probability Matching* (PM) [16,18,22]. PM implements a roulette wheel-like selection process, where the operator rate is proportional to its reward. Some care is however exercised in order to enforce a sufficient amount of exploration, through keeping the operator rate above some threshold p_{min}. Otherwise, an operator which is inefficient in the early stages of evolution would never be considered again, even though it might become the best operator later on. A side effect however is to keep the best operator rate below $p_{max} = 1 - (K - 1)p_{min}$ being K the number of operators. In practice, all mildly relevant operators keep being selected, slowing down evolution [23].

This drawback is partly addressed by *Adaptive Pursuit* (AP) [23], a method originally proposed for learning automata, which implements a winner-takes-all strategy. The main difference compared to PM is that the rate of the best rewarded operator goes to p_{max} whereas all the others go to p_{min}; an additional β parameter controls the greediness of the winner-take-all update.

Others, such as APGAIN [24], use a sequence of exploration/exploitation phases. During each exploration phase, operators are uniformly selected and their rewards are estimated; during the following exploitation phase, operators are selected according to their reward. The fraction of generations devoted to exploration phases (circa 25 % in [24]) is meant to catch up with the changes in the reward distribution; unfortunately, it severely harms the population and the progress of evolution whenever disruptive operators are considered [20].

3 Extreme Dynamic Multi-Armed Bandit

The Extreme Dynamic Multi-Armed Bandit (*Ex-DMAB*) AOS combines Dynamic-Multi Armed Bandit as *Operator Selection* rule and Extreme Value Based *Credit Assignment*. For the sake of self-containedness, this section summarizes both heuristics, referring the interested reader respectively to [4] and [5] for more details.

3.1 Dynamic Multi-Armed Bandit

The choice of an operator within an Evolutionary Algorithm can be viewed as yet another instance of the *Exploration vs. Exploitation* (EvE) dilemma: on the one hand, one wishes to select the operator with best empirical behavior (exploitation); on the other hand, other operators should also be selected in order to check whether the best empirical operator so far truly is the best one (exploration). This dilemma has been intensively studied in the context of *Game Theory* within the so-called Multi-Armed Bandit (MAB) framework [6,25].

The MAB framework involves a set of N arms; the i-th arm, when selected, gets reward 1 with probability p_i and 0 otherwise. A MAB algorithm is a decision making algorithm, selecting an arm at every time step with the goal of maximizing the cumulative reward gathered along time. The widely studied Upper Confidence Bound (UCB) algorithm devised by Auer *et al.* [6], provably maximizing the cumulative reward with optimal convergence rate, proceeds as follows. Let $n_{i,t}$ denote the number of times the ith arm has been played up to time t, and let $\hat{p}_{i,t}$ denote the average empirical reward received from arm i. UCB1 selects in each time step t the arm maximizing the following quantity:

$$\hat{p}_{j,t} + C * \sqrt{\frac{\log \sum_k n_{k,t}}{n_{j,t}}} \tag{1}$$

The left term in Eq. (1) favors the option with best average empirical reward (exploitation). The right term ensures that each arm is selected infinitely often (exploration); the lapse of time between two selections of under-optimal arms however increases exponentially. The scaling factor C controls the exploration/exploitation trade-off.

The operator selection problem can indeed be formalized as a MAB problem, taking each operator as an arm [4], with two caveats. Firstly, arms are assumed to be independent, which is definitely not the case in AOS as operators apply on the same population. Secondly, and even more importantly, the reward probabilities are fixed in standard MAB settings, whereas the operator rewards depend on the current population and the evolution stage. In other words, AOS corresponds to a dynamic MAB problem. It must be emphasized that although UCB keeps exploring all arms, it would need quite some time to detect that the best operator has changed. Therefore, a statistical change detection test was coupled with UCB in [4], defining the Dynamic MAB (DMAB) algorithm. Specifically, the Page-Hinkley (PH) test [7] is used to detect whether the empirical rewards collected for the best current operator undergo an abrupt change. Upon triggering the PH test (suggesting that the current best operator is no longer the best one), the MAB algorithm is restarted from scratch.

Formally, the PH test considers $\bar{r}_t$, the empirical average of the instant rewards $r_1, \ldots r_t$. Let e_t denote the difference $r_t - \bar{r}_t + \delta$, where δ is a tolerance parameter, and let m_t be the sum of e_i for $i = 1$ to t. The PH test is triggered when the difference between the maximum of $|m_i|$ for $i = 1$ to t, and the current $|m_t|$ is greater than a user-specified threshold γ. The PH test is thus controlled from two parameters, γ governing the trade-off between false alarms and unnoticed changes, and δ enforcing the test robustness when dealing with slowly varying distributions. Following initial experiments [4], δ is set to 0.15 in all experiments in this paper.

3.2 Extreme Value Based Credit Assignment

The second AOS component measures the operator impact on the progress of evolution 2.2. Letting $\mathcal{F}$, o and x respectively denote the fitness function (to be maximized), a variation operator and an element of the current population, the standard instant reward is set to the current fitness improvement of the offspring $(\mathcal{F}(o(x)) - \mathcal{F}(x))^+$ (the $^+$ superscript indicates the positive part of the fitness difference).

The main originality of the *Credit Assignment* proposed in [5] is to consider the *extreme* as opposed to the *average* instant reward. Let us compare an operator bringing frequent small improvements, and an operator bringing rare but large improvements. Even though both operators might have the same expected impact on evolution, with high probability an average-reward based AOS would only consider the former one: after the first trials, the former operator dominates the latter one, which is thus hardly selected thereafter, and thus prevented from gathering any further rewards. In other words, average reward-based AOS is risk-adverse. Another strategy, first investigated by [21], thus is to consider extreme rewards. Notably, the role of extreme events in design has long been acknowledged in numerical engineering (e.g. taking into account rogue waves when dimensioning an oil rig); it receives an ever growing attention in the domain of complex systems, as extreme events govern diffusion-based processes ranging from epidemic propagation to financial markets.

The *Extreme Value Based* (EVB) *Credit Assignment* first presented in [5] proceeds as follows. To each operator o is associated a register storing the last W (positive) instant rewards collected by o. The operator reward used within the DMAB *Operator Selection* is the maximum instant reward in the operator register. This *Credit Assignment* mechanism thus involves the window size W as single parameter. W is meant to reflect the time scale of the process; if too large, operators will be applied after their optimal epoch and the switch from the previous best operator to the new best one will be delayed. If W is too small, operators causing large but infrequent jumps will be ignored (as successful events will not be observed at all in the first place) or too rapidly forgotten.

4   Experimental Results

This section reports on the empirical validation of the Extreme - Dynamic Multi-Armed Bandit (*Ex-DMAB*) AOS, combining Extreme-Value-Based *Credit Assignment* and DMAB *Operator Selection*, first described in [5].

Previous results on the One-Max problem, the "Drosophila of EC", are recalled in section 4.2, and comparatively discussed with respect to some new results obtained with Average-Value-Based *Credit Assignment*. A different family of problems, the Long k-paths [26], is considered in section 4.3. Both sets of experiments have been conducted using the same experimental setting, described in section 4.1.

4.1   Experimental Setting

All experiments consider a standard $(1 + \lambda)$-EA, where λ offspring are created from a single parent, and the best individual among the current offspring and parent becomes the parent in the next generation. For the sake of reproducibility, the initial individual is set to $(0, \ldots, 0)$.

For the simplicity of assessment, the AOS only considers mutation operators: the standard $1/\ell$ bit-flip operator (every bit is flipped with probability $1/\ell$, where ℓ is the bit-string length), the 1-bit, 3-bit and 5-bit mutation operators (the b-bit mutation flips exactly b bits, uniformly selected in the parent). This setting makes it feasible to compute the optimal mutation operator depending on the stage of evolution (fitness of the current parent), using Monte-Carlo simulations. Two baseline approaches are considered: the first one uniformly selects an operator out of the whole set of operators; the second one selects an operator out of the best two operators.

Two *Credit Assignment* procedures are considered: the Extreme-Value and the Average-Value based reward (out of the last W instant rewards for the operator).

These are combined with three *Operator Selections*: AP, PM and DMAB. The results obtained with PM will be omitted as this *Operator Selection* is found significantly dominated by the other two.

Every AOS is assessed from the average time-to-solution, averaged over 50 independent runs. The ability of each AOS to correctly identify the best operator is also considered.

The best parameters of every considered AOS have been computed offline, in order to compare them at their best level of performance (see [4,5]). For the One-Max scenario, the parameters were determined after a Design of Experiment campaign [4,5]. For the Long k-path, the following set of values was tried for each parameter: for AP and PM, $p_{min} \in \{0, .05, .1, .2\}$; $\alpha|\beta \in \{.1, .3, .6, .9\}$; for DMAB, $C \in \{.1, .5, 1, 5, 10, 50, 100\}$; $\gamma \in \{1, 5, 10, 25, 50, 100, 250, 500, 1000\}$; and for all techniques, concerning the *Credit Assignment*, $W \in \{50, 500\}$. Given the number of possible configurations, the F-Race [11] method was used. Formally, racing techniques proceed by pruning every configuration as soon as it is not going to be the best one after the available statistical evidence. The F-Race was applied using a confidence level of 95%, with 11 runs being done for each configuration before the first elimination round, up to 50 runs done or a single candidate configuration left.

4.2   The One-Max Problem

The One Max problem involves an unimodal fitness function that simply counts the number of "1"s in the individual binary bitstring. The only difficulty comes from the size of the problem; in the presented experiments, the size N of the bitstring is 10,000. This problem is viewed as a "sterile EC-like" environment, where the ideal AOS behavior can be computed. Fig. 2 (bottom) displays the optimal mutation operators for a $(1 + 50)$-EA, depending on the stage of evolution; for each fitness of the current parent, the expected fitness improvement (estimated over 100 independent runs) of a $(1+50)$-EA is computed. The landscape presented in such figure thus serves as a reference to assess the basic skills of an AOS mechanism: the ability to pick up the best operator and stick to it as long as appropriate, to catch up the changes and switch to the next best operator in transition phases, and to remain efficient in desert phases.

Fig. 2 displays the operator rates of *Ex-DMAB* and *Ex-AP* (averaged over 50 runs) against the "oracle"; the vertical grey lines indicate the changes of the current best mutation operator of the oracle.

Table 1 summarizes the performance of all approaches, together with their optimal setting. The Extreme Value-Based *Credit Assignment*, coupled with either AP or DMAB, closely matches the optimal behavior and significantly improves on the baseline approaches (see [5] for more detail). Interestingly, the Extreme Value-Based *Credit Assignment* appears to be more stable than the Average Value-Based one (the performance of the latter is much degraded when combined with DMAB), despite the smoothness of the One-Max landscape (which should thus make little difference between average and extreme rewards).

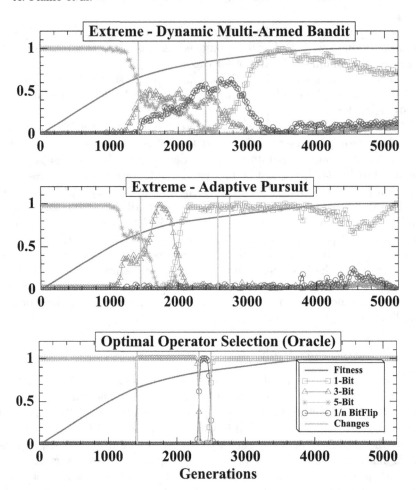

Fig. 2. The *Ex-DMAB* (top) and the *Ex-AP* (middle) AOS compared with the Optimal Operator Selection (bottom) on the 10,000 bit One-Max: operator selection rates averaged on 50 runs

Table 1. Comparative AOS Results on the 10,000 One-Max problem, averaged over 50 runs ($W = 50$)

Credit Assignment	Operator Selection	Configuration	Gens. to Optimum
Extreme	DMAB	$C = 1; \gamma = 250$	**5467** $\pm$ 513
Average		$C = 10; \gamma = 25$	7727 $\pm$ 642
Extreme	Adaptive Pursuit	$p_{min} = 0; \alpha = .3; \beta = .3$	**5478** $\pm$ 299
Average		$p_{min} = .05; \alpha = .1; \beta = .9$	5830 $\pm$ 324
-	Optimal Strategy	Given by "Oracle"	**5069** $\pm$ 292
-	Best Naive	$\mathcal{U}$(1-Bit+5-Bit)	6793 $\pm$ 625
-	Complete Naive	$\mathcal{U}$(4 ops.)	7813 $\pm$ 708

4.3   The Long k-Path Problem

First proposed by [26], Long Paths are unimodal problems designed to challenge local search algorithms. Specifically, the optimum can be found by following a path in the fitness landscape, the length of which increases exponentially w.r.t. the bitstring length ℓ. Solving the Long Path using the 1-bit mutation thus requires a time increasing exponentially with ℓ.

A generalization of Long Path problems was proposed by [8], referred to as Long k-path, where k is the minimal number of bits to be simultaneously flipped in order to take a shortcut on the path. Long k-path problems have the following properties [27]:

- Points that are not on the path have a "Zero-Max" fitness, i.e. their fitness is their number of 0s;
- The first point on the path is $0, 0, \ldots, 0$ and has fitness ℓ;
- Any point on the path has exactly 2 neighbors on the path with Hamming distance 1; two consecutive points on the path have a fitness difference of 1;
- The length of the path is $(k + 1)2^{(l-1)/k} - k + 1$;
- A mutation of $i < k$ bits can only lead to a point which is either off the path (hence with a very low fitness), or on the path but only i positions away from the original point; *shortcuts* (i.e. jumps to very distant points on the path) can only be achieved by mutating at least k bits; using a mutation operator which mutates every bit independently with probability p, the probability of finding a given shortcut is hence $p^k(1 - p)^{l-k}$.

Long k-path problems are defined by recurrence on ℓ. Starting from the Long k-path $P(k, \ell)$, the $P(k, \ell + k)$ path is made of three parts: (i) the first part S_0 is made by concatenating k 0's to each point of $P(k, \ell)$; the third part S_1 is made by concatenating k 1's to each point of $P(k, \ell)$ in reverse order; S_0 and S_1 are linked by a "bridge" containing $(k - 1)$ points, created by concatenating $0 \ldots 01, 0 \ldots 011, \ldots, 001 \ldots 1, 01 \ldots 1$ to the final point of $P(k, \ell)$. The original Long Path problem is a Long k-path with $k = 2$. The path length decreases as k increases, together with the probability of finding a shortcut.

After [27], shortcuts provably speed up the convergence if $k \leq \sqrt{\ell}$ (for higher values of k one should simply follow the path). In such cases, "exceptional properties of operators sometimes reflect EA behavior more accurately than average typical properties".

AOS and Long 3-Path Problems

The reported experiments consider $k = 3$ with ℓ ranging in $\{43, 49, 55, 61\}$. An additional mutation operator, the $3/\ell$ bit-flip (flipping each bit with probability $3/\ell$), has been added to the operator set.

Note that Long k-paths are challenging problems for AOS: when the parent individual belongs to the path, the 1-bit mutation improves the fitness by 1, with probability $1/\ell$ while all other mutation operators will fail to improve the fitness (reward 0) in the vast majority of cases. Experimentally, the *Adaptive Pursuit* AOS does not cope well with Long k-paths and will be omitted in the following; the best results are obtained for $p_{min} = .2$, i.e. for a uniform selection of the operators.

Results

By construction, some Long k-path runs can be "lucky" and discover the shortcuts, thus yielding large standard deviations in the performance. For instance, the optimal result obtained for $\ell = 49$ reaches the solution in 3590 ± 3327 generations (averaged over 50 runs). For this reason, the results will be described in terms of min and median number of generations needed to reach the solution (as opposed to, average and standard deviation).

Table 2 displays the results obtained for *Ex-DMAB*, *Avg-DMAB* and the baseline approaches: the Oracle one always selects the optimal operator (determined in the same way as for the One-Max problem) and the Naive one uniformly selects an operator in the operator set. The AOS setting is the best one found by the F-Race over all considered Long k-path.

Table 3 reports the results obtained for the best AOS setting, found by the F-Race over each considered Long k-path. The optimal setting $(W(C, \gamma))$ is indicated below the min and median number of generations to the solution. As could have been expected, the *Avg-DMAB* AOS goes for a medium window size ($W = 50$) whereas the *Ex-DMAB* needs a much larger window size ($W = 500$). Besides, the F-Race retains many good settings for the AOS parameters, suggesting that the C and γ parameters together control the Exploration vs Exploitation tradeoff, and might be redundant to some extent.

Table 2. Extreme vs Average Reward and DMAB AOS on the Long k-path, $k = 3$, min - median number of generations to the solutions out of 50 runs; the robust optimal AOS parameters $(W, (C, \gamma))$ are indicated below

ℓ	DMAB - $W(C, \gamma)$		Optimal	Uniform
	Extreme	Average		
	500 (100; 100)	50 (50; .5)		
43	11 - 2579	61 - 2342	2 - 1202	50 - 3393
49	17 - 4467	6 - 6397	19 - 2668	5 - 4904
55	161 - 6190	54 - 8222	45 - 3224	344 - 10068
61	251 - 13815	94 - 15304	8 - 5408	12 - 9590

Table 3. Extreme vs Average Reward and DMAB AOS on the Long k-path, $k = 3$, min - median number of generations to the solutions out of 50 runs, using the optimal AOS parameter for each ℓ

ℓ	DMAB - $W(C, \gamma)$		Optimal	Uniform
	Extreme	Average		
43	11 - 2216	66 - 2487	2 - 1202	50 - 3393
	500(50; 50)	50(.5; 100)		
49	17 - 3244	6 - 5321	19 - 2668	5 - 4904
	500(100; 500)	50(.1; 1000)		
55	161 - 6190	54 - 8158	45 - 3224	344 - 10068
	500(100; 100)	50(50; .1)		
61	80 - 10253	94 - 13865	8 - 5408	12 - 9590
	500(50; 25)	50(.5; 50)		

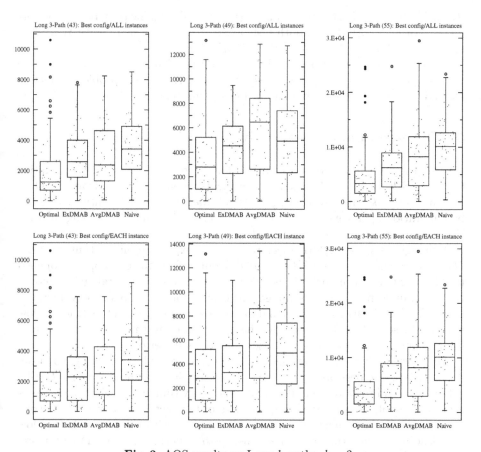

Fig. 3. AOS results on Long k-paths, $k = 3$

The significance of these results is assessed using unsigned Wilcoxon rank sum and Kolmogorov-Smirnov non-parametric tests (thereafter referred to as W and KS).

In the robust scenario (AOS parameters are selected by F-race over all Long k-path problems), *Ex-DMAB* is outperformed by (respectively similar to) the Oracle AOS for $\ell \in \{43, 61\}$ (resp. $\ell \in \{49, 55\}$) with confidence 99% according to both W and KS tests. *Ex-DMAB* outperforms *Avg-DMAB* for $\ell = 49$ (with W at 90% and K at 95%) and for $\ell = 55$ (with W at 90%).

In the fine tuning scenario (AOS parameters are selected by F-race for each Long k-path problem), *Ex-DMAB* obtains better results as could have been expected: no significant difference between *Ex-DMAB* and the Oracle strategy is observed, with confidence 99% according to both W and KS tests. In the meanwhile, *Ex-DMAB* significantly improves on the Naive AOS in all instances ($\ell = 43$, for W at 99% and KS at 95%; $\ell = 49$, for W at 95% and KS at 90%; $\ell = 55$, for W and KS at 99%), except for the $\ell = 61$ one. Comparatively to

Avg-DMAB, *Ex-DMAB* obtains similar (respectively significantly better) performances for $\ell \in \{43, 61\}$ with both tests at 99% (resp. for $\ell = 49$, with W at 99% and KS at 95%; for $\ell = 55$, with W at 90%).

The empirical distributions of all approaches are displayed on Fig. 3. The case $\ell = 61$ is omitted as no strategy was found effective on this problem, which is blamed on the very low probability of finding shortcuts.

5   Discussion and Perspectives

This paper provides a proof of principle for the proposed *Ex-DMAB* Adaptive Operator Selection, based on ample empirical evidence gathered from the One-Max and Long k-path problems. *Ex-DMAB* was found to efficiently detect the best mutation operators during the whole course of evolution, keeping up with the Oracle strategy in the majority of cases.

Although its good performances rely on the expensive offline tuning of *Ex-DMAB* parameters, *Ex-DMAB* was found to outperform the main options opened to the naive EA user, namely (i) using a fixed or deterministic strategy (including the naive, uniform selection, strategy; (ii) using a former AOS strategy. Furthermore, *Ex-DMAB* involves a fixed and limited number of parameters (the window size W, the scaling factor C and the change detection test threshold γ), whereas the number of operator rates increases with the number of operators.

The most challenging situations for *Ex-DMAB* are the last stages of evolution. At this point, the best operator hardly brings any improvement, and *Ex-DMAB* is found to tend toward the uniform naive strategy (uniformly selecting an operator in the pool). A tentative interpretation for this fact is as follows. On the one hand, fitness improvements are more and more rare with respect to the window size, leading to uniform rewards and hence to uniform selection. Furthermore, when a reward occurs after a long wandering period, it is likely to trigger the change detection test, causing the Dynamic Multi-Armed Bandit to restart from scratch, thus increasing the exploration bias.

Further research will aim at addressing the above weaknesses. A first perspective is opened by learning across runs, specifically recording the statistics of fitness improvement in relation with the current average fitness; these statistics will serve to adjust the window length W in the last stages of evolution. A second perspective is to adjust online the Page Hinkley parameter γ, depending on the estimated number of transitions (change of the best operator) in the fitness landscape. Along the same lines, we shall investigate how γ and the scaling factor C relate, as both cooperate to control the exploration vs exploitation trade-off.

Acknowledgments. This work was supported in part by the European STREP EvoTest (IST-33472) and by the European Network of Excellence PASCAL-2. The authors thank Olivier Teytaud, INRIA Saclay - Île-de-France, for fruitful discussions.

References

1. Grefenstette, J.: Optimization of control parameters for genetic algorithms. IEEE Transactions on Systems, Man and Cybernetics 16(1), 122–128 (1986)
2. Lobo, F., Lima, C., Michalewicz, Z. (eds.): Parameter Setting in Evolutionary Algorithms. Studies in Computational Intelligence, vol. 54. Springer, Heidelberg (2007)
3. Davis, L.: Adapting operator probabilities in genetic algorithms. In: Schaffer, J.D. (ed.) Proc. ICGA 1989, pp. 61–69. Morgan Kaufmann, San Francisco (1989)
4. Da Costa, L., Fialho, A., Schoenauer, M., Sebag, M.: Adaptive operator selection with dynamic multi-armed bandits. In: Keijzer, M. (ed.) Proc. GECCO 2008, pp. 913–920. ACM Press, New York (2008)
5. Fialho, A., Da Costa, L., Schoenauer, M., Sebag, M.: Extreme value based adaptive operator selection. In: Rudolph, G., Jansen, T., Lucas, S., Poloni, C., Beume, N. (eds.) PPSN 2008. LNCS, vol. 5199, pp. 175–184. Springer, Heidelberg (2008)
6. Auer, P., Cesa-Bianchi, N., Fischer, P.: Finite-time analysis of the multiarmed bandit problem. Machine Learning 47(2-3), 235–256 (2002)
7. Hinkley, D.: Inference about the change point from cumulative sum-tests. Biometrika 58(3), 509–523 (1971)
8. Rudolph, G.: Convergence Properties of Evolutionary Algorithms. Verlag Dr. Kovac (1997)
9. Eiben, A.E., Hinterding, R., Michalewicz, Z.: Parameter control in Evolutionary Algorithms. IEEE Transactions on Evolutionary Computation 3(2), 124–141 (1999)
10. Eiben, A.E., Michalewicz, Z., Schoenauer, M., Smith, J.E.: Parameter control in Evolutionary Algorithms. In: Lobo, F.G., et al. (eds.) Parameter Setting in Evolutionary Algorithms, pp. 19–46. Springer, Heidelberg (2007)
11. Birattari, M., Stützle, T., Paquete, L., Varrentrapp, K.: A racing algorithm for configuring metaheuristics. In: Langdon, W.B., et al. (eds.) Proc. GECCO 2002, pp. 11–18. Morgan Kaufmann, San Francisco (2002)
12. Yuan, B., Gallagher, M.: Statistical racing techniques for improved empirical evaluation of evolutionary algorithms. In: Yao, X., Burke, E.K., Lozano, J.A., Smith, J., Merelo-Guervós, J.J., Bullinaria, J.A., Rowe, J.E., Tiňo, P., Kabán, A., Schwefel, H.-P. (eds.) PPSN 2004. LNCS, vol. 3242, pp. 172–181. Springer, Heidelberg (2004)
13. Bartz-Beielstein, T., Lasarczyk, C., Preuss, M.: Sequential parameter optimization. In: McKay, B. (ed.) Proc. CEC 2005, pp. 773–780. IEEE Press, Los Alamitos (2005)
14. Nannen, V., Eiben, A.E.: Relevance estimation and value calibration of evolutionary algorithm parameters. In: Veloso, M. (ed.) Proc. IJCAI 2007, pp. 975–980 (2007)
15. De Jong, K.: Parameter Setting in EAs: a 30 Year Perspective. In: Lobo, F.G., et al. (eds.) Parameter Setting in Evolutionary Algorithms, pp. 1–18. Springer, Heidelberg (2007)
16. Lobo, F., Goldberg, D.: Decision making in a hybrid genetic algorithm. In: Porto, B. (ed.) Proc. ICEC 1997, pp. 121–125. IEEE Press, Los Alamitos (1997)
17. Tuson, A., Ross, P.: Adapting operator settings in genetic algorithms. Evolutionary Computation 6(2), 161–184 (1998)
18. Barbosa, H.J.C., Sá, A.M.: On adaptive operator probabilities in real coded genetic algorithms. In: Workshop on Advances and Trends in AI for Problem Solving – SCCC 2000 (2000)

19. Julstrom, B.A.: What have you done for me lately? Adapting operator probabilities in a steady-state genetic algorithm on genetic algorithms. In: Eshelman, L.J. (ed.) Proc. ICGA 1995, pp. 81–87. Morgan Kaufmann, San Francisco (1995)

20. Maturana, J., Saubion, F.: A compass to guide genetic algorithms. In: Rudolph, G., Jansen, T., Lucas, S., Poloni, C., Beume, N. (eds.) PPSN 2008. LNCS, vol. 5199, pp. 256–265. Springer, Heidelberg (2008)

21. Whitacre, J.M., Pham, T.Q., Sarker, R.A.: Use of statistical outlier detection method in adaptive evolutionary algorithms. In: Keijzer, M. (ed.) Proc. GECCO 2006, pp. 1345–1352. ACM Press, New York (2006)

22. Goldberg, D.E.: Probability matching, the magnitude of reinforcement, and classifier system bidding. Machine Learning 5(4), 407–425 (1990)

23. Thierens, D.: An adaptive pursuit strategy for allocating operator probabilities. In: Beyer, H.G. (ed.) Proc. GECCO 2005, pp. 1539–1546. ACM Press, New York (2005)

24. Wong, Y.Y., Lee, K.H., Leung, K.S., Ho, C.W.: A novel approach in parameter adaptation and diversity maintenance for genetic algorithms. Soft Computing 7(8), 506–515 (2003)

25. Lai, T., Robbins, H.: Asymptotically efficient adaptive allocation rules. Advances in Applied Mathematics 6(1), 4–22 (1985)

26. Horn, J., Goldberg, D.E., Deb, K.: Long path problems. In: Davidor, Y., Männer, R., Schwefel, H.-P. (eds.) PPSN 1994. LNCS, vol. 866, pp. 149–158. Springer, Heidelberg (1994)

27. Garnier, J., Kallel, L.: Statistical distribution of the convergence time of evolutionary algorithms for long-path problems. IEEE Transactions on Evolutionary Computation 4(1), 16–30 (2000)

Comparison of Coarsening Schemes for Multilevel Graph Partitioning

Cédric Chevalier[1] and Ilya Safro[2]

[1] Sandia National Laboratories, Albuquerque, NM, USA
ccheval@sandia.gov[*]
[2] Argonne National Laboratory, Argonne, IL, USA
safro@mcs.anl.gov[**]

Abstract. Graph partitioning is a well-known optimization problem of great interest in theoretical and applied studies. Since the 1990s, many multilevel schemes have been introduced as a practical tool to solve this problem. A multilevel algorithm may be viewed as a process of graph topology learning at different scales in order to generate a better approximation for any approximation method incorporated at the uncoarsening stage in the framework. In this work we compare two multilevel frameworks based on the geometric and the algebraic multigrid schemes for the partitioning problem.

1 Introduction

Graph partitioning is a computing technique used in many fields of computer science and engineering. Applications include VLSI design, minimizing the cost of data distribution in parallel computing, optimal tasks scheduling, etc. The goal is to partition the vertices of a graph into a certain number of disjoint sets of approximately the same size, so that a cut metric is minimized. Because of the NP-hardness [1] of the problem and its practical importance, many heuristics of different nature (spectral [2], combinatorial [3,4], evolutionist [5], etc.) have been developed to provide an approximate result in a reasonable (and, one hopes, linear) computational time. However, only the introduction of the multilevel methods during the 1990s has really provided a breakthrough in efficiency and quality.

During the past two decades many attempts have been made to use multilevel strategies for solving combinatorial optimization problems [6,7]. The most frequent branches on which the multilevel algorithms have been applied are VLSI design [8,9,10], graph optimization problems [11] (with special attention to the partitioning problem [12,13,14,15,16,17,18]), and several others [19,20,21].

---

[*] Sandia is a multiprogram laboratory operated by Sandia Corporation, a Lockheed Martin company, for the U.S. Department of Energy's National Nuclear Security Administration under contract DE-AC04-94AL85000.

[**] This work was supported by the Office of Advanced Scientific Computing Research, Office of Science, U.S. Department of Energy, under Contract DE-AC02-06CH11357.

T. Stützle (Ed.): LION 3, LNCS 5851, pp. 191–205, 2009.

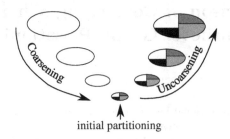

initial partitioning

Fig. 1. Example of multilevel framework for a 4-partitioning problem. Three empty ellipses represent the three levels of the coarsening. The smallest colored by four colors ellipse corresponds to the coarsest level graph. Four colors of the graphs through the uncoarsening stage correspond to the 4-partitioning approximation.

The main objective of a multilevel based algorithm is to create a hierarchy of problems (*coarsening*), each representing the original problem, but with fewer degrees of freedom. For the partitioning and other graph modeled problems, this hierarchy may be viewed as a process of learning of a graph topology prior to applying any approximation method. The construction of hierarchies at different scales ends up at the level with a very small number of degrees of freedom (*coarsest level*) that allows to get a first approximation to the original problem at very large scale within an insignificant running time (even for exact algorithm) in comparison to the size of original problem. Then, the obtained approximation is sequentially projected along all levels of the hierarchy (*interpolation* or *projection*) until it reaches the original problem with some approximation for it. The projection stage can be reinforced at each level by some *refinement* algorithm that improves the quality of approximation before further projection. The projection reinforced by a refinement method is called *uncoarsening*. In terms of graph partitioning problem, the hierarchy of coarse graphs is constructed for different scales, and at each scale the approximation algorithm for this problem is applied in order to project and improve current approximations. In Figure 1, we present a small example of a multilevel framework (called V-cycle) for the 4-partitioning problem. For needed references and background on multilevel techniques, we refer the reader to [6].

Almost all previously developed multilevel schemes for simple graphs possess exactly the same strict coarsening. It is carried out by matching groups (usually pairs) of vertices together and representing each group with a single vertex in the coarsened space (e.g., matching [17,18], first choice [10]). Another class of multilevel schemes used for several combinatorial optimization problems is based on an *algebraic multigrid* (AMG) method [22,23,15,11]. The principal difference between these two approaches is explained in graph model terms in [11]. Because of the difficulties in performing a rigorous analysis of multilevel schemes for discrete problems, the empirical judgment of all these algorithms is usually based on the best achieved results on some test set. Multilevel algorithms consist of many algorithmic parts, and it is not easy to realize which part plays the crucial role. This paper is about the role of a coarsening scheme in a multilevel framework.

The main goal of this paper is a systematic comparison of the AMG-based scheme versus strict scheme based on heavy edge matching (HEM, adopted since 1995 and implemented in many multilevel packages) for the partitioning problem while having the uncoarsening parts (based on the popular sequential algorithm called Fiduccia-Mattheyses (FM) [4]) exactly the same in both cases. This issue still has not been studied empirically, in contrast to many other works in which a number of a more or less successful uncoarsening and postprocessing procedures have been suggested. The framework used for these experiments is SCOTCH [24], since it provides an open architecture to easily plug in different algorithms and choose between them at the runtime with the strategy string, a powerful way to dynamically choose the methods and the parameters we want to use. The AMG-based coarsening procedure was taken from [11].

2   Definitions and Notation

Consider a simple weighted graph $G = (V, E)$, where $V = [1, n]$ is the set of vertices (nodes) and E is the set of edges. Denote by w_{ij} the non-negative weight of the undirected edge $ij \in E$; if $ij \notin E$, then $w_{ij} = 0$. Let v_i be a positive weight of vertex $i \in V$ and $v(A) = \sum_{i \in A} v_i$, where $A \subseteq V$.

The goal of the general graph k-partitioning problem is to find a partition of V into a family of k disjoint nonempty subsets $(\pi_p)_{1 \leq p \leq k}$, while enforcing the following:

1) $\sum_{i \in \pi_p \Rightarrow j \notin \pi_p} w_{ij}$ is minimized (called *interface size* or *edgecut*) and

2) $\max_{p \in [1,k]} \left| v(\pi_p) - \dfrac{v(V)}{k} \right|$ is minimized (called *balanced objective*).

It is accepted to call one subset π_p as a *part* and a family $(\pi_p)_{1 \leq p \leq k}$ as a *partition* of V. In general, two minimization objectives can often be in conflict. Thus, in most of the partitioning formulations the balance objective is restrained to be a constraint

$$\forall p \in [1, k], v(\pi_p) \leq (1 + \alpha) \cdot \frac{v(V)}{k} ,$$

where α is a given *imbalance factor*. In this paper, we refer to the constrained version of the problem as the graph k-partitioning problem.

A common method of solving the k-partitioning problem when $k > 2$ is to adopt a divide and conquer approach [25] that uses *recursive bisection* (or bipartitioning). To simplify the explanation, without loss of generality, we will talk about bipartitioning rather than k-partitioning.

3   Coarsening Schemes

In general, any coarsening can be interpreted as a process of *aggregation* of graph nodes to define the nodes of the next coarser graph. In this paper we compare

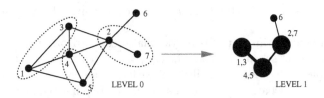

Fig. 2. Schematic demonstration of the SAG scheme. The dashed ovals correspond to the pairs of vertices at the fine level that form aggregates at the coarse level. For example, vertices "1" and "3" are aggregated into one coarse node "1,3".

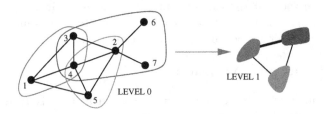

Fig. 3. Schematic demonstration of the WAG scheme. The closed curves at the left graph correspond to the subsets of vertices that form aggregates at the coarse level. These subsets are not disjoint; in other words, vertices in intersection are divided among several aggregates.

two coarsening schemes: strict and weighted aggregations (SAG and WAG). For completeness we briefly review their description. In SAG (also called edge contraction or matching of vertices) the nodes are blocked in small disjoint subsets, called aggregates. Two nodes i and j are usually blocked together if their coupling is *locally strong*, meaning that w_{ij} is comparable to $\min\{\max_k w_{ik}, \max_k w_{kj}\}$ (see Figure 2).

In WAG, each node can be divided into *fractions*. Different fractions belong to different aggregates (see Figure 3); that is, V will be covered by (presumably) small intersecting subsets of V. The nodes that belongs to more than one subset will be divided among corresponding coarse aggregates. In both cases, these aggregates will form the nodes of the *coarser level*, where they will be blocked into larger aggregates, forming the nodes of a *still coarser level*, and so on.

As AMG solvers have shown, *weighted*, instead of *strict*, aggregation is important in order to express the *likelihood* of nodes to belong together; these likelihoods will then accumulate at the coarser levels of the process, indicating tendencies of larger-scale aggregates to be associated to each other. SAG, in contrast, may run into a conflict between the local blocking decision and the larger-scale picture.

For both aggregation schemes, the construction of a coarse graph is divided into three stages: (a) a subset of the fine nodes is chosen to serve as the *seeds* of the aggregates (which form the nodes of the coarser level), (b) the rules for interpolation are determined, and (c) the weights of the edges between the aggregates are calculated. For simplicity, we will unify stages (a) and (b) into one stage in case of SAG. Here are the basic steps of these aggregation schemes.

SAG: Coarse Nodes. Visit the vertices according to some order [18] and choose an appropriate (heaviest, lightest, random, etc., see [18]) edge for making a coarse aggregate from its two endpoints i and j. The weight of a coarse aggregate will be $v_i + v_j$.

WAG: Coarse Nodes-(a). The construction of the set of seeds $C \subset V$ and its complement $F = V \setminus C$ is guided by the principle that each F-node should be "strongly coupled" to C. Starting from $C = \emptyset$ and $F = V$, transfer nodes from F to C until all remaining $i \in F$ satisfy

$$\sum_{j \in C} w_{ij} / \sum_{j \in V} w_{ij} \geq \Theta ,$$

where Θ is a parameter (usually $\Theta \approx 0.5$).

WAG: Coarse Nodes-(b). Define for each $i \in F$ a coarse neighborhood N_i consisting of C-nodes to which i is connected. Let $I(j)$ be the ordinal number in the coarse graph of the node that represents the aggregate around a seed whose ordinal number at the fine level is j. The classical AMG interpolation matrix P is defined by

$$P_{iI(j)} = \begin{cases} w_{ij} / \sum_{k \in N_i} w_{ik} & \text{for } i \in F, \ j \in N_i \\ 1 & \text{for } i \in C, \ j = i \\ 0 & \text{otherwise} \end{cases} .$$

$P_{iI(j)}$ thus represents the likelihood of i to belong to the $I(j)$th aggregate. The volume of the pth coarse aggregate is $\sum_j v_j P_{jp}$. Note that $|N_i|$ is controlled by the parameter called *interpolation order*.

SAG: Coarse Edges. Introduce a weighted coarse edge between aggregates p and q created from fine pairs of vertices (i_1, i_2) and (j_1, j_2), respectively. Then w_{pq} will accumulate all possible connections between different components of these pairs.

WAG: Coarse Edges. Assign the edge connecting two coarse aggregates p and q with the weight $w_{pq} = \sum_{k \neq l} P_{kp} w_{kl} P_{lq}$.

In general, both processes might be reformulated as a single algorithm. Note that, given two consecutive levels l and L, in both cases

$$\sum_{i \in G_l} v_i = \sum_{i \in G_L} v_i.$$

In contrast to the widely used SAG scheme, we are aware of two partitioning solvers [23] and [15] in which an AMG-based scheme was employed. In contrast to [23] we employed the AMG-based coarsening only once directly on the original graph. In [23] two coarsening schemes were employed: PMIS and CLJP. Some parts of these schemes were adapted for their purposes. This solver is very successful, however, in that work was discussed a newly introduced refinement only. The process of coarsening in [15] is reinforced by compatible Gauss-Seidel relaxation [26], which improves the quality of the set of coarse-level variables, prior to deriving the coarse-level equations. The quality measure of the set of coarse-level variables is the convergence rate of F-nodes with respect to C. However, work on improving the quality of coarse-level variables and equations is in progress, and currently it is not clear wheather this relaxation plays an important role for the partitioning problem.

4   Uncoarsening

In this section we will provide the details about the uncoarsening stage and several recommendations of relatively easy improvement of it.

4.1   Disaggregation

To compare the different coarsening methods described in the preceding section, we have chosen to use the same refinement techniques for all the schemes and not to develop one specifically designed for WAG.

The uncoarsening phase typically consists of two steps: the projection of the partition from the coarser graph to the finer graph and the refinement, a local optimization of the partition using the available topological information at the current level.

The projection phase is simple in the case of a strict coarsening scheme. It consists only of assigning the same part number for a fine vertex as the one assigned to its associated coarse vertex.

For a nonstrict coarsening scheme, the projection phase is more complex. We can directly project only the seeds exactly as with a strict coarsening; but for fine vertices that are not seeds, we have to do an interpolation to compute their assignments with respect to the assignments of their neighbors.

In this paper, we have focused on two simple interpolation methods. Both require computing the probability that a fine vertex belongs to a specific part. In the case of bipartitioning, only the knowledge of the probability to be in the part 0 (or 1) matters. Let us call $\mathcal{P}_0(i)$ the probability that vertex i is in part 0, i.e.,

$$\mathcal{P}_0(i) = \sum_{k \in N_i, I(k) \in \pi_0} P_{iI(k)} \ .$$

The first strategy assigns a vertex i to the part 0 (1) if the probability $\mathcal{P}_0(i)$ is greater (lower) than $\frac{1}{2}$. The second strategy assigns the part number proportionally to the probability $\mathcal{P}_0$.

In these two schemes, the projection and interpolation involve two consecutive loops. The first loop browses all the seeds and set their assignments to be those of their corresponding coarse vertices. The second loop scans all the fine nonseed vertices and fixes them in their parts by computing $\mathcal{P}_0$. The cost in time is thus $\Theta(|C| + |F| \cdot io)$, where io is the interpolation order, instead of $\Theta(|F|)$ for SAG.

The next phase consists of the optimization of current assignment using a relaxation (or refinement) methods. In our experiments, we use one of the most popular refinement techniques, Fiduccia-Mattheyses (FM) [4]. This algorithm is popular because it is fast and allows randomized optimization for the cost function. Its principle is simple: Order the vertices according to the gain in edgecut obtained if the vertex is moved to another part; then move the vertex of highest gain, and update the gain for the neighbors; then loop. It is possible that the gain can signify a decrease in the partition quality, but one hopes it can lead to a better local minimum for the edgecut. The number of degradation moves is a parameter and by default set in SCOTCH to 80.

However, we have also chosen to try a poorer refinement, for two reasons. First, to really compare the coarsening schemes, we have to avoid a too powerful refinement because it can hide some artifacts caused by the coarsening. The second reason is that a hill-climbing refinement is sometimes not available, especially in parallel algorithms. To do this poorer refinement, we continue to use FM but with a limitation during its execution: we force FM to stop if the best move will degrade the partition quality. Thus, we obtain only a gradient-like refinement.

4.2 Further Improvements

As mentioned, this paper compares of two coarsening schemes given a significantly simplified, common uncoarsening stage that can be easily parallelized. However, we would like to include in this paper a list of possible further improvements of the AMG-based algorithm. These improvements were tested on the partitioning and linear ordering problems, and all have a good chance of exhibiting superior results to a basic algorithm.

Prolongation by Layers. In classical AMG schemes the initialization of fine level variables is done by a prolongation operator that is equal to the transpose of the restriction operator. In several multilevel algorithms the initialization process depends on the already-initialized variables [11,15], while the order of the initialization is determined by the strength of connection between variables and the set of already initialized variables.

Compatible Relaxation. This type of strict minimization was introduced in [26] as a practical tool for improving the quality of selecting the coarse variables and consequently the relations between the fine and coarse variables. In general, this relaxation minimizes the local energy contribution of fine variables while keeping coarse variables invariant (see [11,15]).

Generating Many Coarse Solutions. Solving the problem exactly at the coarsest level may be reinforced by producing many solutions that differ from each other and involve a lowcost partitioning simultaneously.

Cycling and Linearization. One complete iteration of the algorithm is called a V-cycle, because of the order of visiting the coarse levels. Other patterns of visiting the coarse levels are also possible. A W_ν-cycle was tested in the multilevel scheme for the linear ordering problems [11] and exhibited an improvement proportionally to the amount of work units it spent in comparison to the V-cycle. For both V- and W_ν-cycles the *linearization* technique [27] was used to provide a current approximated solution as an initial point for further approximation.

5   Computational Results

To prevent possible unexpected problems of implementation and to make a fair comparison of the two methods, we implemented full WAG multilevel algorithm with two separate software packages: [11] for the coarsening stage and SCOTCH for the uncoarsening. The entire strict aggregation (HEM) multilevel partitioning algorithm was taken from the SCOTCH package. The combination of two separate packages limited us in performing the bisection experiments only, since the general k-way partitioning might be produced by SCOTCH by applying a bisection method recursively. However, this limitation does not play a crucial role in understanding the general process. Usually, the quality of the k-way partitioning strongly depends on the quality of the bisectioning algorithm incorporated into the general scheme, as a small bias on the first dissection has consequence on all the next levels of bisection.

As is done in most multilevel graph partitioning implementations, the coarsening is continued until the size of the coarsest graph is more than 100 vertices. Then, an aggressive heuristic is applied to get an initial partitioning. The exact partitioning of the coarse graph does not influence the final quality if it is not used in the context of a multiprojection of different partitions.

The comparison is based on the set of real-world graphs presented in Table 1. The imbalance ratio was kept at 1% during all experiments. In order to estimate the algorithmic stability, each test was executed twenty times with different random seeds and initial reshuffling of V and E. Experiments with 100 executions per test did not provide a better estimation of a general statistical view (minimum, maximum, average, and standard deviation).

5.1   Discussion

A frequent weakness of the classical matching-based coarsening schemes may be formulated as the following observation: the results are quite unpredictable. This can be characterized by high standard deviation of the edgecuts, undesirable sensitivity to the parameters, random seed dependence, and other factors that can influence the robustness of the heuristic. In terms of the coarsening stage, this weakness can be heuristically explained by conflicts between local decisions

Table 1. Some of the graphs on which we ran our experiments

Graph name	$|V|(\times 10^3)$	$|E|(\times 10^3)$	Avg. degree	Type
4elt	15	46	5.88	2D finite element mesh
altr4	26	163	12.5	Mesh, CEA-CESTA
oilpan	74	1762	47.78	3D stiffness matrix
ship001	35	2304	132	Parasol matrix
tooth	78	452	11.5	3D finite element mesh
m14b	215	1679	15.6	3D finite element mesh
ocean	143	410	5.71	3D finite element mesh
fe_rotor	100	662	13.3	3D finite element mesh
598a	111	742	13.37	3D finite element mesh
144	144	1074	15	3D finite element mesh
Peku01-25	13	112	17.86	Placement graph
bcsstk32	45	985	44.16	3D stiffness matrix
thread	30	2220	149.32	Parasol matrix
plgr_2500_2	2.5	24	19.61	Power-law graph
plgr_5000_1	4	6.2	3.03	Power-law graph
plgr_5000_2	3.8	5.2	2.72	Power-law graph
plgr_5000_3	4.1	6.2	3.03	Power-law graph
fxm4_6	19	239	25.3	Optimization problem
p2p_1	11	31	5.72	p2p network
p2p_2	11	31	5.62	p2p network

(at the fine scale) and the global solution. In other words, by matching two vertices we assume that, according to some argument, they will share a common property (belonging to the same π_p) and this property will be assigned to each of them at the interpolation stage as initial solution. Unfortunately, because of the NP-hardness of the partitioning problem (still) no argument can provide enough pairwise local (even with high probability) decisions for the vertices to belong to the same part. Thus, making a local decision without collecting enough global information regarding the graph can lead to the unexpectedness.

In contrast to SAG, WAG consists of two ways to prevent itself from making local decisions before collecting the global information: (a) each vertex must be connected to enough seeds and (b) the nonseed vertex might be divided between several seeds (when $io > 1$). Thus, the obtained covering of a fine graph by aggregates (like those depicted in Figure 3) is smoother and the connectivity of a coarse aggregate is better than a pair matching can ensure. However, by increasing the number of possible F-vertices divisions, the connectivity of an aggregate may be too high and can cause an increased coefficient in linear running time. Thus, it must be controlled by the *interpolation order*. Moreover, it was never observed that too high an interpolation order (more than twenty) has improved the final results significantly. It can lead to the global averaging process which result can be far from an optimality.

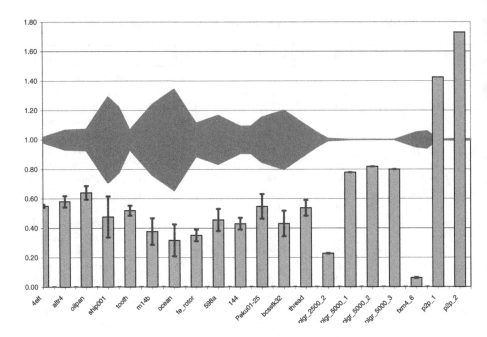

Fig. 4. Edgecut between HEM and WAG with interpolation order 4 when using a gradient refinement. The HEM average is 1. The area colored by dark-gray represents the standard deviation, counted positively and negatively. The light-gray bars are the WAG average, and the dark-gray boxes the standard deviation.

In Figure 4 we compare HEM and WAG with $io = 4$ (which still has a low complexity and exhibits superior results); no randomized optimization was applied at the refinement in either case. Except int two peer-to-peer graphs, WAG clearly outperforms HEM, producing two times better average cuts.

This significant improvement is explained by a better conservation of a graph topology during the WAG coarsening. The main reason is that one can expect a good AMG coarsening of the graph Laplacian when the problem is associated with, or approximated by, the problem of minimizing the quadratic functional given by $\sum_{i,j} w_{ij}(x_i - x_j)^2$, which is, in general, a natural problem that can be solved better by AMG than by geometric multigrid approaches [6]. The partitioning problem yields such an approximation while, for example, considering spectral methods [2] or quadratic programming [28].

In particular, this improvement is interesting in light of designing *parallel* graph partitioners, since many efforts are needed to obtain an efficient parallel refinement. Another important observation is that the WAG standard deviation is lower; that is, the quality of the partitions is more predictable. It can certainly reduce the number of executions of the algorithm as it works in several tools and by default in SCOTCH.

The second experiment consists of applying the same WAG and HEM reinforced by FM *with hill-climbing* capabilities. The results are presented in

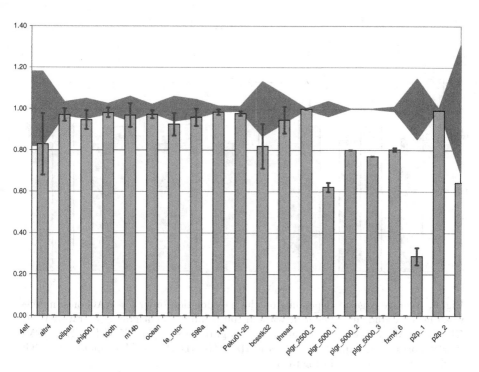

Fig. 5. Edgecut between HEM and WAG with interpolation order 4 when using a FM refinement with hill climbing. The symbols are the same as in Figure 4.

Figure 5. For all test graphs in this case, WAG remains superior to HEM while having a lower standard deviation. More aggressive uncoarsening allows us to better exploit a graph topology and leads to the better partitions. Note that WAG clearly outperforms HEM on power-law graphs. However, in the current (not fully optimized) WAG version poorest refinement can give better results.

Another experiment was performed to determine the influence of the interpolation order in WAG. In the previous tests, only HEM was matching-based; that is, the size of one aggregate was limited by two. However, a study on graph partitioning of power-law graphs [12] shows that the size of the aggregate could be important. In Figure 6, WAG with interpolation order of 1 corresponds to a generic SAG, and we can observe that increasing the interpolation order usually leads to better results.

A method of *increased interpolation orders* (marked here by "inc_io") was proposed in [27]. According to this method the interpolation order must be increased as the coarse graphs become smaller. This hardly affects the total complexity of the algorithm, but it does systematically improve the results since it helps to learn better a graph topology before the uncoarsening stage.

Average edgecuts for partitions computed by standard HEM and by WAG are summarized in Table 3. On our set of test graphs, WAG is on average 15% better than HEM, and worse only for the graph *p2p_1*. However, we explain

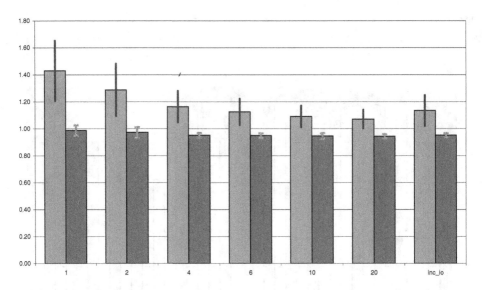

Fig. 6. Average edgecut depending on the interpolation order for WAG on 9 meshes, when using a gradient refinement without hill climbing (light-gray bars, ratios WAG-GRAD/HEM-FM), or a classical FM (dark-gray bars, ratios WAG-FM/HEM-FM). It is important to emphasize that ratio HEM-GRAD/HEM-FM is about 2.61 . The symbols are the same as in Figure 4.

Table 2. Minimum edgecut obtained on 20 runs. HEM is the standard SCOTCH matching and the *Best known* results are from Chris Walshaw's database. All the HEM results and all the WAG results are obtained by using FM refinement with at most 80 unproductive moves for hill climbing.

Graph name	Best known	HEM	WAG io=1	io=2	io=4	io=6	io=10	io=20	inc_io	Delta (%) best	HEM
4elt	138	140	138	138	138	138	138	138	138	0.00	-1.43
tooth	3823	4029	4100	3971	3921	3949	3947	3894	3987	1.86	-3.35
m14b	3826	3915	3888	3882	3860	3877	3858	3871	3863	0.84	-1.46
ocean	387	406	388	387	387	387	387	387	387	0.00	-4.68
fe_rotor	2045	2104	2085	2072	**2041**	**2039**	2053	2056	2070	-0.29	-3.09
598a	2388	2451	2428	2429	2414	2418	2402	2398	2415	0.42	-2.16
144	6479	6688	6638	6622	6596	6576	6556	6575	6600	1.19	-1.97
bcsstk32	4667	5009	4740	4788	4776	4757	4938	5013	4743	1.56	-5.37

this particular problem as a lack of compatible relaxation, which has to be a natural part of any AMG-based algorithm. Usually increasing the interpolation order gives better results, but it seems to be a problem in some cases when too-simplified projection and refinement are applied, since they are not designed to deal efficiently with the gain of precision of a high interpolation order.

Table 3. Average edgecut over 20 run with hill-climbing FM for HEM and WAG with various interpolation orders. Delta is the difference between HEM and the average of WAG, in percent. The numbers in bold correspond to the best average edgecut for a graph.

Graph name	HEM	WAG							Delta (%) HEM
		io=1	io=2	io=4	io=6	io=10	io=20	inc_io	
4elt	170	152	154	141	142	**140**	142	142	-14.70
altr4	1656	1638	1619	1607	1593	**1587**	1593	1619	-2.88
oilpan	9433	9533	9153	8923	8974	**8744**	8771	8854	-4.66
ship001	17052	16787	16754	16733	16720	**16616**	16562	16770	-2.03
tooth	4346	4418	4289	4211	4152	**4129**	4114	4133	-3.21
m14b	4029	4043	3968	3920	3920	3913	**3907**	3914	-2.20
ocean	427	420	417	395	393	392	**391**	395	-6.33
fe_rotor	2205	2147	2129	2115	2112	2160	**2107**	2123	-3.51
598a	2484	2478	2458	2444	2438	2424	**2416**	2440	-1.67
144	6826	6909	6763	6671	6637	6632	**6621**	6672	-1.83
Peku01-25	8305	7008	7049	6802	6798	6742	**6711**	6819	-17.55
bcsstk32	5588	5576	5425	5286	5067	5154	5176	**5066**	-6.06
thread	55933	55966	55899	55917	55970	**55898**	55943	55990	0.01
plgr_2500_2	6908	2308	**1605**	1920	2117	2588	3294	2018	-67.22
plgr_5000_1	946	771	775	757	751	752	**740**	762	-19.84
plgr_5000_2	627	491	496	483	478	466	**466**	482	-23.41
plgr_5000_3	939	772	774	754	751	**747**	**747**	753	-19.37
fxm4_6	1639	511	478	**471**	**471**	484	518	**471**	-70.33
p2p_1	1951	**1944**	1985	1936	2326	2447	2221	2254	10.65
p2p_2	3762	**1981**	2123	2423	2451	2261	2663	2240	-38.71

Although the goal of this work was not to obtain the best known results, we present a comparison against the best known results obtained at Walshaw's database (The graph partitioning archive. http://staffweb.cms.gre.ac.uk/~c.walshaw/partition/) in Table 2. Observe that the best WAG result is on average less than 0.6% from the best known partitioning. We can note that the best WAG partition is always better than the best of HEM, on average by 3% on meshes and 15% on all of our test cases. Another interesting point is that all the best results are obtained with normal FM refinement, except those for the power-law graphs. Thus, at least for this class of graphs, the improvements of uncoarsening (suggested in Section 4.2) may be interesting.

6 Conclusions

This paper compares two coarsening schemes in the context of graph partitioning. As a main result of this work, we recommend the adoption of WAG instead of classical HEM because of its higher ability of graph topology learning prior to the uncoarsening stage. In general, WAG improves the quality of the partitions and thus provides a better chance of finding a good approximation. In particular, the robustness of WAG was better than that of SAG coarsening when a poor refinement was employed and WAG provided still good-quality results. Since parallel implementations in algebraic multigrid solvers [29,30] are very scalable and since the refinement is often poor in parallel, WAG appears to be an ideal candidate to design highly scalable efficient parallel graph partitioning tools.

The framework we use allowed us to combine different coarsenings with different uncoarsenings. For example, we have done several experiments with a band-FM refinement [31], which, despite the simplicity of our projection, worked well with similar results. WAG allows one to obtain superior results with several different tested methods that are of great interest for parallel implementation [32].

The partitionings obtained during our experiments certainly may be improved by using more sophisticated projection and relaxation methods at the refinement, as mentioned in 4.2.

Acknowledgments. This work was funded by the CSCAPES institute, a DOE project, and in part by DOE Contract DE-AC02-06CH11357. We express our gratitude to Dr. Erik Boman for useful advice.

References

1. Garey, M.R., Johnson, D.S., Stockmeyer, L.: Some simplified NP-complete graph problems. Theoretical Computer Science 1, 237–267 (1976)
2. Pothen, A., Simon, H.D., Liou, K.P.: Partitioning sparse matrices with eigenvectors of graphs. SIAM Journal of Matrix Analysis 11(3), 430–452 (1990)
3. Kernighan, B.W., Lin, S.: An efficient heuristic procedure for partitioning graphs. BELL System Technical Journal, 291–307 (1970)
4. Fiduccia, C.M., Mattheyses, R.M.: A linear-time heuristic for improving network partitions. In: Proceedings of the 19th Design Automation Conference, pp. 175–181. IEEE, Los Alamitos (1982)
5. Bui, T.N., Moon, B.R.: Genetic algorithm and graph partitioning. IEEE Trans. Comput. 45(7), 841–855 (1996)
6. Brandt, A., Ron, D.: Ch. 1: Multigrid solvers and multilevel optimization strategies. In: Cong, J., Shinnerl, J.R. (eds.) Multilevel Optimization and VLSICAD. Kluwer, Dordrecht (2003)
7. Walshaw, C.: Multilevel refinement for combinatorial optimisation problems. Annals Oper. Res. 131, 325–372 (2004)
8. Chan, T.F., Cong, J., Romesis, M., Shinnerl, J.R., Sze, K., Xie, M.: mpl6: a robust multilevel mixed-size placement engine. In: Groeneveld, P., Scheffer, L. (eds.) ISPD, pp. 227–229. ACM, New York (2005)
9. Chang, C., Cong, J., Pan, D., Yuan, X.: Multilevel global placement with congestion control. IEEE Trans. on Computer-Aided Design of Integrated Circuits and Systems 22, 395–409 (2003)
10. Cong, J., Shinnerl, J.R. (eds.): Multilevel Optimization and VLSICAD. Kluwer, Dordrecht (2003)
11. Safro, I., Ron, D., Brandt, A.: Multilevel algorithms for linear ordering problems. Journal of Experimental Algorithmics 13, 1.4–1.20 (2008)
12. Abou-Rjeili, A., Karypis, G.: Multilevel algorithms for partitioning power-law graphs. In: IPDPS (2006)
13. Alpert, C.J., Huang, J.H., Kahng, A.B.: Multilevel circuit partitioning. In: Design Automation Conference, pp. 530–533 (1997)

14. Banos, R., Gil, C., Ortega, J., Montoya, F.: Multilevel heuristic algorithm for graph partitioning. In: Raidl, G.R., Cagnoni, S., Cardalda, J.J.R., Corne, D.W., Gottlieb, J., Guillot, A., Hart, E., Johnson, C.G., Marchiori, E., Meyer, J.-A., Middendorf, M. (eds.) EvoIASP 2003, EvoWorkshops 2003, EvoSTIM 2003, EvoROB/EvoRobot 2003, EvoCOP 2003, EvoBIO 2003, and EvoMUSART 2003. LNCS, vol. 2611, pp. 143–153. Springer, Heidelberg (2003)
15. Ron, D., Wishko-Stern, S., Brandt, A.: An algebraic multigrid based algorithm for bisectioning general graphs. Technical Report MCS05-01, Department of Computer Science and Applied Mathematics, The Weizmann Institute of Science (2005)
16. Barnard, S.T., Simon, H.D.: A fast multilevel implementation of recursive spectral bisection for partitioning unstructured problems. Concurrency: Practice and Experience 6, 101–107 (1994)
17. Hendrickson, B., Leland, R.W.: A multi-level algorithm for partitioning graphs. In: Supercomputing (1995)
18. Karypis, G., Kumar, V.: A fast and high quality multilevel scheme for partitioning irregular graphs. Technical Report 95-035, University of Minnesota (1995)
19. Catalyurek, U., Aykanat, C.: Decomposing irregularly sparse matrices for parallel matrix-vector multiplications. In: Saad, Y., Yang, T., Ferreira, A., Rolim, J.D.P. (eds.) IRREGULAR 1996. LNCS, vol. 1117, pp. 75–86. Springer, Heidelberg (1996)
20. Devine, K., Boman, E., Heaphy, R., Hendrickson, B., Vaughan, C.: Zoltan data management services for parallel dynamic applications. Computing in Science and Engineering 4(2), 90–97 (2002)
21. Walshaw, C.: A multilevel approach to the travelling salesman problem. Tech. Rep. 00/IM/63, Comp. Math. Sci., Univ. Greenwich, London SE10 9LS, UK (2000)
22. Hu, Y.F., Scott, J.A.: A multilevel algorithm for wavefront reduction. SIAM J. Scientific Computing 23, 2000–2031 (2001)
23. Meyerhenke, H., Monien, B., Sauerwald, T.: A new diffusion-based multilevel algorithm for computing graph partitions of very high quality. In: Proc. 22nd International Parallel and Distributed Processing Symposium (IPDPS 2008). IEEE Computer Society, Los Alamitos (2008); Best Algorithms Paper Award
24. SCOTCH: Static mapping, graph partitioning, and sparse matrix block ordering package, http://www.labri.fr/~pelegrin/scotch/
25. Simon, H.D., Teng, S.H.: How good is recursive bisection. SIAM J. Sci. Comput. 18, 1436–1445 (1997)
26. Brandt, A.: General highly accurate algebraic coarsening. Electronic Trans. Num. Anal. 10(2000), 1–20 (2000)
27. Safro, I., Ron, D., Brandt, A.: Graph minimum linear arrangement by multilevel weighted edge contractions. Journal of Algorithms 60(1), 24–41 (2006)
28. Hager, W.W., Krylyuk, Y.: Graph partitioning and continuous quadratic programming. SIAM J. Discret. Math. 12(4), 500–523 (1999)
29. Henson, V.E., Yang, U.M.: Boomeramg: a parallel algebraic multigrid solver and preconditioner. Appl. Numer. Math. 41(1), 155–177 (2002)
30. Gee, M., Siefert, C., Hu, J., Tuminaro, R., Sala, M.: ML 5.0 smoothed aggregation user's guide. Technical Report SAND2006-2649, Sandia National Laboratories (2006)
31. Chevalier, C., Pellegrini, F.: Improvement of the efficiency of genetic algorithms for scalable parallel graph partitioning in a multi-level framework. In: Nagel, W.E., Walter, W.V., Lehner, W. (eds.) Euro-Par 2006. LNCS, vol. 4128, pp. 243–252. Springer, Heidelberg (2006)
32. Chevalier, C., Pellegrini, F.: Pt-scotch: A tool for efficient parallel graph ordering. Parallel Comput. 34(6-8), 318–331 (2008)

Cooperative Strategies and Reactive Search: A Hybrid Model Proposal

Antonio D. Masegosa[1,*], Franco Mascia[2,**], David Pelta[1], and Mauro Brunato[2]

[1] Universidad de Granada, Granada, Spain
{admase, dpelta}@decsai.ugr.es
[2] Università degli Studi di Trento, Trento, Italy
{mascia,brunato}@disi.unit.it

Abstract. Cooperative strategies and reactive search are very promising techniques for solving hard optimization problems, since they reduce human intervention required to set up a method when the resolution of an unknown instance is needed. However, as far as we know, a hybrid between both techniques has not yet been proposed in the literature. In this work, we show how reactive search principles can be incorporated into a simple rule-driven centralised cooperative strategy. The proposed method has been tested on the Uncapacitated Single Allocation p-Hub Median Problem, obtaining promising results.

1 Introduction

Given limited time and space resources, metaheuristic algorithms are very effective in providing good quality solutions. Given a problem instance, it is generally impossible to determine *a priori* the best heuristics to solve it, and the performance of a metaheuristic depends strongly on parameters whose adjustment is not trivial, and is usually done by experts or extensive experimentation.

The first issue can be handled by cooperative strategies [1], where a set of potentially good heuristics for the optimization problem are executed in parallel, sharing information during the run. The second problem is successfully addressed by reactive strategies [2], and the use of sub-symbolic machine learning to automate the parameter tuning process, making it an integral part of the algorithm. Both techniques will be described more in detail in the following section.

Despite the success of reactive strategies in other fields, to the best of our knowledge, these principles have not yet been incorporated in cooperative strategies. In this work we propose a hybrid between reactive search and cooperative

---

* A.D. Masegosa is supported by the program FPI from the Spanish Ministry of Science and Innovation (MSI). This work has been partially promoted by the project TIN-2008-01948 from the MSI and P07-TIC-02970 from the Andalusian Government.

** Work by F. Mascia and M. Brunato was partially supported by project BIONETS (FP6-027748) funded by the FET program of the European Commission.

T. Stützle (Ed.): LION 3, LNCS 5851, pp. 206–220, 2009.

strategies. For this purpose, we will use the simple centralised cooperative strategy presented in [3]. The aim of this paper is to evaluate the performance of this new hybrid model. The benchmark-problem used is the Uncapacitated Single Allocation p-Hub Median Problem.

The paper is structured as follows. Section 2 outlines the principles behind reactive search and cooperative strategies. Section 3 focuses on the description of the cooperative strategy and the hybrid model. In Section 4, the benchmark used is described. In Section 5 we present the results of the empirical performance assessment of the various techniques, and we draw conclusions in Section 6.

2   Cooperative Strategies and Reactive Search

When we talk about cooperative strategies, we refer to a set of heuristics that share any type of information among them. The cooperation can be centralised, if the information flow is managed by a coordinator, or decentralised, when the data exchange is directly done among the components of the strategy. These ideas have been applied in several fields both explicitly and implicitly: *Multi-agents systems*, where a set of very simple entities (ants in ACO [4] or particles in PSO [5]) proceed in a cooperative way to progressively improve a population of solutions; *Hyper-heuristics* [6], where a set of low-level heuristics are managed by a higher-level heuristic that determines which one should be applied at each iteration; *Multi-thread parallel meta-heuristics* [1], where several heuristics, which are executed in parallel, cooperate sharing performance information to reach promising regions of the search space, to avoid local minima, etc.

The term *reactive search* refers to an algorithmic framework [2] where optimization techniques are coupled with machine learning algorithms. In particular, the machine learning component analyzes the behavior of the optimization algorithm and provides feedback by fine tuning its parameters, thus adapting it to the properties of the instance being solved. Parameter tuning can be performed: *Offline*: the machine learning component analyzes the behavior of the optimization algorithm after a series of runs on different instances. The purpose of this method is to learn a mapping between some instance features and a satisfactory value of the algorithm's parameters. In this case, the algorithm simply replaces the researcher in performing offline adjustments when he applies an algorithm to a new domain, with a trial-and-error approach; *Online*: the machine learning component operates alongside with the optimization algorithm and tries to detect hints of bad performance, such as repeated visits to the same configuration or low improvement rate. By performing online adjustments to the optimization component, the system can adapt to the local features of the search landscape.

The latter approach, which dates back to the seminal paper on Reactive Tabu Search [7], will be followed in this paper.

3   General Scheme and Strategies

The cooperative strategy described in [3,8] consists of a set of solvers/threads, each one implementing the same or a different resolution strategy for the problem at hand. These threads are controlled by a coordinator, which processes the information received from them. It produces subsequent adjustments of the solver behaviours by making decisions based on fuzzy rules, and sending instructions to the threads. The information exchange process is done through a blackboard architecture [9].

An important part of this strategy is the information flow, which is divided in three steps: 1) performance data (reports are sent from the solvers to the coordinator, 2) the data are stored and processed by the coordinator and 3) the coordinator sends orders to the solvers. Each report in the first step contains:

- solver identification;
- a time stamp t;
- the current solution of the solver at that time s^t;
- the best solution reached until that time by this solver s_{best};
- a list with the local minima found by the method.

The coordinator stores the last two reports from each solver. From these two reports, the coordinator calculates the improvement rate as $\Delta_f = \frac{f(s^t) - f(s^{t'})}{t - t'}$, where $t - t'$ represents the elapsed time between two consecutive reports, $s^{t'}$ is the current solution sent by the solver in the last report and f is the objective function. The values Δ_f and $f(s^t)$ are then stored in two fixed length ordered data structures, one for the improvements and the other for the costs, whose sizes are set to $4 \times$ *number of solvers*. The list of local minima is processed by the coordinator that keeps the history of all local optima in a hash table. Each entry of this table has also a collision counter with the number of times that a solution has been visited by any search thread. The behaviour of the solvers is controlled by a set of rules. These rules allow the coordinator to determine if a solver is behaving correctly, as well as the action that should be performed to correct such behavior. These rules are of the type:

$$\textbf{if } condition \textbf{ then } action.$$

The action consists of sending a solution to the solver (the best solution ever seen by the coordinator C_{best}, a random one, etc.). This solution is used by the solver as a restart point for its search. In this way, the coordinator controls the location of the solver in the search space.

3.1   Fuzzy Control Rule

In the past, we proposed a fuzzy rule that yielded good results on different problems [3]. The rule was designed following the principle: "If a solver is working well, keep it; but if a solver seems to be trapped, do something to alter its behaviour". Its precise definition is as follows:

> **if** the quality of the current solution reported by *solver$_i$* is *low* and the improvement rate of *solver$_i$* is *low* **then** send perturbed C_{best} to *solver$_i$*

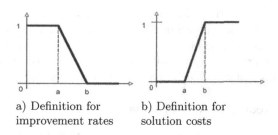

a) Definition for
improvement rates

b) Definition for
solution costs

Fig. 1. Membership function of the two sets used in the fuzzy control rule

The labels *low* are defined as fuzzy sets whose membership functions $\mu(x)$, for both improvement rates and solution costs, are shown in Figure 1. The variable x corresponds to the relative position (resembling the notion of percentile rank) of a value (an improvement rate or a cost) in the samples stored in the respective fixed length data structure. The parameters (a, b) are fixed to $(80, 100)$ and $(0, 20)$ for the data structure of costs and improvements, respectively. This way of measure the quality of the improvement rates and the solution is independent of the problem, instance or scale. C_{best} denotes the best solution ever recorded by the coordinator.

What the rule says, is that, if the values reported by a solver are among the worst stored in the two sets, then the coordinator sends the solution C_{best} (with a small perturbation) to the solver. By doing so, it relocates the solver to a more promising region of the search space, trying to increase the chances of finding better solutions. The details of the perturbation applied to C_{best} will be explained in Section 4.2.

The fuzzy rule shares some ideas with reactive search: in both cases, the antecedent of the rule aims at discovering of the thread/solver stagnation. However the action coded in the consequent of the fuzzy rule is fixed, while in the case of reactive search, the intensity of the perturbation is adapted dynamically.

3.2 Reactive Control Rule

Following the principles in [7,10], this rule uses the history of visited local minima to detect search stagnation. In other words, the hash table kept by the coordinator is used to determine if a solver is visiting an already explored area of the search space. If this is the case, the coordinator drives the searchers diversification by restarting the stagnated algorithm from a perturbation of the best configuration found across all solvers. The definition of this rule is the following:

> **if** the collision counter cc of the last local minima visited by $solver_i$ is bigger than $\lambda_{reaction}$, **then** the coordinator sends C_{best} to $solver_i$ perturbed by degree ϕ.

The threshold $\lambda_{reaction}$ regulates the activation of the rule and ϕ is defined as:

$$\phi = \begin{cases} cc - \lambda_{reaction}, & \text{if } cc - \lambda_{reaction} < \phi_{max} \\ \phi_{max}, & \text{if } cc - \lambda_{reaction} \geq \phi_{max} \end{cases}$$

The idea of this rule is that the more often a local minimum is visited, the higher the probability that it belongs to a large attraction basin, and therefore the perturbation needs to be higher in order to escape from that optimum. The strength of the perturbation is controlled by the application of different mutation operators. The description of such operators will be seen in Section 4.2.

3.3   Fuzzy + Reactive Control Rule

From the previous description, one can observe that the fuzzy rule tries to drive the intensification of the strategy relocating the threads to promising areas, whereas the reactive rule guides the diversification of the solvers to avoid their stagnation in local minima. As a consequence, it is expected that if both rules are used simultaneously, the exploration and the exploitation phases of the cooperative strategy will be better balanced.

From an operational point of view, if just one of the rules is activated (its antecedent evaluates to true), then the corresponding consequent is applied. When both rules are activated, we need to decide which consequent would be applied. If the fuzzy alternative is taken, then we would move a thread towards the C_{best} solution. As all the threads implement the same algorithm with the same parameters, then the opportunity to improve may be low. If the threads were heterogeneous, then it would make sense to search the same region using different search strategies. Under this situation, we consider that when both rules are activated, then the best alternative is to apply the reactive rule's consequent as a way to promote diversification.

4   Case Study Details

In this section, we describe all the aspects related to the case study used for assessing the performance of the methods: Uncapacitated Single Allocation Problem, and the concrete implementation of the strategies.

4.1   The Uncapacitated Single Allocation p-Hub Median Problem

The objective of hub location problems is composed of two steps: (1) **Hub location:** it decides which nodes should be the hubs and the number of them, to distribute the flow across them. (2) **Non-hub to hub allocation:** it assigns the rest of the nodes to the hubs. Generally, these tasks are performed by minimising an objective function that depends on the exchanged flow and its cost. A general survey of this kind of problems can be seen in [11]. Here we will focus on the Uncapacitated Single Allocation p-Hub Median Problem (USApHMP), a specific case of these problems where number of hubs is fixed to p.

In this paper, we will use the formulation given by O'Kelly in [12]. Let N be a set of n nodes. We define W_{ij} as the amount of flow from node i to j, and C_{ij} the cost of transporting a unit between nodes i and j. Each flow W_{ij}, has three different components: collection, transfer and distribution. Collection corresponds to the transport from i to its hub. Transfer represents the movement between the hubs of i and j. Distribution occurs from this second hub to j. Each of these activities can modify the cost per unit flow by a determined constant. Such constants are denoted, respectively, as χ (generally $\chi = 1$), α (generally $\alpha < 1$) and δ (generally $\delta = 1$).

Let X_{ij} be a decision variable whose value is 1 if node i is allocated to hub j and 0, otherwise. The USApHMP can be formulated as:

$$min \sum_{i,j,k,l \in N} W_{ij}(\chi C_{ik} X_{ik} + \alpha C_{kl} X_{ik} X_{jl} + \delta C_{jl} X_{jl}) \tag{1}$$

subject to

$$\sum_{j=1}^{n} X_{jj} = p \tag{2} \qquad \sum_{j=1}^{n} X_{ij} = 1, \forall i = 1, \dots, n \tag{3}$$

$$X_{ij} \leq X_{jj}, \forall i, j = 1, \dots, n \tag{4} \qquad X_{ij} \in \{0,1\}, \forall i, j = 1, \dots, n \tag{5}$$

The objective function (1) minimises the sum of the origin-hub, hub-hub and hub-destination flow costs. Constraint (2) ensures that exactly p hubs exist. Constraint (3) indicates that a node can only be allocated to a unique hub. Condition (4) guarantees that a non-hub point can only be allocated to a hub and not to another non-hub node. Finally, (5) is the classical binary constraint.

The instances used in this work were obtained from ORLIB [13]. Concretely, we used the AP data set. The instances are divided in two groups, those with 20, 25, 40 and 50 nodes, and those with 100 and 200. For the first set, $p \in \{3, 4, 5\}$, while for the second one, we have $p \in \{3, 4, 5, 10, 15, 20\}$. The value of the constants χ, α and δ were fixed to 3, 0.75 and 2 respectively. The stopping condition for all experiments was set to 50000 evaluations for instances with 50 nodes or less, and to 500000 for those with more than 50 nodes. The optimum for the instances with less than 50 nodes was provided by ORLIB, and for the other instances we considered the best solution found by one of the state-of-art algorithms for this problem, presented in [14]. The quality of the solutions is measured as $percent\ error = 100 \times \frac{obtained\ value - optimum}{optimum}$.

4.2   Strategy Implementation

When implementing multi-threaded cooperative strategies, one can resort to real parallel implementations, or simulate the parallelism in a single-processor computer. The later is the strategy adopted here: we construct an array of solvers and we run them using a round-robin scheme. In this way, each of them is run for a certain number of evaluations of the objective function. This number is randomly generated from the interval [500;560] which has been chosen following previous studies. Once a solver is executed, the communication with the coordinator takes place. These steps are repeated until the stop condition is fulfilled.

Another important parameter for this study is the number of solvers as well as which metaheuristic they implement. The cooperative strategy has three solvers that implement the same algorithm, which is a standard version of Simulated

Annealing that follows the basic guidelines described in [15]. The neighborhood operator used by this method is composed of two distinct mechanisms: change of the assignment of non-hub nodes and change of the location of hubs. More precisely, the steps for the assignment of non-hub nodes are the following:

1. Choose randomly a group G_j, being $G_j = \{i | X_{ij} = 1, i \neq j\}$ the group of those nodes that are assigned to hub j.
2. Select randomly a node $i \in G_j$
3. Choose randomly another group G_k, $k \neq j$
4. Allocate the selected node to the new group: $X_{ij} \leftarrow 0, X_{ik} \leftarrow 1$

The second neighborhood operator changes the location of a hub j to another node that is currently allocated to hub i. If there are no nodes allocated to j, a different node is selected as hub and j is assigned to another group. More in detail, the following steps have to be performed:

1. Choose randomly a group G_j
2. If there is at least one node in the group ($|G_j| > 0$) then:

   (a) Select a random node $i \in G_j$
   (b) Allocate all nodes in G_j and its hub node j to the new hub node i: $\forall k \in G_j : X_{kj} \leftarrow 0, X_{jj} \leftarrow 0, X_{ki} \leftarrow 1, X_{ji} \leftarrow 1$ and $X_{ii} \leftarrow 1$
3. If the group has no nodes ($|G_i| = 0$), then:

   (a) Choose randomly another group $G_k, k \neq j$ with at least one node.
   (b) Select a random node $i \in G_k$.
   (c) Make a new group with the selected node i. $X_{ii} \leftarrow 1$
   (d) Allocate the last hub j as a normal node to another hub selected randomly. $X_{jr} \leftarrow 1$ where r is a random hub.

When the cooperative strategy is run, each solver starts from a different initial solution. As for the Fuzzy rule, its activation takes place when the output value of the antecedent is higher than 0.9. The modification done to C_{best}, when it is sent to the solvers, corresponds with one assignment change and one location change. For the Reactive rule, $\lambda_{reaction}$ was set to 10. The parameter ϕ_{max} is set to 3, that is, 3 different degrees of perturbation for C_{best} are considered. This degree of perturbation is determined by the number of changes of assignment ($\#_{assignment}$) and location ($\#_{location}$) applied to this solution. When the degree $\phi = i$, then $\#_{assignment} = \#_{location} = number\ of\ hubs/(\phi_{max} - i - 1)$, $i = 1, \ldots, \phi_{max}$.

5   Experimental Comparison

The aim of our experimentation is to assess the performance of the hybrid strategy proposed. Specifically, we will compare the following models:

- The independent strategy (I), which is the baseline case, where the solvers do not exchange information.

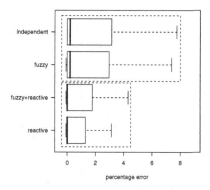

Fig. 2. Box diagram for the percentage error obtained by I, F, R and F+R strategies over all instances. The algorithms grouped by a dotted rectangle do not perform differently at significance level 0.05 (non-parametric Mann-Whitney U-test).

- The cooperative strategy that uses the Reactive rule (R).
- The cooperative strategy utilising the Fuzzy rule (F).
- The cooperative strategy using both rules (F+R).

Each strategy is run 30 times, starting from different initial random solutions, for each instance. The performance is assessed in terms of: quality of the solutions; convergence speed; and how the rules modifies the threads.

5.1   Benefits of the Cooperative-Reactive Hybrid

Looking at the general picture, we show in Figure 2 a box diagram with information about the percentage error obtained by each method over all instances, where the boxes are sorted by the median. The Kruskal-Wallis non-parametric test for multiple comparisons has been used to asses the differences between the performances of the four strategies. The null hypothesis could not be rejected at significance level 0.01. The statistical information about the comparison between pairs of algorithms is also shown in Figure 2: when two boxes are inside a dotted rectangle, a statistically significant difference could not be shown between the percentage error distributions of the corresponding algorithms (non-parametric Mann-Whitney U-test at significance level $\alpha = 0.05$). Looking at the medians there is no strong differences between the algorithms. However, if we take into account the upper quartile represented by the whisker, the differences are more noticeable, and show that the R strategy obtains the best performance. The second position is occupied by F+R, followed by F and I in that order. Furthermore, the statistical non-parametric test shows that R and F+R perform significantly better that the two worst methods.

The following analysis focuses on detailed per-instance results. Figure 3 shows a number of scatter plots providing pairwise comparisons among the four methods evaluated. Every instance is plotted as a pair (x, y) where x (y) is the normalised mean percentage error obtained by the strategy named in the X (Y)

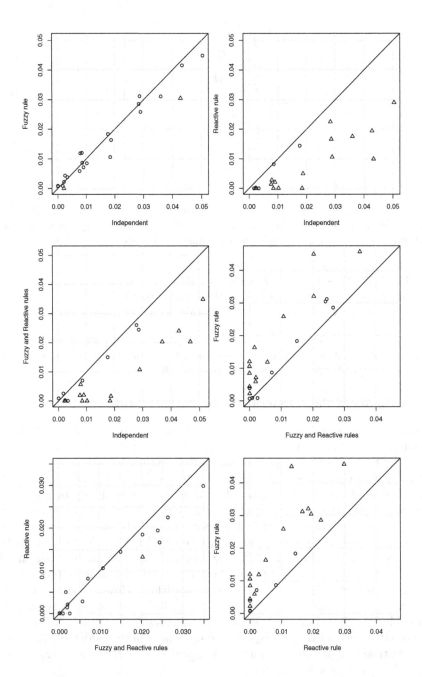

Fig. 3. Comparison of the mean relative deviation from optimum (smaller values are better). Triangles represent the instances on which the algorithms being compared perform differently at significance level 0.05 (Wilcoxon's unpaired rank sum test).

axis. If the marker is above the diagonal, then the algorithm in X is better than the one in the Y axis. A triangle is used when there is a statistically significant difference between the performance of the two algorithms for the corresponding instance (non-parametric Mann-Whitney U-test at significance level $\alpha < 0.05$). A circle is used in the other case.

The first plot shows how F hardly improves over I. Only in two instances the difference between both methods is statistically significant, which means that this rule does not obtain good results for this problem. When the coordinator uses R, it outperforms the I strategy. As we can see in the plot, the performance of this coordination type is significantly better on seventeen instances. When the control is carried out by F+R, the results are similar to those obtained by R, although now the number of cases where this control rule achieves a significant difference over I is reduced to fourteen.

If we compare the different rules among them, we observe that F+R improves upon F. The contribution of F+R is significantly evident in thirteen instances. However, when F+R is compared with R, apart from one instance, the null hypothesis of the equivalence of the performances could not be rejected. The comparison between F and R confirms the intuition on the importance of the reactive rule, since on fifteen cases there is a significant improvement over F.

The next step is to identify the instances where there is a significant difference between the different cooperation schemes. Table 1 shows the mean percentage error for the different strategies as well as the std. deviation. From the table, we can notice that the differences of R and F+R with respect to F and I are larger when the size of the instances increases, i.e., for a fixed number of nodes, when the number of hubs increases. If we consider the best solution found, something similar happens. The difference in terms of both the number of times the optimum is reached and the quality of the best solution found is bigger for instances with a higher number of hubs, with the only exception of F in 200-{20}.

5.2   Study of the Dynamic Behaviour

Two different aspects of the dynamic behaviour of the strategy will be investigated now. The first is the performance of the strategies during the search process. For this, we studied how the percentage error of the best solution found for each method evolves over time. We will focus on the six hardest instances, which are 100-{10,15,20} and 200-{10,15,20}. The results are shown in Figure 4.

The first issue we want to highlight is the early stagnation in local minima experienced by I and F for the instances with one hundred nodes. This behaviour can also be observed when the instances have two hundred nodes, although in these cases it takes places in the last stages of the search. Unlike these methods, R and F+R do not suffer from this problem and are able to improve the quality of the solutions during the whole search process. Just on instance 100-{10}, a small stagnation is noticeable. This fact confirms that the reactive control rule is capable of driving the diversification of the strategy in such a way that the solvers can escape from the different local minima.

Table 1. Mean, std deviation and best solution for I, F, R and F+R. In the column best solution, when the value is an integer it indicates the number of times that the optimum was reached. Otherwise, the cost of the best solution found by that strategy.

N	H	mean				std. dev				best solution			
		I	F	R	F+R	I	F	R	F+R	I	F	R	F+R
20	3	0.25	0.44	0.00	0.00	0.74	1.00	0.00	0.00	25	23	30	30
	4	1.87	1.63	0.50	0.16	1.54	1.58	1.02	0.53	11	13	24	27
	5	1.02	0.85	0.01	0.00	1.04	1.33	0.04	0.00	10	16	29	30
25	3	0.90	0.72	0.21	0.20	1.38	1.29	0.77	0.77	20	21	27	28
	4	0.76	0.59	0.14	0.19	1.26	1.26	0.17	0.17	4	7	16	13
	5	1.84	1.06	0.00	0.00	2.79	2.14	0.01	0.00	11	12	29	30
40	3	0.00	0.00	0.00	0.00	0.00	0.00	0.00	0.00	30	30	30	30
	4	0.33	0.38	0.00	0.00	1.29	1.26	0.00	0.00	28	27	30	30
	5	0.79	1.19	0.28	0.57	1.08	1.18	0.77	1.02	7	5	24	21
50	3	0.17	0.08	0.00	0.25	0.64	0.46	0.00	0.77	28	29	30	27
	4	0.84	1.20	0.00	0.00	2.08	2.45	0.00	0.00	20	19	30	30
	5	2.86	3.12	1.66	2.44	2.05	2.29	2.11	2.16	8	8	18	12
100	3	0.00	0.09	0.00	0.09	0.00	0.47	0.00	0.47	30	29	30	29
	4	0.21	0.00	0.00	0.00	1.14	0.00	0.00	0.00	25	30	30	30
	5	1.76	1.83	1.44	1.50	0.81	0.76	1.11	1.08	2	2	11	10
	10	2.89	2.58	1.06	1.08	1.74	1.23	0.98	0.98	0.49	0.49	3	2
	15	5.12	4.94	1.75	2.47	1.56	1.61	1.16	1.44	1.48	1.27	0.75	0.42
	20	4.35	3.87	2.50	2.68	1.39	1.64	0.96	0.79	2.38	1.14	0.73	0.64
200	3	0.00	0.07	0.00	0.00	0.00	0.38	0.00	0.00	30	29	30	30
	4	0.22	0.22	0.00	0.00	1.13	1.13	0.01	0.01	20	16	28	26
	5	0.77	0.79	0.74	0.62	0.43	0.46	0.53	0.52	0.17	1	3	1
	10	2.94	2.97	2.36	2.76	1.18	1.12	1.39	1.68	1.66	1.30	0.11	0.15
	15	6.68	6.12	4.51	5.03	1.58	1.76	1.49	2.17	3.15	1.89	1.56	1.48
	20	6.81	5.54	4.41	4.88	1.69	2.16	1.13	1.82	3.24	2.42	2.77	2.82

Another interesting behaviour can be found in 200-{20}. In this instance, F achieves a faster convergence than I, which leads to better results. This can be considered as an indication that the use of this control rule is useful for larger instances where the concentration of the solvers in the same region of the search space is needed to, at least, obtain high quality local optima. More experimentation is needed in order to confirm this fact.

The other aspect of the dynamic behaviour we have studied is the evolution of the rule activation with time. Figure 5 shows the evolution of the mean number of times the control rule is fired for I, F, R and F+R. For this last strategy, the disaggregated behaviour for each rule (F+R:F and F+R:R) is shown. As in the former case, we focus on the six hardest instances.

Looking at the left-side plots of Figure 5, we can observe a much higher number of activations of R and F+R:R with respect to F and F+R:F. However, that order is inverted in the second column, that is, when the number of nodes is increased to 200. This effect is due to an important drop in the number of triggers for R and F+R:R. This behaviour variation is explained by the different

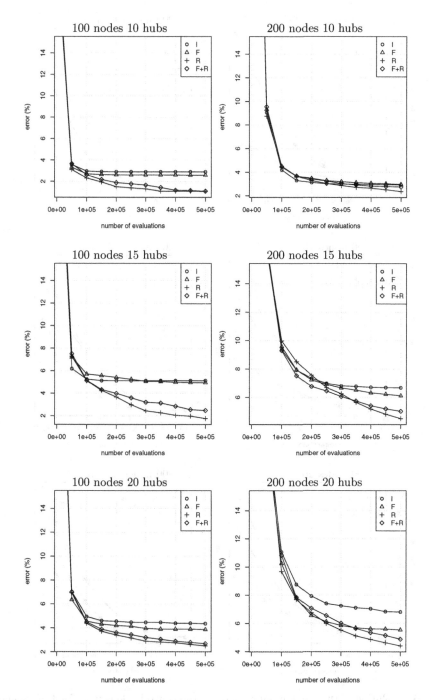

Fig. 4. Evolution of the mean percentage error of the best solution found by I, F, R and F+R during the execution time

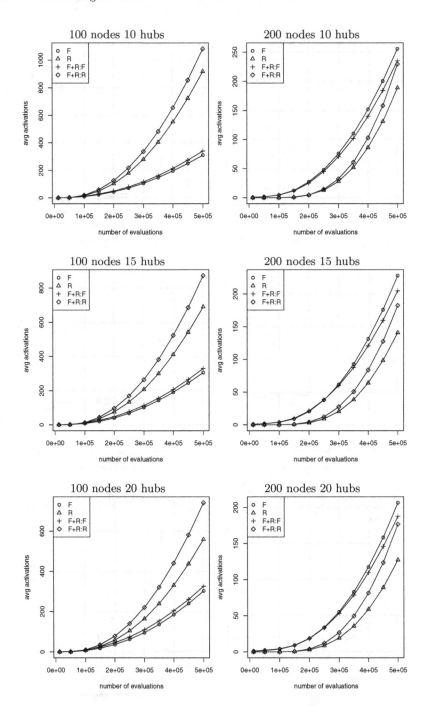

Fig. 5. Evolution of the mean number of times the control rule is fired for I, F, R and F+R. For this last strategy, the disaggregated behaviour for each rule (F+R:F and F+R:R) is shown.

performance of the solvers in the two cases. As we have seen before, the independent strategy stagnates in the first phases of the search process for instances with one hundred nodes. This situation is captured by the reactive rule, which increases the number of triggers since the early stages of the search. On the contrary, for 200-10,15,20, the independent solvers experience a slower convergence, and stop improving at the end of the process. This is also reflected in the behaviour of the reactive rule, that delays its activation. In this way, we see that the reactive rule is able to detect the difference between both instance sizes, adjusting its behaviour to each case.

We can also observe that using the two rules simultaneously produces an increase in their number of activations. This is due to the difference between the objectives of both rules. Since the fuzzy rule tries to reallocate the threads around promising regions of the search space, the probability of solvers to find big local minima is higher, which leads the reactive rule to be triggered more times. Although in a lesser degree, that interaction in the opposite sense also happens. When the cooperation is driven by the Reactive rule, the coordinator attempts to bring the solvers away from the local minima, and put them in regions of the search space relatively far from the best solution. If this distance is big enough and the solver does not achieve a good improvement rate, then the fuzzy rule will be fired for this thread.

6   Conclusions

In this work we have presented a hybrid composition of cooperative strategies and reactive search. Concretely, we have incorporated a rule based on the reactive search framework into a simple centralised cooperative strategy. Experiments on instances of the Uncapacitated Single Allocation p-Hub Median problem have shown that the proposed hybridization (cooperative strategy + reactive rule) achieves better results with respect to a strategy based on independent solvers (where solvers do not exchange information), and with respect to the same cooperative strategy based on a fuzzy control rule previously tested with success on other problems. We have also tested the cooperative strategy using both rules, reactive and fuzzy, at the same time. In this case, the bigger complexity does not pay off because the reactive rule alone performs at least as well as both rules together. Moreover, the reactive rule is able to adapt its behaviour according to the characteristics of the instance, and is effective for detecting stagnation and for driving diversification strategies.

The behavior of the cooperative strategy coupled with the fuzzy rule is somehow deceptive as it can not perform better than the independent strategy. Further research is needed in order to assess if this is strictly related with the p-hub problem, since previous work showed the benefit of such combination [3,8].

Our next step will be a comparison with state of the art algorithms in order to have a good reference for the performance of the proposed technique. We plan also to test the proposed schemes with solvers based on other optimization algorithms such as Tabu Search or Variable Neighborhood Search, possibly with a heterogeneous set of solvers.

Finally, the behaviour of the proposed strategy can be analysed according to the chosen reaction mechanism: while this paper dealt with the increase/decrease of the perturbation applied to a solution, a critical solver parameter can be varied, for instance, the annealing schedule for simulated annealing, or the tabu tenure for tabu search.

References

1. Bouthillier, A.L., Crainic, T.G.: A cooperative parallel meta-heuristic for the vehicle routing problem with time windows. Computers & Operations Research 32(7), 1685–1708 (2005)
2. Battiti, R., Brunato, M., Mascia, F.: Reactive Search and Intelligent Optimization. Operations Research/Computer Science Interfaces, vol. 45. Springer, New York (2008)
3. Pelta, D., Sancho-Royo, A., Cruz, C., Verdegay, J.L.: Using memory and fuzzy rules in a co-operative multi-thread strategy for optimization. Information Sciences 176(13), 1849–1868 (2006)
4. Dorigo, M., Stützle, T.: Ant Colony Optimization. The MIT Press/Bradford Books, Cambridge (2004)
5. Kennedy, J., Eberhart, R.C.: Swarm intelligence. Morgan Kaufmann Publishers Inc., San Francisco (2001)
6. Burke, E., Kendall, G., Newall, J., Hart, E., Ross, P., Schulenburg, S.: Hyper-Heuristics: An Emerging Direction in Modern Search Technology. In: Handbook of Metaheuristics, pp. 457–474 (2003)
7. Battiti, R., Tecchiolli, G.: The reactive tabu search. ORSA Journal on Computing 6(2), 126–140 (1994)
8. Cruz, C., Pelta, D.: Soft computing and cooperative strategies for optimization. Applied Soft Computing 9(1), 30–38 (2009)
9. Ferber, J.: Multi-Agent Systems: An Introduction to Distributed Artificial Intelligence. Addison-Wesley Longman Publishing Co., Inc., Boston (1999)
10. Battiti, R., Mascia, F.: Reactive local search for maximum clique: A new implementation. Technical Report DIT-07-018, Department of Information and Communication Technology, University of Trento (May 2007)
11. Campbell, J., Ernst, A., Krishnamoorthy, M.: Hub location problems. In: Facility Location: Applications and Theory, pp. 373–406. Springer, Heidelberg (2002)
12. O'Kelly, M., Morton, E.: A quadratic integer program for the location of interacting hub facilities. European Journal of Operational Research 32(3), 393–404 (1987)
13. Beasley, J.: Obtaining test problems via internet. Journal of Global Optimization 8(4), 429–433 (1996)
14. Kratica, J., Stanimirović, Z., Duščan Tovšić, V.F.: Two genetic algorithms for solving the uncapacitated single allocation p-hub median problem. European Journal of Operational Research 182(1), 15–28 (2007)
15. Henderson, D., Jacobson, S., Johnson, A.: The Theory and Practice of Simulated Annealing. In: Handbook of Metaheuristics. International Series in Operations Research & Management Science, vol. 57, pp. 287–320. Kluwer, Norwell (2003)

Study of the Influence of the Local Search Method in Memetic Algorithms for Large Scale Continuous Optimization Problems

Daniel Molina[1], Manuel Lozano[2], and Francisco Herrera[2]

[1] Department of Computer Languages and Systems,
Universidad de Cádiz, Cádiz, Spain
daniel.molina@uca.es
[2] Department of Computer Science and Artificial Inteligence,
Universidad de Granada, Granada, Spain
{lozano,herrera}@decsai.ugr.es

Abstract. Memetic algorithms arise as very effective algorithms to obtain reliable and high accurate solutions for complex continuous optimization problems. Nowadays, high-dimensional optimization problems are an interesting field of research. Its high dimension introduces new problems for the optimization process, making recommendable to test the behavior of optimization algorithms to large-scale problems. In memetic algorithms, the local search method is responsible of exploring the neighborhood of the current solutions; therefore, the dimensionality has a direct influence over this component. The aim of this paper is to study this influence. We design different memetic algorithms that only differ in the local search method applied, and they are compared using two sets of continuous benchmark functions: a standard one and a specific set with large-scale problems. The results show that high dimensionality reduces the differences among the different local search methods.

1 Introduction

It is now well established that hybridization of *evolutionary algorithms* (EAs) with other techniques can greatly improve the efficiency of search [1,2]. EAs that have been hybridized with local search (LS) techniques are often called *memetic algorithms* (MAs) [3,4,5]. One commonly used formulation of MAs applies LS to members of the EA population after recombination and mutation, with the aim of exploiting the best search regions gathered during the global sampling done by the EA.

MAs comprising efficient local improvement processes on continuous domains (*continuous LS methods*) have been presented to address the difficulty of obtaining reliable solutions of high precision for complex continuous optimization problems [6,7,8,9,10]. In this paper, they will be named MACOs (MAs for continuous optimization problems).

Nowadays, high-dimensional optimization problems arise as a very interesting field of research, because they appear in many important new real-world

T. Stützle (Ed.): LION 3, LNCS 5851, pp. 221–234, 2009.

problems (bio-computing, data mining, etc.). Unfortunately, the performance of most available optimization algorithms deteriorates rapidly as the dimensionality of the search space increases [11]. Thus, the ability of being scalable for high-dimensional problems becomes an essential requirement for modern optimization algorithm approaches.

In recent years, it has been increasingly recognized that the influence of the continuous LS algorithm employed has a major impact on the search performance of MACOs [8]. There exists a group of continuous LS algorithms that stand out as brilliant local search optimizers. They include the Solis and Wets's algorithm [12], the Nelder and Mead's simplex method [13], and the CMA-ES algorithm [14]. However, on some occasions, they may become very expensive, because of the wa y they exploit local information to guide the search process. In this paper, they are called *intensive* continuous LS methods. Given the potential of these LS methods, a specific MACO model, called *MACO based on LS chains*, has been designed that can effectively use them as local search operators [10]. An instance of this MACO was experimental studied, which employed CMA-ES as local optimizer. The results showed that it was very competitive with state-of-the-art on both MACOs and EAs for continuous optimization problems. Particularly, significant improvements were obtained for the problems with the highest dimensionality among the ones considered for the empirical study (in particular, $D = 30$ and $D = 50$), which suggests that the application of this MACO approach to optimization problems with higher dimensionalities is indeed worth of further investigations.

In this paper, we undertake an extensive study of the performance of three MACOs based on LS chains that apply different instances of intense LS procedures on a specific set of benchmark functions with large scale problems, which was defined for the *CEC'2008 Special Session on Large Scale Global Optimization* ($D = 100$ and $D = 200$). In addition, since we are interested on investigating the ways these MACOs respond as dimensionality increases, we have evaluated them, as well, on the benchmark functions defined for the *CEC'2005 Special Session on Real-Parameter Optimization* ($D = 10$, 30, and 50).

The paper is set up as follows. In Section 2, we describe the different continuous LS methods analyzed. In Section 3, we review some important aspects of the MACO used for our study. In Section 4, we present the experimental study of three MACO instances based on continuous LS mechanisms on test problems with different dimensionalities. Finally, in Section 5, we provide the main conclusions of this work.

2   Continuous LS Methods

In this section, we present a detailed description of the continuous LS methods used in our empirical study. They are three well-known continuous local searchers: the Solis and Wets's algorithm [12], the Nelder and Mead's simplex method [13], and the CMA-ES algorithm [14]. Next, we present a detailed description of these procedures.

2.1   Solis and Wets' Algorithm

The classic *Solis and Wets' algorithm* [12] is a randomised hill-climber with an adaptive step size. Each step starts at a current point x. A deviate d is chosen from a normal distribution whose standard deviation is given by a parameter ρ. If either $x+d$ or $x-d$ is better, a move is made to the better point and a success is recorded. Otherwise, a failure is recorded. After several successes in a row, ρ is increased to move quicker. After several failures in a row, ρ is decreased to focus the search. It is worth noting that ρ is the strategy parameter of this continuous LS operator. Additionally, a bias term is included to put the search momentum in directions that yield success. This is a simple LS method that can adapt its size search very quickly. More details about this procedure may be found in [12].

2.2   Nelder-Mead Simplex Algorithm

This is a classical and very powerful local descent algorithm. A simplex is a geometrical figure consisting, in n dimensions, of $n+1$ points $s_0, \cdots, s_n$. When a points of a simplex is taken as the origin, the n other points are used to describe vector directions that span the n-dimension vector space. Thus, if we randomly draw an initial starting point s_0, then we generate the other n points s_i according to the relation $s_i = s_0 + \lambda e_j$, where the e_j are n unit vectors, and λ is a constant that is typically equal to one.

Through a sequence of elementary geometric transformations (reflection, contraction, expansion and multi-contraction), the initial simplex moves, expand or contracts. To select the appropriate transformation, the method only uses the values of the function to be optimized at the vertices of the simplex considered. After each transformation, a better vertex replaces the current worst one. A complete picture of this algorithm may be found in [13].

This method has the advantage of creating initially a simplex composed by movements in each direction. This is a very useful characteristic to deal with high-dimensional spaces.

2.3   CMA-ES Method

The *covariance matrix adaptation evolution strategy* (CMA-ES) [14,15] was originally introduced to improve the LS performances on evolution strategies. Even though CMA-ES reveals competitive global search performances [16], it has exhibited effective abilities for the local tuning of solutions; in fact, it was used as continuous LS algorithm to create *multi-start LS metaheuristics*, L-CMA-ES [17], and G-CMA-ES [18]. At the 2005 congress of evolutionary computation, these algorithms were ones of the winners of the real-parameter optimisation competition [19].

In CMA-ES, not only is the step size of the mutation operator adjusted at each generation, but so too is the step direction in the multidimensional problem space, i.e., not only is there a mutation strength per dimension but their combined update is controlled by a covariance matrix whose elements are updated as the search proceeds. In this paper, we use the (μ_W, λ) CMA-ES model. For every generation, this algorithm generates a population of λ offspring by sampling a multivariate normal distribution:

$$x_i \sim N\left(m, \sigma^2 C\right) = m + \sigma N_i(0, C) \text{ for } i = 1, \cdots, \lambda,$$

where the mean vector m represents the favourite solution at present, the so-called step-size σ controls the step length, and the covariance matrix C determines the shape of the distribution ellipsoid. Then, the μ best offspring are recombined into the new mean value using a *weighted intermediate recombination*: $\sum_{i=1}^{\mu} w_i x_{i:\lambda}$, where the positive weights sum to one. The covariance matrix and the step-size are updated as well following equations that may be found in [14] and [16]. The default strategy parameters are given in [16]. Only the initial m and σ parameters have to be set depending on the problem.

3   MACOs Based on LS Chains

In this section, we describe a MACO approach proposed in [10] that employs continuous LS methods as LS operators. It is a steady-state MA model that employs the concept of *LS chain* to adjust the LS intensity assigned to the intense continuous LS method. In particular, this MACO handles LS chains, throughout the evolution, with the objective of allowing the continuous LS algorithm to act more intensely in the most promising areas represented in the EA population. In this way, the continuous LS method may adaptively fit its strategy parameters to the particular features of these zones.

In Section 3.1, we introduce the foundations of steady-state MAs. In Section 3.2, we explain the concept of *LS chain*. Finally, in Section 3.3, we give an overview of the MACO approach presented in [10], which handles LS chains with the objective of make good use of intense continuous LS methods as LS operators.

3.1   Steady-State MAs

In *steady-state* GAs [20] usually only one or two offspring are produced in each generation. Parents are selected to produce offspring and then a decision is made as to which individuals in the population to select for deletion in order to make room for the new offspring. Steady-state GAs are *overlapping* systems because parents and offspring compete for survival. A widely used replacement strategy is to replace the worst individual only if the new individual is better. We will call this strategy the *standard replacement strategy*.

Although steady-state GAs are less common than generational GAs, Land [21] recommended their use for the design of *steady-state MAs* (steady-state GAs plus LS) because they may be more stable (as the best solutions do not get replaced until the newly generated solutions become superior) and they allow the results of LS to be maintained in the population.

3.2   LS Chains

In steady-state MAs, individuals resulting from the LS invocation may reside in the population during a long time. This circumstance allows these individuals to become starting points of subsequent LS invocations. In [10], Molina et al. propose *to chain* an LS algorithm invocation and the next one as follows:

> *The final configuration reached by the former (strategy parameter values, internal variables, etc.) is used as initial configuration for the next application.*

In this way, the LS algorithm may continue under the same conditions achieved when the LS operation was previously halted, providing an *uninterrupted connection between successive LS invocations*, i.e., forming a *LS chain*. Two important aspects that were taken into account for the management of LS chains are:

- Every time the LS algorithm is applied to refine a particular chromosome, a fixed LS intensity should be considered for it, which will be called *LS intensity stretch* (I_{str}).

  In this way, a LS chain formed throughout n_{app} LS applications and started from solution s_0 will return the same solution as the application of the continuous LS algorithm to s_0 employing $n_{app} \cdot I_{str}$ fitness function evaluations.
- After the LS operation, the parameters that define the current state of the LS processing are stored along with the reached final individual (in the steady-state GA population). When this individual is latter selected to be improved, the initial values for the parameters of the LS algorithm will be directly available. For example, if we employ the Solis and Wets' algorithm (Section 2.1) as LS algorithm, the stored strategy parameter may be the current value of the ρ parameter. For the more elaborate CMA-ES (Section 2.3), the state of the LS operation may be defined by the covariance matrix (C), the mean of the distribution (m), the size (σ), and some additional variables used to guide the adaptation of these parameters.

3.3   A MACO Model That Handles LS Chains

In this section, we introduce a MACO model that handles LS chains (see Figure 1) with the following main features:

1. It is a steady-state MA model.
2. It ensures that a fixed and predetermined local/global search ratio is always kept. With this policy, we easily stabilise this ratio, which has a strong influence on the final MACO behaviour. Without this strategy, the application of intense continuous LS algorithms may induce the MACO to prefer super exploitation.
3. It favours the enlargement of those LS chains that are showing promising fitness improvements in the best current search areas represented in the steady-state GA population. In addition, it encourages the activation of innovative LS chains with the aim of refining unexploited zones, whenever the

---

1. Generate the **initial population**.
2. Perform the **steady-state GA** throughout n_{frec} evaluations.
3. Build the set S_{LS} with those individuals that **potentially may be refined by LS**.
4. **Pick the best individual** in S_{LS} (Let's c_{LS} to be this individual).
5. If c_{LS} belongs to an **existing LS chain** then
6.   Initialise the LS operator with the **LS state stored** together with c_{LS}.
7. Else
8.   Initialise the LS operator with the **default** LS state.
9. Apply the LS algorithm to c_{LS} with an LS intensity of I_{str} (Let's c_{LS}^r to be the resulting individual).
10. Replace c_{LS} by c_{LS}^r in the **steady-state GA population**.
11. Store the **final LS state** along with c_{LS}^r.
12. If (*not termination-condition*) go to step 2.

Fig. 1. Pseudocode algorithm for the MACO based on LS chains

current best ones may not offer profitability. The criterion to choose the individuals that should undergo LS is specifically designed to manage the LS chains in this way (Steps 3 and 4).

This MACO scheme defines the following relation between the steady-state GA and the intense continuous LS method (Step 2): *every n_{frec} number of evaluations of the steady-state GA, apply the continuous LS algorithm to a selected chromosome, c_{LS}, in the steady-state GA population.* Since we assume a fixed $\frac{L}{G}$ ratio, $r_{L/G}$, n_{frec} may be calculated using the following equation:

$$n_{frec} = I_{str} \frac{1 - r_{L/G}}{r_{L/G}}. \tag{1}$$

where n_{str} is the LS intensity stretch (Section 3.2) and $r_{L/G}$ is defined as the percentage of evaluations spent doing local search from the total assigned to the algorithm's run.

The following mechanism is performed to select c_{LS} (Steps 3 and 4):

1. Build the set of individuals in the steady-state GA population, S_{LS} that fulfils:
   (a) They have never been optimized by the intense continuous LS algorithm, or
   (b) They previously underwent LS, obtaining a fitness function improvement greater than δ_{LS}^{min} (a parameter of our algorithm).
2. If $|S_{LS}| \neq 0$, then apply the continuous LS algorithm to the best individual in this set. If this condition is not accomplished, the LS operator is applied to the best individual in the steady-state GA population.

With this mechanism, when the steady-state GA finds a new best so far individual, it will be refined immediately. In addition, the best performing individual in the steady-state GA population will always undergo LS whenever the fitness improvement obtained by a previous LS application to this individual is greater than the δ_{LS}^{min} threshold. The last condition is very important in order to avoid the overexploitation of search zones where the LS method may not make substantial progresses any more.

4   Experiments

In this section, we present the experimental study carry out with three MACOs based on LS chains that use the Solis and Wets's algorithm, the Nelder and Mead's simplex method, and the CMA-ES algorithm, respectively, on two different test suites:

- *CEC'2005 test suite.* Benchmark functions recommended for the *Special Session on Real Parameter Optimization organized in the 2005 IEEE Congress on Evolutionary Computation.* It is possible to consult in [19] the complete description of the functions, furthermore in the link the source code is included. The set of test functions is composed of 5 unimodal functions, 7 basic multimodal functions, 2 expanded multimodal functions, and 11 hybrid functions. Three dimension values were considered for these problems: $D = 10$, $D = 30$, and $D = 50$.
- *CEC'2008 test suite.* A set of benchmark functions specifically designed as large-scale problems, which were defined for *CEC'2008 Special Session on Large Scale Global Optimization.* This test suite is used to study the behavior of the MACOs on problems with high dimensionality. It consists in 7 test functions with two different dimension values: $D = 100$ and $D = 200$.

This section is structured in the following way. In Section 4.1, we describe the three instances of MACO used for the experiments. In Section 4.2, we detail the experimental setup and statistical methods that were used for this experimental study. In Section 4.3, we analyze the results on the *CEC'2005* and, in Section 4.4, the ones on the *CEC'2008 test suite.* In Section 4.5, we study the influence of the LS method when the dimensionality is increased. In Section 4.6, we compare our best model with two algorithms based on CMA-ES, L-CMA-ES [17] and G-CMA-ES [18], which give very good results in continuous optimization.

4.1   Three Instances of MACO Based on LS Chains

In this section, we build three instances of the MACO model described in Figure 1, which apply the Solis and Wets's algorithm (Section 2.1), the Nelder and Mead's simplex method (Section 2.2), and the CMA-ES algorithm (Section 2.3), respectively, as intense continuous LS algorithms.

Next, we list their main features:

Steady-State GA. It is a real-coded steady-state GA [22] specifically designed to promote high population diversity levels by means of the combination of the BLX-α crossover operator (see [22]) with a high value for its associated parameter ($\alpha = 0.5$) and the *negative assortative mating* strategy [23]. Diversity is favored as well by means of the BGA mutation operator (see [22]).

Continuous LS Algorithms. The three instances follow the MACO approach that handles LS chains, with the objective of tuning the intensity of the three LS algorithms considered, which are employed as intense continuous LS operators. In particular, the application of CMA-ES for refining an individual, C_i, is carried out following the next guidelines:

- We consider C_i as the initial mean of distribution (m).
- The initial σ value is half of the distance of C_i to its nearest individual in the steady-state GA population (this value allows an effective exploration around C_i).

CMA-ES will work as local searcher consuming I_{str} fitness function evaluations. Then, the resulting solution will be introduced in the steady-state GA population along with the current value of the covariance matrix, the mean of the distribution, the step-size, and the variables used to guide the adaptation of these parameters (B, BD, D, p_c and p_σ). Latter, when CMA-ES is applied to this inserted solution, these values will be recovered to proceed with a new CMA-ES application. When CMAE-ES is performed on solutions that do not belong to existing chains, default values, given in [16], are assumed for the remaining strategy parameters.

The Solis and Wets's algorithm and the Nelder and Mead's simplex method are used in a similar fashion.

Parameter Setting. For the experiments, the three MACO instances apply BLX-α with $\alpha = 0.5$. The population size is 60 individuals and the probability of updating a chromosome by mutation is 0.125. The n_{ass} parameter associated with the negative assortative mating is set to 3. They use $I_{str} = 500$ and the value of the $\frac{L}{G}$ ratio, $r_{L/G}$, was set to 0.5, which represents an equilibrated choice. Finally, a value of 10^{-8} was assigned to the δ_{LS}^{min} threshold. All these parameter values are recommended in [10].

4.2   Experimental Setup and Statistical Analysis

The experiments have been carried out following the instructions indicated in the documents associated to each set of benchmark functions. The main characteristics are:

- Each algorithm is run 25 times for each test function, and the error average of the best individual of the population is computed.

Table 1. Parameters used for the experiments

Parameters	Functions Bench-mark CEC'2005	Functions Benchmark CEC'2008
Execution Numbers	25	25
Dimensions	10, 30, 50	100, 200
Maximum Evaluations Number (NE)	$10,000 * D$	$5,000 * D$
Stopping Criterion	NE achieved or $error < 1e - 8$	NE achieved

- The study has been made with dimensions $D = 10, D = 30$, and $D = 50$ for the *CEC'2005* and $D = 100$ and $D = 200$ for the *CEC'2008 test suite*.
- The maximum number of fitness evaluations is $10,000 \cdot D$ for the *CEC'2005* and $5,000 \cdot D$ for the *CEC'2005 test suite*.
- Each run stops either when the error obtained is less than 10^{-8}, or when the maximal number of evaluations is achieved.

To analyse the results we have used *non-parametric tests*, because it has been shown that parametric tests can not be applied with security for these test suites [24]. We have applied the non-parametric recommended in [24], thus it can be consulted this paper to obtain a detailed explanations of them. Next, these tests are briefly explained:

- The *Iman-Davenport's* test. This non-parametric test is used for answering this question: *In a set of k samples (where $k \geq 2$), do at least two of the samples represent populations with different median values?*. It is a non-parametric procedure employed in a hypothesis testing situation involving a design with two or more samples; therefore, it is a multiple comparison test that aims to detect significant differences between the behaviour of two or more algorithms.
- The *Holm's* test as a post-hoc procedure, to detect whose algorithms are worse than the algorithm with best results. This test only can be applied if the Iman-Davenport's test detects a significant difference. It sequentially checks the hypotheses ordered according to their significance. If p_i is lower than $\alpha/(k - i)$, the corresponding hypothesis is rejected and the process continues. In other case, this hypothesis and all the remaining hypotheses are maintained as supported.

4.3 Results for the CEC'2005 Test Suite

Firstly, we applied the *Iman-Davenport's* test to see if there is a significant difference between the different MACO instances, considering the three different dimension values. Table 2 shows the results of this statistical test.

We may observe in Table 2 that for dimension 30 and 50 there exist significant differences among the rankings of the algorithms (the statistical value is greater

Table 2. Results of the Iman-Davenport's test comparing the instances for $D = 10$, 30, and 50

Dimension	Iman-Davenport value	Critical value	Sig. differences?
10	2,2991	3,19	No
30	9,2012	3,19	Yes
50	5,5276	3,19	Yes

Table 3. Comparison using the Holm's test of the MACO instances with respect the best one (based on CMA-ES), with $D = 30, 50$

Dimension	LSMethod	z	p-value	α/i	Sig. differences?
30	Simplex	3,2527	0,00114	0,0250	Yes
	Solis Wets	2,4749	0,01333	0,0500	Yes
50	Simplex	2,9698	0,00298	0,0250	Yes
	Solis Wets	2,1213	0,03389	0,0500	Yes

than the critical one, 3,19). Then, attending on these results, we compare the MACO instances by means of the *Holm's* test. Table 3 shows the results.

From Table 3, we can see that the election of the continuous LS method is crucial for $D = 30$ and $D = 50$. For $D = 10$, the problems become easy and then, all instances achieve similar results. In this case, the MACO based on CMA-ES is the best algorithm.

4.4   Results for CEC'2008 Test Suite

In this section, we compare the different instances on the large-scale problems in the CEC'2008 test suite. First, we have applied the *Iman-Davenport's* test. Table 4 shows the results.

We may see in Table 4 that for high dimensionality (greater or equal than 100) there are no differences between the results achieved by MACO instances based on the different LS methods. Thus, when MACOs based on LS chains are applied on problems with high dimensionality, the election of the continuous LS method does not provide significant statistical differences. Table 5 shows the results for each instance and function.

4.5   Remarks from the Results

In the previous sections, we have used non-parametric test to determine when differences between the MACO instances are statistically significant. Figure 2

Table 4. Results of the Iman-Davenport's test for $D = 100$ and $D = 200$

Dimension	Iman-Davenport value	Critical value	Sig. differences?
100	0,2222	3,89	No
200	1,0000	3,89	No

Table 5. Results of each instance for the CEC'2008 test suite

LS Method	Dimension 100						
	F1	F2	F3	F4	F5	F6	F7
CMA-ES	9,50E-9	**5,89E-3**	1,65E+2	**2,31E+0**	**9,73E-9**	9,76E-9	-1,42E+3
Simplex	**9,43E-9**	5,33E+0	**8,21E+1**	2,83E+0	6,90E-4	8,93E-8	-1,45E+3
Solis Wets	9,92E-9	1,71E+1	1,50E+2	2,55E+0	5,91E-4	9,94E-9	-1,45E+3
LS Method	Dimension 200						
	F1	F2	F3	F4	F5	F6	F7
CMA-ES	9,69E-9	**1,57E+0**	2,31E+2	1,28E+1	2,96E-4	**9,87E-9**	-2,70E+3
Simplex	**9,66E-9**	1,74E+1	**1,22E+2**	**3,26E+0**	**9,82E-9**	9,88E-9	**-2,76E+3**
Solis Wets	9,95E-9	4,15E+1	2,85E+2	5,29E+0	2,07E-3	9,98E-9	-2,73E+3

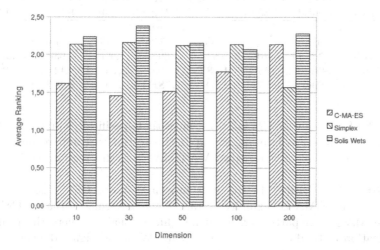

Fig. 2. Mean ranking of each MACO instance on the problems with different dimensionalities

shows the average rankings of these algorithms for the different dimension values considered in this study, with the aim of showing, graphically, the differences between them. We can obtain the following conclusions from this table:

- The improvements obtained from the best LS method, CMA-ES, decrease as dimensionality increases (for $D = 10$, the difficulty is not enough to obtain a clear difference), achieving, finally, a situation ($D = 200$), where it might not be the best LS method. The difference between the Solis Wets' method and simplex method is also reduced.
- For $D = 100$, the simplex method outperforms the others continuous LS methods.

4.6 Comparisons with Other CMA-ES Algorithms

In the *CEC'2005 Special Session on Real-Parameter Optimization* two algorithms arise as the best algorithms [25]: the L-CMA-ES [17] and G-CMA-ES

Table 6. Results of the Iman-Davenport's test, comparing MA-CMAES,L-CMA-ES, and G-CMA-ES for $D = 100$ and $D = 200$

Dimension	Iman-Davenport value	Critical value	Sig. differences?
100	1,0000	3,89	No
200	0,3913	3,89	No

Table 7. Results of each CMA-ES based algorithms for the CEC'2008 test suite

LS	Dimension 100						
Method	F1	F2	F3	F4	F5	F6	F7
G-CMA-ES	1,04E-13	3,41E-10	7,97E-1	2,29E+2	1,72E-12	1,71E+1	-8,63E+2
L-CMA-ES	**4,49E-14**	**3,98E-11**	**6,38E-1**	2,01E+2	**9,44E-13**	2,13E+1	-8,58E+2
MA-CMA-ES	9,50E-9	5,89E-3	1,65E+2	**2,31E+0**	9,73E-9	**9,76E-9**	**-1,42E+3**

LS	Dimension 200						
Method	F1	F2	F3	F4	F5	F6	F7
G-CMA-ES	9,72E-14	**6,96E-10**	9,56E-1	4,69E+2	5,06E-12	1,63E+1	**-3,96E+8**
L-CMA-ES	**4,99E-14**	2,53E-9	**4,78E-1**	5,39E+2	**2,74E-12**	2,14E+1	-3,12E+7
MA-CMA-ES	9,69E-9	1,57E+0	2,31E+2	**1,28E+1**	2,96E-4	**9,87E-9**	-2,70E+3

[18], both of them invoke CMA-ES instances that specifically emphasise the local refinement abilities of this algorithm. In this section we compare our instance with the CMA-ES algorithm (called MA-CMA-ES in this section) with them.

CEC'2005 Test Suite. In [10] it is proven that, for CEC'2005 test suite, our model with CMA-ES exhibits overall better performance than L-CMA-ES and G-CMA-ES, in particular, at higher dimensionality, where the proposed hybridisation method outperforms the others. With the higher dimension, it is the best algorithm, and statistically better than the pure restart local search strategy (L-CMA-ES). In [10] this analysis is explained in detail.

CEC'2008 Test Suite. Table 6 shows the results of Iman-Davenport's test comparing the two CMA-ES algorithms (G-CMA-ES, L-CMA-ES) with the instance with CMA-ES, using the CEC'2008 test suite, for dimension 100 and 200. Table 7 shows the results for each one of these algorithms for CEC'2008.

Although we can observe than there is no statistically differences between them, from Table 7 we can obtain several conclusions. First, in unimodal functions (F1 and F2), L-CMA-ES and G-CMA-ES obtain the best results. For more complex functions (F4, F6, and F7 in dimension 100) our model with CMA-ES obtain clearly the best results. In general, for easy functions where every algorithm obtains a good result (a error lower than 10^{-8}) the restart models achieve the best results, due to a greater exploitation in the search. For other more complex functions, our model can avoid local optima and obtain better results, achieving a clear improvement over the multistart models.

5   Conclusions

In this paper, we have built several instances of MACO based on LS chains that differ in the intense continuous LS method applied: the Solis and Wets's algorithm, the Nelder and Mead's simplex method, and the CMA-ES algorithm. We have compared their performance on two set of test problems including problems with different dimensionality, the *CEC'2005* and *CEC'2008* test suites. The main conclusions obtained are:

- Our instance with CMA-ES gives better results than other restart algorithms that use CMA-ES.
- The dimensionality strongly affects the performance of the MACO instances based on the different LS methods; not only the differences among them, but also, which algorithm results the best one.
- We have observed that, while there are significant differences for medium and low dimension values (and the election of the LS method is crucial), when the dimensionality increases, the differences between them are reduced.
- The simplex method improves as the dimension of the problems increase. Maybe the reason is that this is an algorithm capable of exploring better the changes in reduced group of dimensions. This is an aspect to consider for future LS methods specifically designed for high-dimension problems.

References

1. Davis, L.: Handbook of Genetic Algorithms. Van Nostrand Reinhold, New York (1991)
2. Goldberg, D.E., Voessner, S.: Optimizing global-local search hybrids. In: Banzhaf, W., et al. (eds.) Proceedings of the Genetic and Evolutionary Computation Conference 1999. Morgan Kaufmann, San Mateo (1999)
3. Moscato, P.: On evolution, search, optimization, genetic algorithms and martial arts: Towards memetic algorithms. Technical report, Technical Report Caltech Concurrent Computation Program Report 826, Caltech, Pasadena, CA (1989)
4. Moscato, P.: Memetic algorithms: a short introduction, pp. 219–234. McGraw-Hill, London (1999)
5. Merz, P.: Memetic Algorithms for Combinational Optimization Problems: Fitness Landscapes and Effective Search Strategies. PhD thesis, Gesamthochschule Siegen, University of Siegen, Germany (2000)
6. Hart, W.: Adaptive Global Optimization With Local Search. PhD thesis, Univ. California, San Diego, CA (1994)
7. Lozano, M., Herrera, F., Krasnogor, N., Molina, D.: Real-coded Memetic Algorithms with Crossover Hill-climbing. Evolutionary Computation 12(2), 273–302 (2004)
8. Ong, Y.S., Keane, A.J.: Meta-Lamarckian Learning in Memetic Algorithms. IEEE Transactions on Evolutionary Computation 4(2), 99–110 (2004)
9. Noman, N., Iba, H.: Accelerating differential evolution using an adaptive local search. IEEE Transactions on Evolutionary Computation 12(1), 107–125 (2008)

10. Molina, D., Lozano, M., García-Martínez, C., Herrera, F.: Memetic algorithms for continuous optimization based on local search chains. Evolutionary Computation (in press, 2009)
11. van den Bergh, F., Engelbrencht, A.: A cooperative approach to particle swarm optimization. IEEE Transactions on Evolutionary Computation, 225–239 (2004)
12. Solis, F.J., Wets, R.J.: Minimization by Random Search Techniques. Mathematical Operations Research 6, 19–30 (1981)
13. Nelder, J., Mead, R.: A simplex method for functions minimizations. Computer Journal 7(4), 308–313 (1965)
14. Hansen, N., Ostermeier, A.: Adapting Arbitrary Normal Mutation Distributions in Evolution Strategies: The Covariance Matrix Adaptation. In: Proceeding of the IEEE International Conference on Evolutionary Computation (ICEC 1996), pp. 312–317 (1996)
15. Hansen, N., Müller, S., Koumoutsakos, P.: Reducing the time complexity of the derandomized evolution strategy with covariance matrix adaptation (cma-es). Evolutionary Computation 11(1), 1–18 (2003)
16. Hansen, N., Kern, S.: Evaluating the CMA evolution strategy on multimodal test functions. In: Yao, X., Burke, E.K., Lozano, J.A., Smith, J., Merelo-Guervós, J.J., Bullinaria, J.A., Rowe, J.E., Tiño, P., Kabán, A., Schwefel, H.-P. (eds.) PPSN 2004. LNCS, vol. 3242, pp. 282–291. Springer, Heidelberg (2004)
17. Auger, A., Hansen, N.: Performance Evaluation of an Advanced Local Search Evolutionary Algorithm. In: 2005 IEEE Congress on Evolutionary Computation, pp. 1777–1784 (2005)
18. Auger, A., Hansen, N.: A Restart CMA Evolution Strategy with Increasing Population Size. In: 2005 IEEE Congress on Evolutionary Computation, pp. 1769–1776 (2005)
19. Suganthan, P., Hansen, N., Liang, J., Deb, K., Chen, Y., Auger, A., Tiwari, S.: Problem definitions and evaluation criteria for the CEC 2005 special session on real parameter optimization. Technical report, Nanyang Technical University (2005)
20. Sywerda, G.: Uniform crossover in genetic algorithms. In: Schaffer, J. (ed.) Proc. of the International Conference on Genetic Algorithms, pp. 2–9 (1989)
21. Land, M.S.: Evolutionary Algorithms with Local Search for Combinational Optimization. PhD thesis, Univ. California, San Diego, CA (1998)
22. Herrera, F., Lozano, M., Verdegay, J.L.: Tackling Real-coded Genetic Algorithms: Operators and Tools for the Behavioral Analysis. Artificial Intelligence Reviews 12(4), 265–319 (1998)
23. Fernandes, C., Rosa, A.: A Study of non-Random Matching and Varying Population Size in Genetic Algorithm using a Royal Road Function. In: Proc. of the 2001 Congress on Evolutionary Computation, pp. 60–66 (2001)
24. García, S., Molina, D., Lozano, M., Herrera, F.: A study on the use of non-parametric tests for analyzing the evolutionary algorithms' behaviour: A case study on the cec 2005 special session on real parameter optimization. Journal of Heuristics (2008) (in press)
25. Hansen, N.: Compilation of Results on the CEC Benchmark Function Set. Technical report, Institute of Computational Science, ETH Zurich, Switerland (2005),
http://www.ntu.edu.sg/home/epnsugan/index_files/CEC-05/
compareresults.pdf

Neural Network Pairwise Interaction Fields for Protein Model Quality Assessment

Alberto J.M. Martin, Alessandro Vullo, and Gianluca Pollastri

School of Computer Science and Informatics and
Complex and Adaptive Systems Laboratory
University College Dublin, Belfield, Dublin, Ireland
{albertoj,alessandro.vullo,gianluca.pollastri}@ucd.ie

Abstract. We present a new knowledge-based Model Quality Assessment Program (MQAP) at the residue level which evaluates single protein structure models. We use a tree representation of the C_α trace to train a novel Neural Network Pairwise Interaction Field (NN-PIF) to predict the global quality of a model. We also attempt to extract local quality from global quality. The model allows fast evaluation of multiple different structure models for a single sequence. In our tests on a large set of structures, our model outperforms most other methods based on different and more complex protein structure representations in both local and global quality prediction. The method is available upon request from the authors. Method-specific rankers may also built by the authors upon request.

1 Introduction

In order to use a 3D model of a protein structure we need to know how good it is, as its quality is proportional to its utility [1]. Moreover most protein structure prediction methods produce many reconstructions for any one protein and being able to sift out good ones from bad ones is often the key to the success of a method.

Several different potential or (pseudo-)energy function have been developed with the aim of mapping a three-dimensional structure into its "goodness" or "native-likeness". These can be roughly divided into physics-based and knowledge-based, with some amount of overlap. The former are based on physico-chemical calculations [2, 3, 4, 5, 6], while the latter are based on statistics obtained from a training set of known structures (also termed "decoys") typically generated by computational methods. These structures or decoys can be represented in different ways, embedded or not into a 3D lattice, in order to reduce the conformational space, and potentials can rely on more or less detailed representations of a structure. As simplicity in the representation increases (e.g. if each residue is represented as a single point or sphere), evaluation speed increases since fewer interactions have to be taken into account. By contrast the reliability of the evaluation tends to decrease when details are stripped from a structure [7, 8, 9, 10].

T. Stützle (Ed.): LION 3, LNCS 5851, pp. 235–248, 2009.

However, there may be other advantages in the use of simplified representations, such as reduced sensitivity to small perturbations in conformations and the ability to take into account complex effects that cannot be described separately [11], plus a reduced sensitivity to inaccuracies and uncertainties in the data (crystal contacts, high R-factors, etc.).

Knowledge-based potentials are not only used to rank models, but for other aims such as to drive 3D predictions and for model refinement [5, 6, 12, 13, 14]; fold recognition [15]; to place side chain and backbone atoms [16, 17]; as mentioned, to select native structures from sets of decoys [7, 8]; to predict stability of proteins [18, 19, 20]; or even to predict residue residue contact maps [21, 22, 23].

Several different representations of protein structures have been used in knowledge-based potentials. They can be classified in: 1) single point/sphere representations of each residue; 2) two or more points for each residue; 3) full atom models. In the first group each residue is usually represented by its C_α atom [24, 25, 26, 27, 28, 29] or by their C_β atom [7, 8, 15, 18, 29, 30, 31], pseudo-C_β, side chain (SD) centre of mass or SD centroids [32]. In group 2) each residue is represented by two or more points, but fewer than its number of atoms; these points range from only two (e.g. $C_\alpha + C_\beta$, C_α+pseudo-C_β [33]), backbone atoms and some sort of representation of side chains like the C_β [19], different rigid-body blocks/fragments [10]. In group 3) are many different potentials [7, 19, 20, 30, 34, 35, 36, 37]. These depend on different reference states and on different physico-chemical assumptions, whereas others are learnt from examples, typically via Machine Learning algorithms [28, 35, 38, 39].

There is a further group of potential functions - those based on clustering of many different structure predictions, and consensus methods. These usually outperform single model evaluation methods [1, 39, 40, 41]. Consensus methods make use of several different potentials to know the quality of the prediction [14, 38, 39]. The potentials are joined in a (possibly weighted) average. Clustering methods use similarities among high numbers of models obtained either from different methods (in this case they are hard to apply outside the CASP experiment environment(*http* : *//predictioncenter.org/*), as many of the methods and their predictions are not easily available outside CASP), or to rank high numbers of reconstructions made by a single method [35, 42, 43].

The method presented here only uses information obtained from the C_α trace and the sequence of residues associated to it. The main advantage is that the quality of a model is assessed based on its overall topology, rather than based on local details. Moreover, there is no need to model backbone and side chain atoms before evaluating a structure, which allows many more C_α traces with different conformations to be produced. From C_α traces it is possible to model backbone and side chain atoms fairly accurately [16, 17], but this may be more computationally expensive than predicting several conformations of the protein structure as simple C_α. If C_α traces can be evaluated effectively, backbone and side chains may be modelled only for those that are deemed to be accurate.

2   Methods

Protein model quality is often measured as the scaled distance between C_αs of models to their postions in the native structure after optimal superimposition of the structures. Here we encode information obtained solely from the C_α trace. First we represent the C_α trace of each structure model as a directed acyclic graph (rooted tree), in which the outer nodes are pairwise interactions. Each residue in the C_α is encoded into a vector describing its environment. Interactions among C_αs are simply characterised by their distance and angles between pseudo-C_β , alongside the two vectors encoding the residues involved. Environments are described by several angles, distances among neighbours, pseudo-Solvent Accessibility (SA), and coarse packing information. Both interactions and environment descriptors are described in section 2. All these numerical descriptors are computed from the C_α trace and are fed into a model (Neural Network Pairwise Interaction Field, NN-PIF) trained to predict global quality. In the NN-PIF each C_α (i.e. its interactions with all the other C_αs) is mapped into a hidden state, which contains the contribution of that residue to the global quality of the structure. Two C_αs are considered as interacting if they are closer than a fixed distance threshold (we use 20Å, see 2.3). The hidden vectors for all C_α are then combined and mapped to a global quality measure. The NN-PIF allows us to evaluate all the interactions at the same time, whereas other knowledge based potentials generally evaluate interactions separately. To train the NN-PIF we use models submitted to previous CASP editions [44], as the main purpose of this MQAP is to rank models from different prediction systems. No native structures are included in the training set.

Incidentally, the NN-PIF is able to evaluate all the 3D server models submitted to CASP7 [44], as it only depends on the C_α trace and a number of predictors only submitted this. Our system runs on all the CASP7 server models (about 24000, including AL models - see section 2.6) in approximately 1 hour on a PentiumIV 3 GHz processor. However, given the very large amount of weight sharing, we found that training has to proceed very slowly, and training times can be dire (in the region of months on a single CPU).

2.1   Graph Representation of a Protein Structure and NN-PIF

Ways to represent structured information (in the form of a DAG) by recursive neural networks have been described in the past (e.g. [45, 46, 47]), and training can proceed by extensions of the backpropagation algorithm. In our case the complex of interactions among all C_α atoms may be naturally represented as an undirected graph in which nodes are atoms and labelled edges describe the nature of the interactions. However, in order to represent a structure as a DAG, we consider interactions themselves as nodes (see figure 1). If the identity (as in the type of residue it belongs to) and environment of the ith C_α atom is encoded by a vector a_i, and the interaction between the ith and j^{th} atom is described by vector d_{ij}, then each pair of atoms is mapped into a hidden state X_{ij} as:

$$X_{ij} = F(a_i, a_j, d_{ij}) \tag{1}$$

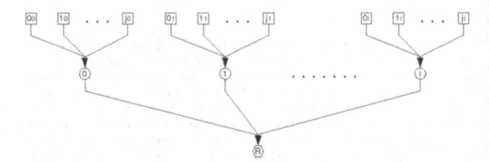

Fig. 1. Tree representation of a C_α trace. Squares, neighbour nodes, are connected to the inner nodes, circles. All the inner nodes, one for each residue in the C_α trace, are connected to the root node represented as an hexagon.

The function $F()$ is implemented by a feed-forward neural network with a single hidden layer and a linear output. The hidden states are then combined together for each residue i, yielding a hidden state for the whole residue:

$$Y_i = \sigma(K \sum_{j \in Ci} X_{ij}) \tag{2}$$

where $Ci = \{\forall j | d(a_i, a_j) < 20\text{Å}\}$, $d()$ is the Euclidian distance, and K is a normalisation constant. Finally, the hidden states for all residues are averaged into a single output, which represents a single property for the whole structure:

$$O = \frac{1}{L} \sum_{i=1}^{L} Y_j \tag{3}$$

The function $F()$ is assumed to be stationary, hence the same network is replicated for all the interactions. The overall NN-PIF architecture is trained by gradient descent. We assume the error to be the squared difference between the network output and the desired property (in our case the "goodness" of the structure). The gradient can be easily computed in closed form, via the backpropagation algorithm. It should be noted that during training the gradient is computed for each replica of the network $F()$, hence there will be as many partial derivatives of the global error with respect to each free parameter in $F()$ as there are interacting pairs of atoms in a model. The contributions to the gradient from each replica of $F()$ are added up component by component to yield the final gradient.

Also notice how here the states describing a pairwise interaction and all the interactions of a residue (X_{ij} and Y_i, respectively) are mapped into the output O through a fixed function with no free parameters. It is also possible to devise a model in which X_{ij} and Y_i are vectors and the average, or sum, of all Y_i is mapped into the desired output through a further feed-forward network. We are currently implementing such model.

2.2   NN-PIF Configuration and Training

The NN-PIF we train here has 10 hidden neurons in the hidden layer of the network implementing $F()$. The learning rate is set to 3.0. During training, the weights are updated after the gradient for all pairs of residues of a single complete structure model has been computed. As we were participating to the 7th edition of the CASP experiment during the preparation of this work, it was possible to run comparisons against other methods in real time. Our performances were evaluated on the models submitted by servers at CASP (see below for details).

2.3   NN-PIF Inputs

To describe two residues in contact (i and j), their environments and the interaction between them, the following descriptors are used:

- Local backbone conformation. We do not explicitly use secondary structure, unlike other MQAP methods [30, 35]. Instead we use several structure descriptors computed from the C_α trace:

  - Distances between all C_α in $[i-2, i+2]$ to each other. A smaller set of distances was used in [48] and to validate protein models in [24].
  - $C\alpha_{i-1} - C\alpha_i - C\alpha_{i+1}$ angles and dihedral angles (angles between vectors formed by $\overrightarrow{C\alpha_{i-1} - C\alpha_i}$ and $\overrightarrow{C\alpha_i - C\alpha_{i+2}}$, and between vectors formed by $\overrightarrow{C\alpha_i - C\alpha_{i+1}}$ and $\overrightarrow{C\alpha_i - C\alpha_{i+2}}$ for both i and j, as in [24, 49, 50].
  - Distance to sequence neighbours: distances between each C_α in $[i-2, i+2]$ and each C_α in $[j-2, j+2]$. A smaller set of distances was used in [48] to assign β Sheets.
  - Relative spatial orientation with respect to sequence neighbours: angles between pseudo-C_β vectors (placed in the direction of the vector formed by the sum of $\overrightarrow{(i-1, i)}$ and $\overrightarrow{(i, i+1)}$ - this vector is also used to compute pseudo-solvent accessibility, see below). Angles between pseudo-C_β of each residue in $[i-1, i+1]$ and in $[j-1, j+1]$ against all the other pseudo-C_β vectors of the residues in the same ranges.

- Residue identities: both residues in contact (i and j) are one-hot encoded (20 inputs/residue).
- Pseudo-Solvent Accessibility (SA) as HSE measure [51]. Briefly, a sphere of radius 6.5Å centred on a C_α is divided into two hemispheres by the plane whose normal vector is the sum of vectors $\overrightarrow{C\alpha_{i-1} - C\alpha_i}$ and $\overrightarrow{C\alpha_i - C\alpha_{i+1}}$, then the number of other atoms falling in either hemisphere is counted.
- Coarse Packing (HSE8): we further split the a sphere, this time into 8 slices induced by three perpendicular planes and count the number of C_αs within each slice. The normal vectors of these three planes are: the pseudo-β vector; the vector obtained from the cross-product of $\overrightarrow{C\alpha_{i-1} - C\alpha_i}$ and $\overrightarrow{C\alpha_i - C\alpha_{i+1}}$; the cross-product of the first two. This time we use a sphere of 13Å .

Some of these inputs can not be computed for residues at the protein termini, in which case they are set to 0. If chain breaks occur (C_αs separated by more than 4.7Å), residues at the edges of the break are treated as termini.

2.4   Model Quality Measurement: Structural Distance

All the measures we consider are based on the Euclidean distance between atoms in the model and in the native protein structure after structural superimposition. Those described below are or may be computed using C_α traces alone. We choose the last in the list (TM score) as our target function.

- RMSD: Root Mean Square Distance. Widely used in the field. Not very suitable to be used as desired output because above a certain threshold it becomes insensitive to highly similar substructures when many atoms of the model are wrongly modelled; also, it is not constrained to a fixed range.
- GDT_TS [52]: it identifies maximum common substructures based on several distance thresholds (e.g. 1, 2, 4 and 8 Å as used in CASP). It may miss fine details because all the atoms within a range (e.g. $(4, 8]$) contribute the same to the scoring function and all the residues without coordinates (atoms not present in the model), contribute the same as those residues further than the highest distance threshold.
- MaxSub score: the average scaled distance using a maximum distance threshold of 3.5 Å (residues further than 3.5 Å do not contribute to the score [53]).
- Average S score: the average scaled distance using a distance threshold of 5Å [54].
- TM-Score [55]: based on scaled RMSD but TM-Score is scaled by the protein length - each residue in the model and native structure contributes to it. It is a number in the $[0, 1]$ range allowing its direct use as desired output.

2.5   Performance Test and Measures

Several different potential quality measures can been used, and here we will use the following ones:

- Enrichment ($E15\%$): the number of top 15% models found among the top 15% top ranked models, divided by the number obtained in a random selection (15% x 15% x number of structures in the model set [56].
- Pearson correlation coefficient (r) for both global quality and local quality.
- Spearman's rank correlation coefficient (ρ).
- Recall ($R = \frac{TP}{TP+FN}$, where TP are true positives, FN false negatives). We use Recall to measure the quality of the output after quantisation into categories (e.g. good vs. bad, defined as greater or smaller than a given threshold).
- Precision ($P = \frac{TP}{TP+FP}$, where FP are false positives). We use Precision on quantised output/target as for Recall.

2.6 Dataset

The main purpose of this MQAP is to know how good protein structure models are (how close are they to the native structure), sorting them by their quality. Because of this we train the NN-PIF only on models submitted to CASP5, 6 and 7, while no native structure is used for training.

To train the NN-PIF we use all human and server predictions of targets for which all 3D predictions were assessed. We randomly divide these into 5 folds, each of which is used to train a different model. To avoid redundancies in the training data (i.e. many models with almost exactly the same C_α trace), TM-Score is divided into 0.001 size bins and only one model per bin for each target is included into each training fold.

All the methods are evaluated on the CASP7 [44] server predictions. We present results on all targets, and on two slightly smaller sets of targets, in order to allow direct comparisons with other methods on the same sets. When testing on one protein, we ensemble those networks that were trained on folds not containing it. This is similar to an n-fold cross validation, except that CASP5 and CASP6 proteins are only included in training sets.

3 Results and Discussion

We compare our MQAP with other methods for global model quality assessment, and also show results for single model quality. We refer to our method as DISTILLF (its identifier at CASP8). To gauge other methods' performances we look at published results. A number of methods were tested on CASP7 targets, and for these the results are extracted from the CASP web site. The results for other methods are quoted from their respective publications.

Table 1 is extracted from [39], and shows average per target Spearman's rank correlations for a number of methods. The table is computed on 87 CASP7 targets. ModFOLD is a consensus method and 3D-Jury is a clustering method, i.e. they are not primary methods, but rather rely on multiple other methods for their predictions. As such, they are in a category of their own and it is not fair to compare them with single model methods. All the other methods evaluate single models, and DISTILLF clearly outperforms all of them, has a rank correlation 7.4% higher than the next method based on TM-Score, and 2.2% based on GDT_TS. The smaller gain is expected as DISTILLF is trained to predict TM-Score.

Table 2 is extracted from [30], and compares QMEAN with other single model evaluation methods available. QMEAN, as DISTILLF, was developed after CASP7. This table is computed on the 95 targets evaluated at CASP7 and on all those server predictions for which all the programs compared were able to make a prediction (22427 models in total). It should be noted that the table is based on GDT_TS, while DISTILLF is trained to predict TM-score, and as such it is at an obvious disadvantage. In the table we also report DISTILLF results based on TM-Score. Not unexpectedly, its performance increases in this case. DISTILLF performs well on correlation measures, and is only outperformed by

Table 1. Spearman's rank correlation on CASP7 targets server submitted models for the 87 targets evaluated in [39], both TS and AL models used

Method	TM − Score	GDT_TS
3D-Jury [39]	0.87	0.857
ModFOLD [39]	0.732	0.754
PROQ [35]	0.574	0.587
Pcons [38]	0.557	0.58
ProQ-MX [34]	0.55	0.556
ModSSEA [39]	0.506	0.52
MODCHECK [15]	0.412	0.444
ProQ-LG [34]	0.289	0.326
DISTILLF	0.647	0.609

Table 2. All CASP7 server models for the 95 targets evaluated. Average per target. All the methods were evaluated on 22427 models of the 95 targets used in [30]. Other methods extracted from [30].

Method	r^2	ρ	$E_{15\%}$
Modcheck [15]	0.64	0.59	2.7
RAPDF [36]	-0.5	0.5	2.44
DFIRE [19]	-0.39	0.53	2.59
ProQ [34]	0.36	0.26	1.22
ProQ_SSE [34]	0.54	0.43	1.71
FRST [57]	-0.57	0.53	2.36
QMEAN3 [30]	-0.65	0.58	2.57
QMEAN4 [30]	-0.71	0.63	2.76
QMEAN5 [30]	-0.72	0.65	2.9
DISTILLF_GDT	0.65	0.59	2.34
DISTILLF_TMS	0.68	0.64	2.53

Table 3. DISTILLF results on ab initio (AI) vs. template-based (TBM) models at CASP7. GDT and TM-score results.

	r^2	ρ	$E_{15\%}$
DISTILLF_GDT_AI	0.39	0.41	2.00
DISTILLF_GDT_TBM	0.71	0.64	2.42
DISTILLF_TMS_AI	0.49	0.49	2.34
DISTILLF_TMS_TBM	0.73	0.67	2.57

the most complete QMEAN potentials, which are based on full-atom models, while DISTILLF only relies on C_α traces, or on a number of interactions smaller by two orders of magnitude. However DISTILLF is less than perfect at selecting the best models ($E_{15\%}$ measure), indicating that it is better at estimating the absolute quality of a model, than at ranking models that are very similar. On all CASP7 targets we obtain an average $E_{15\%}$ (enrichment over random choice)

Table 4. Single residue correlation coefficient on all the CASP7 targets server models (4.55 million residues, both AL and TS models)

Method	r
DISTILLF_TMS	0.71
DISTILLF_MaxSub	0.70
DISTILLF_S	0.69

Table 5. Ability to identify correctly modelled residues. Correct residues are those with scaled distance ≥ 0.7. TP, TN, FP and FN are, respectively, true positives, true negatives, false positives and false negatives, in millions of residues. True values obtained with TM-Score package with default options, and setting D_0 to 3.5Å and 5Å for the MaxSub and S scores respectively. $4.55 \cdot 10^6$ residues in total.

ScaledDistance	TP	FP	TN	FN	P	R
$TM-Score$	0.963	0.396	1.777	1.417	0.708	0.405
$MaxSub$	0.926	0.433	1.846	1.348	0.681	0.407
$Sscore$	0.815	0.544	2.073	1.121	0.600	0.421

Table 6. Ability to identify badly modelled residues. Badly modeled residues are those with scaled distance ≤ 0.3. TP, TN, FP and FN are, respectively, true positives, true negatives, false positives and false negatives, in millions of residues. True values obtained with TM-Score package with default options, and setting D_0 to 3.5Å and 5Å for the MaxSub and S scores respectively. $4.55 \cdot 10^6$ residues in total.

ScaledDistance	TP	FP	TN	FN	P	R
$TM-Score$	1.107	1.543	1.522	0.381	0.418	0.744
$MaxSub$	1.144	1.506	1.492	0.411	0.432	0.736
$Sscore$	1.361	1.29	1.376	0.526	0.513	0.721

of 2.3-2.5 - this allows us to sift out most unfolded or poorly folded models, but is rarely sufficient to pick out the best of all models. This is not surprising, as very similar models (e.g. good predictions based on homology to known structures) often have to be distinguished based on local atomic details, which DISTILLF does not rely on. Results on ab initio CASP7 targets are less good than on template-based targets (Table 3). This is not only a characteristic of DISTILLF and is probably caused by two reasons: ab initio results have a wider distribution; especially, there are far fewer ab initio targets than template-based ones.

To measure the performance of DISTILLF on single residue quality, we use all the TS and AL models submitted by automatic servers for all the 98 CASP7 targets. All measures are computed on 4.55 million residues. S and MaxSub scores are generated fixing d_0 in the TM-Score package to 5 and 3.5Å respectively. We take the Y_i hidden value from the NN-PIF (see eq.2) as the local estimate of quality for residue i by DISTILLF. Table 4 shows the Pearson's correlation against TM, S and MaxSub scores. Table 5 reports results on identifying well modelled residues (those with a scaled distance ≥ 0.7). Table 6 reports results on identifying wrongly modelled residues (those with an actual scaled distance ≤ 0.3).

4   Conclusion

In this manuscript we have described a novel predictor of model quality for protein structure prediction. The main novelty of the predictor is the model it is based on, a neural network designed to estimate properties of sets of pairwise interactions, which we have provisionally termed NN-PIF. A single feed-forward network is used to estimate each of the interactions, and the outputs of all replicas of the network are combined without resorting to any free parameter, to yield a single property. In this manuscript we predict the "goodness" or "native-likeness" of a protein model. However, the model can be used more in general to learn about data that can be represented as undirected graphs. While the model we used in this manuscript is fairly simple, it can be easily extended to more expressive versions, for instance one in which each pairwise interaction is mapped into a hidden vector (rather than a single hidden state, as it is in this work), and a combination of all hidden vectors is then mapped to a property of interest via a further network. We are currently working on such model. Another simple extension is one in which properties of single nodes of the undirected graph are predicted. This can be achieved by mapping hidden states describing each single node (Y_i in eqn.2, or a multi-dimensional extension thereof) to the property of the node via a second network.

The DISTILLF predictor which we have described in this manuscript, relies on simple C_α traces as inputs. The fact that we can use such simple representation induces a set of interactions that is two orders of magnitude smaller than that of a full-atom model, allows very large scale processing of protein models, and is a direct consequence of using a model that does not rely on physico-chemical laws, but only on geometrical information and machine learning. In spite of its simplicity, we have shown that DISTILLF is accurate, more so than any of the CASP7 primary algorithms for model quality assessment, and only slightly less accurate than a newer, far more computationally complex system based on full-atom models. Although DISTILLF is meant to predict global model quality, we have also shown that an accurate estimate of local quality can be extracted very simply from it.

It is also important to note that, although much of the appeal of DISTILLF is that it relies on C_α traces, the NN-PIF model is equally suited to deal with full-atom representations of molecules. We are currently testing NN-PIF on the prediction of protein-ligand binding energies based on full-atom models, with encouraging preliminary results.

A further future/current direction of research is whether NN-PIF may be applied directly as potentials for the *ab initio* prediction of protein structures. In this case, rather than on endpoints of structure prediction searches, decoys representing intermediate stages of the search need to be used for training. Although building sets of examples with the correct distribution may be a hard task, and so is training a network on a potentially enormous set, even limited success at this task may yield a fast, flexible predictor which could be input a vast range of non-homogeneous information.

Acknowledgements. This work is supported by Science Foundation Ireland grant 05/ RFP/CMS0029, grant RP/2005/219 from the Health Research Board of Ireland and a UCD President's Award 2004. We would like to thank Liam McGuffin for his benchmarks[39], and Pascal Benkert for kindly providing us lists of CASP targets.

References

1. Cozzetto, D., Kryshtafovych, A., Ceriani, M., Tramontano, A.: Assessment of predictions in the model quality assessment category. Proteins 69(suppl. 8), 175–183 (2007)
2. Cornell, W., Cieplak, P., Bayly, C., Gould, I., Merz, K., Ferguson, D., Spellmeyer, D., Fox, T., Caldwell, J., Kollman, P.: A second generation force field for the simulation of proteins, nucleic acids, and organic molecules. J. Am. Chem. Soc. 117, 5179–5197 (1995)
3. MacKerell, A., Bashford, D., Bellott, M., Dunbrack, R., Evanseck, J., Field, M., Fischer, S., Gao, J., Guo, H., Ha, S., Joseph-McCarthy, D., Kuchnir, L., Kuczera, K., Lau, F., Mattos, C., Michnick, S., Ngo, T., Nguyen, D., Prodhom, B., Reiher, W., Roux, B., Schlenkrich, M., Smith, J., Stote, R., Straub, J., Watanabe, M., Wiorkiewicz-Kuczera, J., Yin, D., Karplus, M.: All-atom empirical potential for molecular modelling and dynamics studies of proteins. J. Phys. Chem. 102, 3586–3616 (1998)
4. Scott, W., Hünenberger, P., Tironi, I., Mark, A., Billeter, S., Fennen, J., Torda, A., Huber, T., Krüger, P., van Gunsteren, W.F.: The gromos biomolecular simulation program package. J. Phys. Chem. 103, 3596–3607 (1999)
5. Krieger, E., Koraimann, G., Vriend, G.: Increasing the precision of comparative models with yasara nova a self-parameterising force field. PROTEINS: Structure, Function, and Bioinformatics 47, 393–402 (2002)
6. Krieger, E., Darden, T., Nabuurs, S., Finkelstein, A., Vriend, G.: Making optimal use of empirical energy functions: Force-field parameterisation in crystal space. PROTEINS: Structure, Function, and Bioinformatics 57, 678–683 (2004)
7. Colubri, A., Jha, A., Shen, M., Sali, A., Berry, R., Sosnick, T., Freed, K.: Minimalist representations and the importance of nearest neighbour effects in protein folding simulations. J. Mol. Biol. 363, 835–857 (2006)
8. Fitzgerald, J., Jha, A., Colubri, A., Sosnick, T., Freed, K.: Reduced c_β statistical potentials can outperform all-atom potentials in decoy identification. Protein Science 16, 2123–2139 (2001)
9. Wu, Y., Lu, M., Chen, M., Li, J., Ma, J.: Opus-c_α: A knowledge-based potential function requiring only c_α positions. Protein Science 16, 1449–1463 (2007)
10. Lu, M., Dousis, A., Ma, J.: Opuspsp: An orientation-dependent statistical all-atom potential derived from side-chain packing. J. Mol. Biol. 376, 288–301 (2008)
11. Leherte, L.: Application of multiresolution analyses to electron density maps of small molecules: Critical point representations for molecular superposition. J. of Math. Chem. 29(1), 47–83 (2001)

12. Simons, K., Kooperberg, T., Huang, E., Baker, D.: Assembly of protein tertiary structures from fragments with similar local sequences using simulated annealing and bayesian scoring functions. J. Mol. Biol. 268, 209–225 (1997)
13. Baú, D., Pollastri, G., Vullo, A.: Distill: a machine learning approach to ab initio protein structure prediction. In: Bandyopadhyay, S., Maulik, U., Wang, J.T.L. (eds.) Analysis of Biological Data: A Soft Computing Approach. World Scientific, Singapore (2006)
14. Wu, S., Skolnick, J., Zhang, Y.: Ab initio modelling of small proteins by iterative tasser simulations. BMC Biology 5, 17 (2007)
15. Pettitt, C., McGuffin, L., Jones, D.: Improving sequence-based fold recognition by using 3d model quality assessment. Bioinformatics 21(17), 3509–3515 (2005)
16. Adcock, S.: Peptide backbone reconstruction using dead-end elimination and a knowledge-based forcefield. J. Comput. Chem. 25, 16–27 (2004)
17. Bower, M., Cohen, F., Dunbrack, R.: Prediction of protein side-chain rotamers from a backbone-dependent rotamer library: A new homology modelling tool. J. Mol. Biol. 267, 1268–1282 (1997)
18. Khatun, J., Khare, S., Dokhlyan, N.: Can contact potentials reliably predict stability of proteins? J. Mol. Biol. 336, 1223–1238 (2004)
19. Zhou, H., Zhou, Y.: Distance-scaled, finite ideal-gas reference state improves and stability prediction structure-derived potentials of mean force for structure selection. Protein Science 11, 2714–2726 (2002)
20. Hoppe, C., Schomburg, D.: Prediction of protein thermostability with a direction- and distance-dependent knowledge-based potential. Protein Science 14, 2682–2692 (2005)
21. Shao, Y., Bystroff, C.: Predicting interresidue contacts using templates and pathways. PROTEINS: Structure, Function, and Bioinformatics 53, 497–502 (2003)
22. Vullo, A., Walsh, I., Pollastri, G.: A two-stage approach for improved prediction of residue contact maps. BMC Bioinformatics 7, 18 (2006)
23. Martin, A., Baú, D., Walsh, I., Vullo, A., Pollastri, G.: Long-range information and physicality constraints improve predicted protein contact maps. Journal of Bioinformatics and Computational Biology 6(5) (2008)
24. Kleywegt, G.: Validation of protein models from c-alpha coordinates alone. J. Mol. Biol. 273, 371–376 (1997)
25. Ngan, S., Inouye, M., Samudrala, R.: A knowledge-based scoring function based on residue triplets for protein structure prediction. Protein Engineering, Desing & Selection 19(5), 187–193 (2006)
26. Feng, Y., Kloczkowski, A., Jernigan, R.: Four-body contact potentials derived from two protein datasets to discriminate native structures from decoys. PROTEINS: Structure, Function, and Bioinformatics 68, 57–66 (2007)
27. Loose, C., Klepeis, J., Floudas, C.: A new pairwise folding potential based on improved decoy generation and side-chain packing. PROTEINS: Structure, Function, and Bioinformatics 54, 303–314 (2004)
28. Heo, M., Kim, S., Moon, E., Cheon, M., Chung, K., Chang, I.: Perceptron learning of pairwise contact energies for proteins incorporating the amino acid environment. Phys. Rev. E Stat. Nonlin. Soft Matter Phys. 72, 011906 (2005)
29. Sippl, M.: Recognition of errors in three-dimensional structures of proteins. PROTEINS: Structure, Function, and Bioinformatics 17, 355–362 (1993)
30. Benkert, P., Tosatto, S., Schomburg, D.: Qmean: A comprehensive scoring function for model quality assessment. PROTEINS: Structure, Function, and Bioinformatics 71(1), 261–277 (2008)

31. Dong, Q., Wang, X., Lin, L.: Novel knowledge-based mean force potential at the profile level. BMC Bioinformatics 7, 324 (2006)
32. Zhang, C., Kim, S.: Environment-dependent residue contact energies for proteins. PNAS 97(6), 2550–2555 (2000)
33. Fogolari, F., Pieri, L., Dovier, A., Bortolussi, L., Giugliarelli, G., Corazza, A., Esposito, G., Viglino, P.: Scoring predictive models using a reduced representation of proteins: model and energy definition. BMC Structural Biology 7(15), 17 (2007)
34. Wallner, B., Elofsson, A.: Can correct protein models be identified? Protein Science 12, 1073–1086 (2003)
35. Wallner, B., Elofsson, A.: Identification of correct regions in protein models using structural, alignment, and consensus information. Protein Science 15, 900–913 (2006)
36. Samudrala, R., Moult, J.: An all-atom distance-dependent conditional probability discriminatory function for protein structure prediction. J. Mol. Biol. 275, 895–916 (1998)
37. Eisenberg, D., Lthy, R., Bowie, J.: Verify 3d: assessment of protein models with three-dimensional profiles. Methods Enzymol. 277, 396–404 (1997)
38. Wallner, B., Fang, H., Elofsson, A.: Automatic consensus-based fold recognition using pcons, proq, and pmodeller. PROTEINS: Structure, Function, and Genetics 53, 534–541 (2003)
39. McGuffin, L.: Benchmarking consensus model quality assessment for protein fold recognition. BMC Bioinformatics 8, 15 (2007)
40. Wallner, B., Elofsson, A.: Prediction of global and local model quality in casp7 using pcons and proq. PROTEINS: Structure, Function, and Bioinformatics 69(suppl. 8), 184–193 (2007)
41. Ginalski, K., Elofsson, A., Fischer, D., Rychlewski, L.: 3d-jury: a simple approach to improve protein structure predictions. Bioinformatics 19(8), 1015–1018 (2003)
42. Qiu, J., Sheffler, W., Baker, D., Noble, W.: Ranking predicted protein structures with support vector regression. PROTEINS: Structure, Function, and Bioinformatics 71, 1175–1182 (2008)
43. Zhou, H., Skolnick, J.: Protein model quality assessment prediction by combining fragment comparisons and a consensus ca contact potential. PROTEINS: Structure, Function, and Bioinformatics 71, 1211–1218 (2008)
44. Battey, J., Kopp, J., Bordoli, L., Read, R., Clarke, N., Schwede, T.: Automated server predictions in casp7. Proteins 69(suppl. 8), 68–82 (2007)
45. Sperduti, A., Starita, A.: Supervised neural networks for the classification of structures. IEEETNN 8(3), 714–735 (1997)
46. Frasconi, P.: An introduction to learning structured information. In: Giles, C.L., Gori, M. (eds.) IIASS-EMFCSC-School 1997. LNCS (LNAI), vol. 1387, pp. 99–120. Springer, Heidelberg (1998)
47. Frasconi, P., Gori, M., Sperduti, A.: A general framework for adaptive processing of data structures. IEEETNN 9(5), 768–786 (1998)
48. Martin, J., Letellier, G., Marin, A., Taly, J., de Brevern, A.G., Gibrat, J.F.: Protein secondary structure assignment revisited: a detailed analysis of different assignment methods. BMC Struct. Biol. 5, 17 (2005)
49. Majumdar, I., Krishna, S., Grishin, N.: Palsse: A program to delineate linear secondary structural elements from protein structures. BMC Bioinformatics 6(202), 24 (2005)
50. Labesse, G., Colloc'h, N., Pothier, J., Mornon, J.: P-sea: a new efficient assignment of secondary structure from c alpha trace of proteins. CABIOS 13(3), 291–295 (1997)

51. Hamelryck, T.: An amino acid has two sides: A new 2d measure provides a different view of solvent exposure. PROTEINS: Structure, Function, and Bioinformatics 59, 38–48 (2005)
52. Zemla, A., Venclovas, C., Moult, J., Fidelis, K.: Processing and analysis of casp3 protein structure predictions. Proteins 37(suppl. 3), 22–29 (1999)
53. Siew, N., Elofsson, A., Rychlewski, L., Fischer, D.: MaxSub: an automated measure for the assessment of protein structure prediction quality. Bioinformatics 16(9), 776–785 (2000)
54. Cristobal, S., Zemla, A., Fischer, D., Rychlewski, L., Elofsson, A.: A study of quality measures for protein threading models. BMC Bioinformatics 2(5), 15 (2001)
55. Zhang, Y., Skolnick, J.: Scoring function for automated assessment of protein structure template quality. PROTEINS: Structure, Function, and Bioinformatics 57, 702–710 (2004)
56. Tsai, J., Bonneau, R., Morozov, A., Kuhlman, B., Rohl, C., Baker, D.: An improved protein decoy set for testing energy functions for protein structure prediction. PROTEINS: Structure, Function, and Bioinformatics 53, 76–87 (2003)
57. Tosatto, S.: The victor/FRST function for model quality estimation. J. Comput. Biol. 12(10), 1316–1327 (2005)

A Graph-Based Semi-supervised Algorithm for Protein Function Prediction from Interaction Maps

Valerio Freschi

ISTI – Information Science and Technology Institute,
University of Urbino, Urbino, Italy
freschi@sti.uniurb.it

Abstract. Protein function prediction represents a fundamental challenge in bioinformatics. The increasing availability of proteomics network data has enabled the development of several approaches that exploit the information encoded in networks in order to infer protein function. In this paper we introduce a new algorithm based on the concept of topological overlap between nodes of the graph, which addresses the problem of the classification of partially labeled protein interaction networks. The proposed approach is tested on the yeast interaction map and compared with two current state-of-the-art algorithms. Cross-validation experiments provide evidence that the proposed method represents a competitive alternative in a wide range of experimental conditions and also that, in many cases, it provides enhanced predictive accuracy.

1 Introduction

The recent development of high-throughput technologies has allowed the generation of massive proteomic data sets for entire organisms. Several experimental protocols (e.g. yeast two-hybrid, mass spectrometry) have been devised to extract information regarding physical interactions between proteins at proteomic-scale [11]. These advancements in post-genomic technologies prompted the need for computational methods that allow to elucidate the complex structure of interactions underlying cellular biochemistry. In particular, the problem of predicting protein functional categories given the set of interactions and the knowledge of the function for a subset of interacting proteins has received increasing attention. Despite a considerable variety of proposed solutions, none of them can be considered fully satisfactory, leaving open research issues to be addressed [3, 6, 10].

Previous approaches can be coarsely classified as *direct methods* or *module-assisted methods* [10]. Direct-method algorithms basically exploit the "guilty-by-association" principle, which transfers annotations among neighbor-nodes in the protein-protein-interaction (PPI) network, assuming that nodes that are located close to each other tend to share the same functional categories [7, 9]. Module-assisted algorithms aim at identifying, as a first computational step, coherent clusters of nodes of the underlying PPI network and, after that, predict functions for all genes in each cluster [1].

T. Stützle (Ed.): LION 3, LNCS 5851, pp. 249–258, 2009.
© Springer-Verlag Berlin Heidelberg 2009

From a machine learning point of view, the problem of classifying nodes in a partially labeled network can be viewed as a graph-based semi-supervised learning problem [17]. In this framework, algorithms exploit both labeled and unlabeled data by leveraging the relationships provided by edges of the graph.

Graph based function prediction has also been tackled as an optimization problem: in [13] the authors predict functional annotations by minimizing the number of times that different functions are annotated in neighbor nodes. The resulting optimization problem has been solved by means of simulated annealing [13] and local search [4]. Finally, a related optimization method has been proposed so as to minimize the sum of costs of edges that connect nodes without any common function, which can be cast as a minimum multiway cut problem that can be exactly solved by means of integer linear programming [7].

In general, classification in partially labeled network is challenging for two reasons. First, the available knowledge of labels is often very sparse w.r.t. the network topology thus potentially impairing methods that only take into account the local topology of the graph in the inference process. Second, data interaction networks are inherently noisy, making methods that exploit the whole topology of the network prone to noise propagation.

In this paper we address these issues by proposing a new approach to function prediction from protein-protein interaction networks based on the concept of topological overlap analysis. Our algorithm takes into account, for each pair of proteins, the level of overlap between the respective sets of neighbors of each interacting protein. This coefficient (hereafter denoted *topological overlap coefficient*), properly quantified, can be used for the derivation of a measure of similarity between nodes that could be used to replace link weights in a weighted-majority setting in order to improve its accuracy, relying on the fact that proteins that share a given number of the same neighbors, probably share some functional role [10, 15]. The topological overlap coefficients computed for all possible pairs of nodes in a network encode a matrix (i.e. the *topological overlap matrix, T*) which has been successfully applied in bioinformatics for the analysis of metabolic networks [8] and the analysis of gene expression networks [14]. Our contribution is the application of this concept to the problem of function prediction in protein interaction networks and its use as building block of a new prediction algorithm.

The paper is organized as follows: in the next section we describe the proposed approach discussing the algorithm and its implementation, in section 3 we introduce the experimental set up and discuss cross-validation, in section 4 we show the results, lastly, in section 5, we conclude with some final remarks.

2   The Proposed Approach

The aim of our approach is the derivation of an algorithm for function prediction (label classification) in PPI networks. A first input of this algorithm is a network of physically interacting proteins that we represent as a graph whose nodes are the proteins and whose (possibly weighted) edges represent the strength of such interactions. A second input is the set of label annotations associated to each protein of a given subset of the nodes. The output of the algorithm is a prediction of the function(s) for each of the proteins whose label is unknown.

In order to achieve this goal we first compute the degree of overlap between sets of nodes that are neighbors of a pair of nodes. Given two nodes i and j we compute the topological overlap coefficient $T(i,j)$ as follows:

$$T(i,j) = \frac{\sum\limits_{k \neq i,j} A(i,k) \cdot A(k,j) + A(i,j)}{\min(deg_i, deg_j) + 1 - A(i,j)} \tag{1}$$

Where A is the adjacency matrix of the graph representing the PPI network and deg_i represents the degree of node i (i.e. the number of interacting proteins directly linked to protein i) in the case of binary unweighted network or its equivalent representative in the case of weighted PPI networks, computed as follows:

$$deg_i = \sum_{k \neq i} A(i,k)$$

It has been demonstrated ([14]) that $0 \leq T(i,j) \leq 1$. Particularly, the topological overlap coefficient $T(i,j)$ takes its minimum value when nodes i and j are not directly linked and do not share common neighbors, while it is maximum when i and j are directly linked and the set of neighbors of i (respectively j) is a subset of the set of neighbors of j (respectively i).

Hence $T(i,j)$ carries information regarding the degree of overlap between nodes i and j: since proteins that are close to each other and tend to form densely connected subgraphs are supposed to be functionally correlated, we hypothesize that also nodes that have significant overlap among their respective neighborhoods could show some degree of correlation in their functional roles.

Finally we compute, for each couple of nodes, a coefficient of similarity $T_m(i,j)$ between nodes:

$$T_m(i,j) = \frac{1}{2}T(i,j) + \frac{1}{2}A(i,j) \tag{2}$$

The final step of our method entails the replacement of the adjacency matrix with the matrix T_m and the application of a majority-vote strategy on the new graph encoded by T_m. The rationale behind our approach is to exploit the capability of the topological overlap matrix of identifying overlapping neighbors (hence, potential clusters of homogeneous functionally linked nodes) while retaining the noise-resilience of a locally-based majority setting. Hence we compute, for each target node i to predict, a score $s(i, f_j)$ that is the sum of the scores contributed by all nodes (not just the interacting ones) that are annotated with a given function f_j (we call this set of nodes $\mathcal{F}_j$):

$$s(i, f_j) = \sum_{k \in \mathcal{F}_j} T_m(i,k)$$

The computation of this score is repeated for every function f_j in the set of possible labels. The overall computation is then repeated for each node of the graph: labels with highest scores above a given threshold are chosen as predicted functional categories for each target node.

Notably, we do not explicitly aim at identifying modules, rather we leverage the information associated to the topological overlap coefficient to reward with a higher score the contribution of those nodes that have a higher degree of overlap between their neighborhoods. On the other side, we do not have a pure local majority approach because we do not look only at immediate neighbors, rather we also exploit the relationships between distant nodes encoded by $T(i, j)$ to improve the effectiveness of the prediction. Hence, the proposed algorithm tries to bridge the gap between local and global algorithms and between direct methods and module-based algorithms by simply re-wiring the network according to the topological overlap analysis and by using this re-wired network as a new starting element for the semi-supervised learning process.

3   Experimental Setup

3.1   Data Sets

We tested the proposed approach on the *Saccharomyces Cerevisiae* PPI network. This organism has been extensively studied and provides a widely accepted benchmark for the validation of network-based prediction methods. In particular we use the reference set compiled by Nabieva et al. for cross-validating their method [7]. It consists of a network of 12531 interactions among 4495 proteins that are known to physically interact. The weights that are assigned to the edges between pair of nodes are computed by evaluating the probability (i.e. the reliability) of the interactions after separation of functional linkage by experimental source of evidence [7]. The evaluation of the probabilities to be used as weights is *per se* an interesting active line of research which particularly involves data integration from multiple sources [12]. It is well known that taking into account probability of interactions instead of binary linkage information, allows more accurate modeling and results in better classification performances [7, 12]. We provide independent confirmation of these findings since (as we will show in the results section) also our algorithm benefits from the usage of probabilistic weights in the PPI graph.

3.2   Functional Annotations

We followed the experimental setup defined in [7] also for what concerns the reference set of labels to be used for annotating the PPI network: in particular we used the MIPS controlled vocabulary for biological processes (second hierarchy level) which consists of 72 labels [5, 7]. 2946 proteins out of the 4495 that take part in the interactome are annotated with MIPS functional categories according to [7]. In the following we also denote the PPI weighted annotated network of Saccharomyces Cerevisiae as SC_w dataset.

3.3 Competing Algorithms

We compared our algorithm with two state-of-the-art approaches: the first one is a standard *Majority Vote* algorithm (MV) which is still considered an effective strategy despite its simplicity [6, 10]; the second one is a flow-based algorithm (called *FunctionalFlow*) that is widely recognized as a high-level accuracy method.

The strategy of MV consists in counting the number of instances of the labels (i.e. functional categories) annotated in the immediate neighbors of the protein node target of the prediction. The most voted categories above a given threshold are taken as predictions [7, 9]. A weighted version of the MV algorithm simply weights the contribution of each neighbor by the weight of the link between the target and the neighbor node.

FunctionalFlow works by propagating functional information according to the whole network topology from sources of functional flow (the annotated nodes) to sinks (the target nodes). At the end of an iteration process the unlabeled nodes are scored according to the algorithm and, once again, the functions corresponding to the highest scores above threshold are considered as candidates for the prediction [7].

3.4 Cross-Validation

Cross-validating semi-supervised protein labeling algorithms is an issue that entails the choice of proper benchmarking evaluations. Apart from differences in measuring the effectiveness of the algorithms, all experimental frameworks need a *ground truth* to compare with. To this aim, known annotations are taken as ground truth and labels corresponding to a given number of proteins are "cleared" (i.e. their functions are supposed to be unknown): algorithms are then tested on their capability of recovering correct labels given the remaining subset of annotated nodes. A first degree of freedom is the number of cleared annotated proteins. Two choices are possible: in the *leave-one-out* cross-validation framework one protein at the time is cleared while in the *leave-a-percentage-out* cross-validation a given fraction of known annotations is cleared and used for testing [6, 7]. We decided to evaluate our approach according to a leave-a-percentage-out method since we believe it could better reflect real problem instances. In fact today's knowledge of protein functional roles is very sparse: for instance we have that the extent of known annotation in reference species ranges from 89.9% for the *C.elegans* interactome to the 23% for the *A.thaliana* (for biological process categories) [10]. Moreover, this setting is also particularly challenging since it is less conservative and provides better insights into the algorithms capability of handling the sparsity of label annotations.

Finally, one needs to adequately measure the effectiveness of function prediction algorithms by means of classification accuracy. We follow in this work a method based on a modification of standard *Receiver Operating Curve* (ROC) analysis that have been proposed by Nabieva et al. to evaluate FunctionalFlow [7] and also recently used for cross-validating a new method based on literature-data integration [2]. Each protein is considered correctly predicted (i.e. is taken as a *true positive*, TP) if the number of correctly predicted functions for that protein is more than half the number of known functions, otherwise the prediction for the protein is considered incorrect (i.e. it is counted as a *false positive*, FP). Both TP and FP are computed for decreasing stringency levels of the algorithm (e.g. properly changing the threshold upon which functional labels are assigned), allowing to explore the tradeoff between predictive accuracy and number of proteins for which a prediction is produced.

4 Results

We implemented our topological-overlap-based method (hereafter called Tom-Pred) and we also re-implemented the MV and FunctionalFlow algorithms in order to compare them on the same benchmark (SC_w dataset). We tested three different levels of sparsity in the annotation of the network to be labeled by clearing, respectively, 20%, 50%, and 80% of the known annotations. We also tested the capability of our algorithm of taking advantage of weighted links, instead of binary unweighted networks, by comparing the performances of TomPred on both the weighted and unweighted version of the same network (hereafter also denoted as SC_u dataset).

Since the proteins to be classified for cross-validation (i.e. the nodes whose labels are cleared) are randomly selected, we repeated each experiment five times with different seeds of the pseudo-random number generator and computed the average of all the results. The consistency of this experimental setup is guaranteed by the small variance of the results within the same type of experiments (with different seeds). Figures 1.a and 1.b report the behavior of the tested algorithms and accounts for variation of the randomly selected input. In particular, Figure 1.a shows a comparison between TomPred and MV for the SC_w dataset when 50% of known proteins are assumed to be unlabeled while Figure 1.b reports the comparison between TomPred and FunctionalFlow under the same experimental conditions. Error bars represent standard deviations which are, as previously stated, limited in their range.

Figures 2, 3 and 4 show the behavior of the three tested algorithms for the SC_w dataset when, respectively, 20%, 50% and 80% of proteins are cleared (error bars are omitted for the sake of clarity, and because of the above mentioned considerations). The proposed algorithm outperforms both MV and FunctionalFlow in almost all the range of the stringency threshold and at different degrees of knowledge of the underlying network. In particular, we can see (Figure 2) that when 20% of nodes are unlabeled FunctionalFlow has a slightly better performance in the first part of the ROC curve while TomPred achieves better results if more than 50 FP are accepted and both perform better than MV. Conversely, when 80% of nodes are cleared TomPred shows higher accuracy except for the rightmost part of the ROC curve (Figure 4). Finally, in the case of 50% of unlabeled nodes, our algorithm results as good as or better than FunctionalFlow and MV in the whole range of the ROC analysis curve (Figure 3).

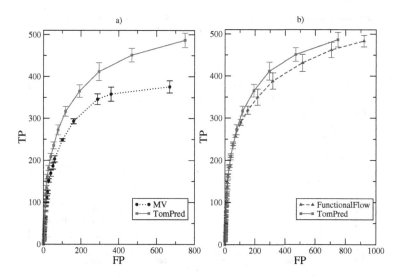

Fig. 1. ROC curve analysis for TomPred, FunctionalFlow and MV (SC_w dataset, unknown proteins: 50%). Error bars report standard deviation computed on 5 different random experiments.

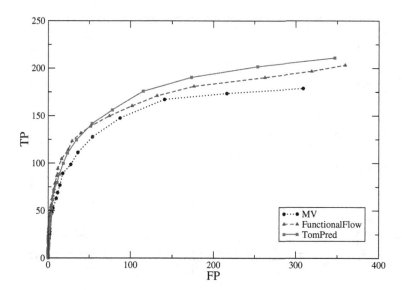

Fig. 2. ROC curve analysis for TomPred, FunctionalFlow and MV (SC_w dataset, unknown proteins: 20%)

As a supplementary result we also present in Figure 5 a comparison between the performance of our algorithm on the same yeast's interactome when unweighted links are used instead of weighted edges (50% of unlabeled nodes): the

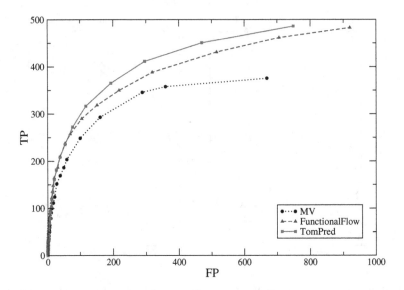

Fig. 3. ROC curve analysis for TomPred, FunctionalFlow and MV (SC_w dataset, unknown proteins: 50%)

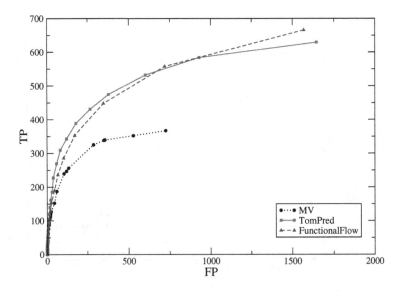

Fig. 4. ROC curve analysis for TomPred, FunctionalFlow and MV (SC_w dataset, unknown proteins: 80%)

improvement of TomPred when run on the SC_w instead of the SC_u dataset is apparent.

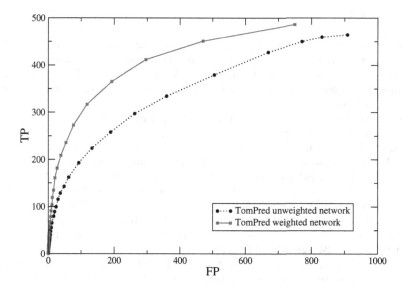

Fig. 5. TomPred performance comparison between unweighted and weighted network (SC_w and SC_u dataset, unknown proteins: 50%)

5   Conclusions

In this paper we have presented a new approach to graph-based semi-supervised protein function prediction from PPI networks. The proposed method relies upon the concept of topological overlap between sets of neighbor nodes of protein pairs. By integrating this information with the linkage between adjacent nodes we are able to exploit local and global properties of the topology of the PPI graph. In fact the algorithm makes use of direct links between nodes but also exploits topological overlap to extract informative data from distant nodes.

We have tested the proposed method on the extensively analyzed interaction map of Saccharomyces Cerevisiae and compared it with two state-of-the-art algorithms. The experimental results provide evidence of the effectiveness of our approach which presents comparable or, in most cases, improved predictive accuracy w.r.t the competing algorithms.

As a subject of future research, we think it could be worthwhile to investigate how different learning algorithms perform within the topological overlap framework: in particular, it could be interesting to check how the re-wiring of the graph according to topological overlap analysis impacts the performance of other classifiers [16, 17].

Acknowledgements. The author thanks Dr. E.Nabieva and Prof. M.Singh (Princeton University) for having kindly provided the PPI network datasets used in their work.

References

1. Bader, G., Hogue, C.W.: An automated method for finding molecular complexes in large protein interaction networks. BMC Bioinformatics 4, 2 (2003)
2. Gabow, A.P., Leach, S.M., Baumgartner, W.A., Hunter, L.E., Goldberg, D.S.: Improving protein function prediction methods with integrated literature data. BMC Bioinformatics 9, 198 (2008)
3. Hu, P., Bader, G., Wigle, D.A., Emili, A.: Computational prediction of cancer-gene function. Nature Reviews Cancer 7(1), 23–34 (2007)
4. Karaoz, U., Murali, T.M., Letovsky, S., Zheng, Y., Ding, C., Cantor, C.R., Kasif, S.: Whole genome annotation by using evidence integration in functional-linkage networks. Proc. Natl. Acd. Sci. USA 101, 2888–2893 (2004)
5. Mewes, H.W., Frishman, D., Guldener, U., Mannhaupt, G., Mayer, K., Mokrejs, M., Morgenstern, B., Munsterkotter, M., Rudd, S., Weil, B.: Mips: a database for genomes and protein sequences. Nucleic Acid Research 30, 31–34 (2002)
6. Murali, T.M., Wu, C.J., Kasif, S.: The art of gene function prediction. Nature Biotechnology 24(12), 1474–1476 (2006)
7. Nabieva, E., Jim, K., Agarwal, A., Chazelle, B., Singh, M.: Whole-proteome prediction of protein function via graph-theoretic analysis of interaction maps. Bioinformatics 21, i302–i310 (2005)
8. Ravasz, E., Somera, A.L., Mongru, D.A., Oltvai, Z.N., Barabasi, A.L.: Hierarchical organization of modularity in metabolic networks. Science 297(5586), 1551–1555 (2002)
9. Schwikowski, B., Uetz, P., Field, S.: A network of protein-protein interactions in yeast. Nature Biotechnology 18, 1257–1261 (2000)
10. Sharan, R., Ulitsky, I., Shamir, R.: Network-based prediction of protein function. Molecular System Biology 3(88), 1–13 (2007)
11. Shoemaker, B.A., Panchenko, A.R.: Deciphering protein-protein interactions. part i. experimental techniques and databases. PLoS Computational Biology 3(3), 337–344 (2007)
12. Srinivasan, B.S., Novak, A.F., Flannick, J.A., Batzoglou, S., McAdams, H.H.: Integrated protein interaction networks for 11 microbes. In: Apostolico, A., Guerra, C., Istrail, S., Pevzner, P.A., Waterman, M. (eds.) RECOMB 2006. LNCS (LNBI), vol. 3909, pp. 1–14. Springer, Heidelberg (2006)
13. Vazquez, A., Flammini, A., Maritan, A., Vespignani, A.: Global protein function prediction from protein-protein interaction networks. Nature Biotechnology 21, 697–700 (2003)
14. Yip, A.M., Horvath, S.: Gene network interconnectedness and the generalized topological overlap measure. BMC Bioinformatics 8, 22 (2007)
15. Yook, S.H., Oltvai, Z.N., Barabasi, A.L.: Functional and topological characterization of protein interaction networks. Proteomics 4, 928–942 (2004)
16. Zhu, X., Ghahramani, Z., Lafferty, J.: Semi-supervised learning using gaussian fields and harmonic functions. In: Proc. of 20th International Conference on Machine Learning, ICML 2003, pp. 912–919 (2003)
17. Zhu, X.: Semi-supervised learning literature survey. Technical Report 1530, Department of Computer Sciences, University of Wisconsin, Madison (2005)

Substitution Matrices and Mutual Information Approaches to Modeling Evolution

Stephan Kitchovitch, Yuedong Song, Richard van der Wath, and Pietro Liò

Computer Laboratory, University of Cambridge, Cambridge, UK
{sk490,ys340,rcv23,pl219}@cam.ac.uk

Abstract. Substitution matrices are at the heart of Bioinformatics: sequence alignment, database search, phylogenetic inference, protein family classification are all based on BLOSUM, PAM, JTT, mtREV24 and other matrices. These matrices provide means of computing models of evolution and assessing the statistical relationships amongst sequences. This paper reports two results; first we show how Bayesian and grid settings can be used to derive novel specific substitution matrices for fish and insects and we discuss their performances with respect to standard amino acid replacement matrices. Then we discuss a novel application of these matrices: a refinement of the mutual information formula applied to amino acid alignments by incorporating a substitution matrix into the calculation of the mutual information. We show that different substitution matrices provide qualitatively different mutual information results and that the new algorithm allows the derivation of better estimates of the similarity along a sequence alignment. We thus express an interesting procedure: generating ad hoc substitution matrices from a collection of sequences and combining the substitution matrices and mutual information for the detection of sequence patterns.

1 Introduction

DNA and amino acid sequences contain both the information of the genetic relationships among species and that of the evolutionary processes that have caused them to diverge. Various computational and statistical methods exist to attempt to extract such information to determine the way in which DNA and protein molecules function. Most of Bioinformatics is based on using substitution matrices: at the core of database search, sequence alignment, protein family classification and phylogenetic inference lies the use of DNA and amino acid substitution matrices for scoring, optimising, and assessing the statistical significance of sequence analysis [11, 12, 19, 33, 37]. Much care and effort has therefore been taken to construct substitution matrices, and the quality of Bioinformatics analysis results depends upon the choice of an appropriate matrix [11, 15, 16, 19, 29, 30, 32, 33, 37]. The substitution matrices can be built empirically, utilising data from comparisons of observed sequences, or parametrically, using the known chemical and biological properties of DNA and amino acids. These models permit the estimation of genetic distance between two homologous sequences, measured by the expected number of nucleotide or amino acid

T. Stützle (Ed.): LION 3, LNCS 5851, pp. 259–272, 2009.

substitutions per site that have occurred. Such distances may be represented as E-values, bit scores (as in BLAST) or branch lengths in a phylogenetic tree where the existent sequences form the leaf nodes of the tree, while their ancestors form the internal branch nodes and are usually unknown. Note that alignments can be also considered as a part of the evolutionary model in inferring the tree or, conversely, that a tree can be used to improve the sequence alignment [42].

It is impractical to generate a single substitution matrix to apply to all tasks: standard substitution matrices are appropriate only for the comparison of proteins with an amino acid composition similar to the one used to derive them. Therefore, for groups of species with biased amino acid compositions, standard substitution matrices are not optimal. It is worth remembering that species may have different (mitochondrial) genetic codes, codon bias, different amino acid composition and gene copy numbers. It is known that close and distant homologous proteins need different PAM (Point Accepted Mutation) or BLOSUM substitution matrices [14]: we use PAM and BLOSUM families for BLAST and Clustal; Dayhoff, JTT and other models for distance, Likelihood and Bayesian methods in phylogenetic inference, etc.

One of the first amino acid substitution matrices was the PAM matrix, developed by Margaret Dayhoff in the 1970s. This matrix was calculated by observing the differences in closely related proteins (with at least 85% similarity between sequences). Dayhoff and coworkers estimated substitution frequencies empirically from alignments of related sequences. From inspection of log odds scores they concluded that amino acids with similar properties indeed tend to form groups that are conserved: members of a group substitute with a high frequency internally compared to substitution frequencies to external amino acids. A matrix for divergent sequences can be calculated from a matrix obtained for closely related sequences by taking the latter matrix to a power. The PAM1 matrix estimates the expected rate of substitution given that 1% of the amino acids have changed. Assuming that future mutations would follow the same rate as those observed so far, we can use PAM1 as the foundation for calculating other matrices. Using this logic, Dayhoff derived matrices as high as PAM250. Whereas the traditional approach considers each amino acid to share an 'average' environment [10, 11, 13], Henikoff and Henikoff [14], using local, ungapped alignments of distantly related sequences, derived the BLOSUM series of matrices. The number after the matrix (BLOSUM62) refers to the minimum percent identity of the blocks used to construct the matrix; as a thumb rule, greater numbers represent lesser distances. It is noteworthy that these matrices are directly calculated without extrapolations. The BLOSUM series of matrices generally perform better than PAM matrices for local similarity searches [14]. Altschul suggested that three matrices should ideally be used: PAM40, PAM120 and PAM250, as the lower PAM matrices will tend to find short alignments of highly similar sequences, while higher PAM matrices will find longer, weaker local alignments [12].

Goldman and collaborators inferred Markov chain models of amino acid replacement for several structural categories and solvent accessibility states, i.e. alpha helix, beta sheets, coils and turns [8, 9, 34]. Here, we show that two substitution matrices derived from two mitochondrial data sets perform better than a standard one, mtREV24 [1], suggesting that a collection of species, while diverging from other life tree branches, represents a special environment in terms of amino acid preferences. For example see Figure 4 in [43] where antarctic fish have undergone a strong fitness bottleneck that has largely affected the amino acid composition. Following on from this we may think of the grid as a distributed system of wells within which evolution occurs, i.e. a collection of sequences mutated according to different parameters. The grid is becoming *the* resource for solving large-scale computing applications in Bioinformatics, system biology and computational medicine.

Furthermore we describe an algorithm which incorporates a substitution matrix in the mutual information formula of an alignment of protein sequences. This refinement of the mutual information provides a better evaluation of the information content and similarity measure in an alignment than using the usual formula. This represents a novel and important refinement of mutual information which is a similarity measure in the sense that small values imply large "distances" in a loose sense. Therefore, given that the scoring matrices provide an effective measure of distances between amino acids, the incorporation of a substitution matrix into a mutual information leads to a more meaningful estimation of the similarity along an alignment. This will have interesting applications in local and global alignment which we will not describe here.

2   Generating and Testing Novel Substitution Matrices Using a Grid Setting

Mitochondrial amino acid sequences were downloaded from GenBank; we focused on two different different collection of species (orders): fish (chondrichthyan and teleosts; 66) and protostomes (molluscs, arthropods, brachiopods, annelids; 42). All the methods for generating mutational data matrices are similar to that described by Dayhoff et al. [13]. The method involves 3 steps:

1. Clustering the sequences into homologous families.
2. Tallying the observed mutations between highly similar sequences.
3. Relating the observed mutation frequencies to those expected by pure chance.

We have generated MtPip - a novel model of evolution for fish phylogenies and evaluated the performances of MtPan, a model of amino acid replacement of insects we have used for phylogenetic inference in [21]. These two models, described in Figure 1 a,b, were built using relative rates of estimated amino acid replacement from pairwise comparisons of mitochondrial fish and insect sequences that are identical by 85% or higher. More precisely the estimates of the relative rates of amino acid replacement were computed by examining the database and recording the number of times that, for each column, amino acid type i is observed in one sequence and type j is observed at the corresponding site in a closely related sequence. The 85% threshold of amino acid identity between aligned sequences ensures that the likelihood of a particular mutation (e.g. L $\rightarrow$ V) being the result of a set of successive mutations (e.g. L $\rightarrow$ x $\rightarrow$ y $\rightarrow$ V) is low.

3   Testing the Matrices

In order to test the models, we performed two different comparisons:

1. We applied the Mantel Test, [44] which computes a correlation between two $n \times n$ distance or similarity matrices. It is based on the normalized cross-product:

$$c_{MT} = \frac{1}{n-1} \sum_{i=1}^{n} \sum_{j=1}^{n} \frac{(a_{ij} - \bar{a})}{s_a} \frac{(b_{ij} - \bar{b})}{s_b}$$

where a_{ij} and b_{ij} are the generic elements of the two matrices A and B we want to compare, $\bar{a}$ and $\bar{b}$ are the corresponding mean values and s_a and s_b are the standard deviations. The null hypothesis is that the observed correlation between the two distance matrices could have been obtained by any random arrangement. The significance is evaluated via permutation procedures. The rows and columns of one of the two matrices are randomly rearranged and the resulting correlation is compared with the observed one.

2. We used Bayesian phylogenetic inference to compute posterior probability of best trees obtained with different models. Although here we are not focusing on phylogenetic trees di per se, the process of testing the substitution matrix involves considering a mathematical model of evolution which describes the instantaneous probabilities transition (P) from one amino acid to another: $dP(T)/dT = QP(T)$ where $\mathbf{Q}$ is the instantaneous rate matrix of transition probabilities.

    The rate matrix for a Markov process is restricted by a mathematical requirement that the row sums are all zero. To calculate $\mathbf{P}(t) = e^{t\mathbf{Q}}$ we need to compute the spectral decomposition (diagonalisation) of $\mathbf{Q}$; if we consider,

$$Q = U \cdot diag\{\lambda_1, \ldots, \lambda_n\} \cdot U^{-1}$$

then

$$P(T) = U \cdot diag\{e^{\lambda_1}, \ldots, e^{\lambda_n}\} \cdot U^{-1}$$

and its component is written as

$$P_{ij}(T) = \sum_k c_{ijk} e^{t\lambda_k}$$

where $i, j, k = 1, .., 20$ for proteins and c_{ijk} is a function of $\mathbf{U}$ and $\mathbf{U}^{-1}$. The row sums of the transition probability matrix over the time t, $\mathbf{P}(T)$ are all ones. Substitution matrices are then incorporated into robust statistical frameworks such as those of Maximum Likelihood or Bayesian.

3.1   Bayesian and Grid Computing for Phylogeny

We have compared the two models of evolution with standard models in use, mtREV24 in particular, using Bayesian inference. A Bayesian analysis combines one's prior beliefs about the probability of a hypothesis with the likelihood which carries the information about the hypothesis contained in the observations. Bayesian inference always produces well-calibrated results on average with respect to the distribution of data and parameter values chosen from the prior. In brief, the Bayesian posterior probability of a phylogenetic tree involves a summation over all possible trees. Given the large number of trees for even moderately-sized problems, for each tree that is considered, the likelihood involves a multidimensional integral over all possible combinations of branch lengths and substitution model parameters (e.g. parameters that allow different rates among the different character states, different stationary character-state frequencies or rate variation across sites). By necessity, posterior probabilities of trees must be approximated.

Markov Chain Monte Carlo [2, 5, 26, 27, 28](MCMC) has been successfully used to approximate the posterior probability distribution of trees. MCMC uses stochastic simulation to obtain a sample from the posterior distribution of trees; inferences are then based on the MCMC sample. The posterior probability distribution of trees can contain multiple peaks. The peaks represent trees of high probability separated from other peaks by valleys of trees with low probability. This is a phenomenon that has been observed for other optimality criteria, such as maximum parsimony and maximum likelihood. Parallel MrBayes is a program that implements a variant of MCMC called "Metropolis-Coupled Markov Chain Monte Carlo"(MCMCMC) [4, 6, 26, 27, 28]. This entails the execution of a certain amount of chains on as many or less processors where all but one of them are heated. By heating a Markov chain the acceptance probability of new states is increased allowing the heated chain to accept more states that a cold chain and consequently crosses valleys more easily in the landscape of probability trees. Integration is improved by attempted state swapping among randomly selected chains. Successful swapping between states allows a chain, that is otherwise stuck on a local maximum in the landscape of trees, to explore other peaks. If the target distribution has multiple peaks, separated by low valleys, the Markov chain may have difficulty in moving from one peak to another. As a result, the chain may get stuck on one peak and the resulting samples will not be representative of the actual posterior density. This is known to occur often in phylogeny reconstruction, where multiple local peaks exist in the tree space during heuristic tree search under maximum parsimony, maximum likelihood, and minimum evolution criteria. The problem same can be expected for stochastic tree search using MCMC. An obvious disadvantage of the MCMCMC algorithm is that m chains are run and only one chain is used for inference. For this reason, it is ideally suited for implementation on parallel machines, since each chain will in general require the same amount of computation per iteration. The grid is not only suitable for Bayesian MCMCMC runs but also to run simulations with different models of evolution.

In the next section we shall review the pertinent properties of mutual information in the Shannon version and the refinement we propose.

4   Incorporating Substitution Matrices into Mutual Information

Assume that one has two random variables X and Y. If they are discrete, we write $p_i(X) = \mathrm{prob}(X = x_i)$, $p_i(Y) = \mathrm{prob}(Y = y_i)$, and $p_{ij} = \mathrm{prob}(X = x_i, Y = y_j)$ for the marginal and joint distributions. Otherwise (and if they have finite densities) we denote the densities by $\mu_X(x), \mu(y)$ and $\mu(x, y)$. Entropies are defined for the discrete case as usual by

$$H(X) = -\sum_i p_i(X) \log p_i(X)$$

and

$$H(X, Y) = -\sum_{i,j} p_{ij} \log p_{ij}$$

Conditional entropies are defined as

$$H(X|Y) = H(X, Y) - H(Y) = -\sum_{i,j} p_{ij} \log p_{i|j}$$

The base of the logarithm determines the units in which information is measured. In particular, taking base two leads to information measured in bits. In the following, we always will use natural logarithms, measured in nats. The mutual information between X and Y is finally defined as

$$I(X, Y) = H(X) + H(Y) - H(X, Y) = \sum_{i,j} p_{ij} \log \frac{p_{ij}}{p_i(X) p_j(Y)}$$

It can be shown to be non-negative, and is zero only when X and Y are strictly independent [20, 22, 23, 40, 41]. In applications, one usually has the data available in form of a statistical sample. To estimate $I(X, Y)$ for a real multiple sequence alignment we have estimated the probability of each amino acid in each column of the alignment as a weighted sum of the amino acid replacement rates according to a substitution matrix. In here $p_i(X)$ refers to the probability of amino acid X occurring at site i. The process of evolution for a specific site k is described by parameters $p^k(X \to Y)$, the relative rate of change from type X to Y.

We can also write $p^k(X \to Y)$ as

$$\pi^k(Y)\, s^k(X \to Y)$$

where $\pi^k(Y)$ represents the equilibrium frequencies for amino acid Y in the column k, and the $s^k(X \to Y) \equiv s^k(Y \to X)$ represent the relative exchangeability of amino acids X and Y in the column k once effects of amino acid frequencies are removed. We have used a substitution matrix to estimate the $p^k(X \to Y)$ for each amino acid, for each alignment column, by averaging all the probabilities of changes from amino acid X to the 19 other amino acids.

5   Results and Discussion

Central to Bioinformatics is the assessment of the statistical relationship among species/genes, which is given by E-values, bit scores, branch lengths etc. We have tested two specie-specific models of mitochondrial sequence evolution. Mitochondrial sequences are widely used in phylogenetic assessment. Mitochondria play a central role in many key aspects of animal physiology and pathophysiology. Their central and ubiquitous task is clearly to be the sites of aerobic respiration and the production of ATP. They also play subtle roles in calcium storage, oxygen sensing, glucose homeostasis [17, 18]. Fig 1a and b show the bubble plots of the substitution matrices of the two models MtPan (insects) and MtPip (fish), a model of evolution specific for fish mitochondrial proteins. Although there are notably similarities with mtREV24, the Mantel tests (not shown) has suggested that these three matrices are indeed different. Figure 2 a,b show the comparison of the performances of the two specie-specific models with respect to mtREV24. The two models perform in average better than mtREV24. In Figure 2a mtPan challenges mtREV24 [29] and another recent model, MtArt [3] based on a smaller data set than our MtPan [21]. One million generations were run, with two MCMCMC chain settings (4-chain version and 8-chain version in Fig 2a; 8-chain version in Fig 2b), and trees were sampled every 100 generations. Most of the runs, regardless of the matrix used, converged to slightly different maxima. This indicates, on one hand, that the resulting topology for each run is highly dependent on the performance of the algorithm to explore the likelihood surface and the starting point of the search, thus suggesting prudence when interpreting the results. On the other hand, this underlines the importance of conducting different parallel runs and comparing the results in order to have a global outlook on these aspects of the analysis. Furthermore, comparing the actual topologies to which each run converges, it becomes evident that while most of the shallow nodes are common to most resulting trees, the deepest nodes tend to

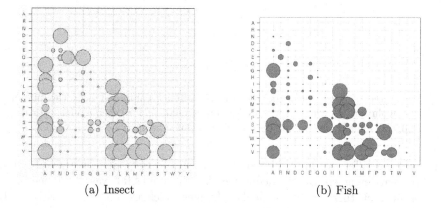

| (a) Insect | (b) Fish |

Fig. 1. A novel amino acid replacement matrix based on a database of Insect and Fish mitochondrial amino acid sequences

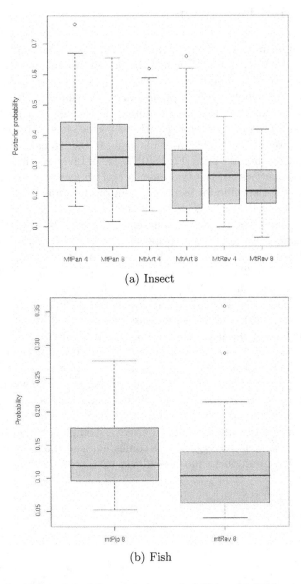

Fig. 2. Box-and-whisker plot of the posterior probabilities obtained using the different models of evolution, showing the MCMCMC implementations of each model separately. Note that the 4 chain versions tended to slightly outperform the 8 chain versions.

vary, and the difference in likelihood observed across runs, little as they are, depend on rearrangements at the deepest nodes. Since mtART was developed on a smaller data set, there are amino acid exchanges which are not exactly estimated affecting the posterior probabilities.

In the analysis of our fish and insect mitochondrial data sets, plots of likelihood versus generations, together with the value of the likelihood towards which each run converges, were used to assess the efficiency of the analysis to explore the likelihood space and reach the best maximum, and the relative performance of the three amino acid substitution matrices. By running multiple analyses, it was possible to show that the MtPip and MtPan matrices generally converge to higher likelihood values than MtREV24. Differences beween MtPan and MtPip are mostly due to the exchange between the following pairs of amino acids: (F,M),(D,N),(Q,E),(T,L),(W,L),(S,L),(V,T); remarkably, these are not exchanges among similar amino acids, i.e. neutral evolution, since they involve charge and volume changes, suggesting adaptive changes under selective pressure. Several authors have found differences in the performances of mitochondrial proteins in recovering expected phylogenies [24, 35, 36]. Liò and Goldman [9] showed that this happens for both short and long range phylogenetic distances and under a variety of models. They found that different models of evolution have small effects on topology but can have large effects on branch lengths and ML scores. Liò has shown that matrices derived from alignments of single mitochondrial proteins are substantially different [25]. Cao and colleagues showed that the phylogenetic relationship based on different vertebrate mitochondrial proteins can suggest wrong trees [24, 34, 35, 36]. Since single proteins often support different trees, mitochondrial phylogeny is generally estimated from all 12 proteins (either concatenated or summing up log-likelihood scores for each gene/protein). This has the effect of making the result less prone to statistical fluctuations, but also less robust to deviations from a uniform model for all portions of the data. Moreover only proteins encoded by the H strand of mtDNA are used, i.e. ND6 is always discarded [43]. The available literature shows that ND5 and ND4 perform generally better than the other mitochondrial proteins. This suggests that single-protein topology deviations from 'reference' or concatenated sequences topologies may not depend on the model of evolution implemented, but may be intrinsic to the protein-specific evolutionary dynamics with respect to the molecular environment.

We have also computed the mutual information of the aligned cytochrome b mitochondrial sequences from a large variety of vertebrate species; we explored several other substitution matrices and found that the incorporation of mtREV24 substitution improves the detection of patterns. Note that our new models are generated from close species sequences so that the small sequence variability result in matrices of accepted point mutations which include a large number of entries equal to 0 or 1. Moreover the presence of a wide range of species in the alignment suggest that mtREV24 should be used. In Figure 3 we compare the standard mutual information (a) and the mutual information which incorporates Dayhoff or mtREV24 substitution rate applied to a mitochondrial cytochrome b alignment. The axes represent site position along the sequence alignment. All figures have undergone the same normalisation procedure. The incorporation of an appropriate substitution matrix results in extracting more information and patterns from the sequence data. We can see that the amount of noise is

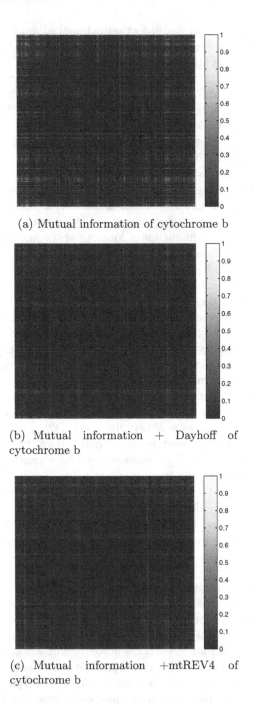

(a) Mutual information of cytochrome b

(b) Mutual information + Dayhoff of cytochrome b

(c) Mutual information +mtREV4 of cytochrome b

Fig. 3. Heatmaps of mutual information of a cytochrome b alignment from a large ensemble of vertebrate mitochondrial species (a); in (b) we have incorporated Dayhoff, in (c) mtREV24 in the mutual information formula

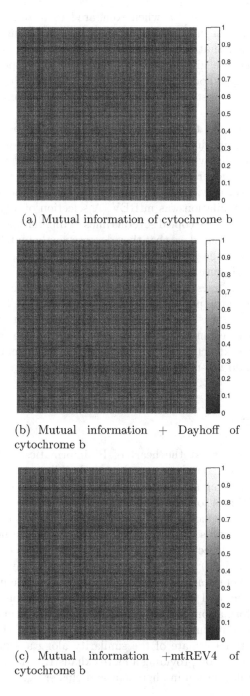

(a) Mutual information of cytochrome b

(b) Mutual information + Dayhoff of cytochrome b

(c) Mutual information +mtREV4 of cytochrome b

Fig. 4. Heatmaps of mutual information of a the Seq-Gen generated alignments (a); with Dayhoff incorporated in the calculation in (b) and mtREV24 in (c)

greatly reduced from Figure 3 a when compared to b and c. Furthermore, a few lines of high mutual information between certain columns become visible in b and c, suggesting that the addition of a substitution matrix has revealed a similarity between the given columns. Therefore we propose an interesting procedure: generating ad hoc substitution matrices from a collection of sequences and using the derived substitution matrices for the detection of patterns using mutual information.

To further demonstrate the benefits of considering a substitution matrix in the calculation of mutual information, we performed an additional test. Using the program Seq-Gen[45], which simulates the evolution of amino acid sequences given a phylogeny, we generated an alignment of amino acids. The phylogenetic tree provided was arbitrary, as we are not concerned with the topology, and the model used for the simulation was mtREV. A selection of adjacent columns in the sequence alignment were copied several times at different positions within the alignment and then permuted, so that they do not exactly match. This alignment was then put through the same process as the aligned sequences above. The results are shown in Figure 4. In 4 c) we have used the mtREV24 substitution matrix and regions or 'lines' of high mutual information can be seen occurring in the positions where the permuted columns were inserted. These regions were not visible without the use of a substitution matrix in a) or when the Dayhoff matrix was incorporated in b). This confirms that the use of an appropriate matrix allows for the extraction of additional information from multiple sequence alignments that would not normally be visible.

6   Conclusions

Substitution matrices are at the heart of Bioinformatics, with many uses in sequence alignment and database search, tree building, protein classification, etc. There is no such thing as a perfect substitution matrix; each matrix has its own limitations. If this is so, then it should be possible to use multiple matrices so that each one complements the limits of the others. The paper is divided in two sections; first we use Bayesian grid setting to derive novel substitution matrices for fish and insects and discuss their performances with respect to commonly used (standard) amino acid replacement matrices. The main result of this paper is a novel refinement of mutual information: the incorporation of a substitution rate generated from the sequence data set or from a larger data set into the mutual information. We show that different substitution matrices give qualitatively different mutual information results and that the new algorithm allows to derive better estimate of the similarity along a sequence alignment. Work in progress focuses on providing a more exhaustive characterisation of this optimised mutual information in alignment and clustering.

Acknowledgment. PL thanks the British Council which is a registered charity 209131 (England and Wales) SC037733 (Scotland) for funding.

References

1. Adachi, J., Hasegawa, M.: Model of amino acid substitution in proteins encoded by mitochondrial DNA. J. Mol. Evol. 42, 459–468 (1996a)
2. Altekar, G., Dwarkadas, S., Huelsenbeck, J.P., Ronquist, F.: Parallel Metropolis coupled Markov chain Monte Carlo for Bayesian phylogenetic inference. Bioinformatics 20, 407–415 (2004)
3. Abascal, F., Posada, D., Zardoya, R.: MtArt: a new model of amino acid replacement for Arthropoda. Mol. Biol. Evol. 24, 1–5 (2007)
4. Huelsenbeck, J.P., Ronquist, F.: MrBayes: Bayesian inference in phylogenetic trees. Bioinformatics 17, 754–755 (2001)
5. Ronquist, F., Huelsenbeck, J.P.: MrBayes3: Bayesian phylogenetic inference under mixed models. Bioinformatics 19, 1572–1574 (2003)
6. Rannala, B., Yang, Z.: Bayes estimation of species divergence times and ancestral population sizes using DNA sequences from multiple loci. Genetics 164, 1645–1656 (2003)
7. Goldman, N., Thorne, J.L., Jones, D.T.: Using evolutionary trees in protein secondary structure prediction and other comparative sequence analyses. J. Mol. Biol. 263, 196–208 (1996)
8. Goldman, N., Thorne, J.L., Jones, D.T.: Assessing the impact of secondary structure and solvent accessibility on protein evolution. Genetics 149, 445–458 (1998)
9. Liò, P., Goldman, N.: Using protein structural information in evolutionary inference: transmembrane proteins. Mol. Biol. Evol. 16, 1696–1710 (1999)
10. Jones, D.T., Taylor, W.R., Thornton, J.M.: The rapid generation of mutation data matrices from protein sequences. CABIOS 8, 275–282 (1992)
11. Jones, D.T., Taylor, W.R., Thornton, J.M.: A mutation data matrix for transmembrane proteins. FEBS Letts 339, 269–275 (1994)
12. Altschul, S.F.: Amino acid substitutions matrices from an information theoretic perspective. J. Mol. Biol. 219, 555–665 (1991)
13. Dayhoff, M.O., Schwartz, R.M., Orcutt, B.C.: A model of evolutionary change in proteins. In: Dayhoff, M.O. (ed.) Atlas of Protein Sequence and Structure, vol. 5(3), pp. 345–352 (1978)
14. Henikoff, S., Henikoff, J.: Amino acid substitution matrices from protein blocks. Proc. Natl. Acad. Sci. USA 89(biochemistry), 10915–10919 (1992)
15. Whelan, S., Liò, P., Goldman, N.: Molecular phylogenetics: State-of-art methods for looking into the past. Trends Genet. 17, 262–272 (2001)
16. Liò, P., Goldman, N.: Models of molecular evolution and phylogeny. Genome Res. 8, 1233–1244 (1998)
17. Chomyn, A.: Mitochondrial genetic control of assembly and function of complex I in mammalian cells. J. Bioenerg. Biomembr. 133, 251–257 (2001)
18. Duchen, M.R.: Mitochondria and calcium: from cell signalling to cell death. J. Physiol. 529, 57–68 (2000)
19. Grantham, R.: Amino acid difference formula to help explain protein evolution. Science 185, 862–864 (1974)
20. Li, M., Badger, J.H., Chen, X., Kwong, S., Kearney, P., Zhang, H.: An information-based sequence distance and its application to whole mitochondrial genome phylogeny. Bioinformatics 17, 149–154 (2001)
21. Carapelli, A., Liò, P., Nardi, F., van der Wath, E., Frati, F.: Phylogenetic analysis of mitochondrial protein coding genes confirms the reciprocal paraphyly of Hexapoda and Crustacea. BMC Evol. Biol. 7(suppl. 2), S8 (2007)

22. Li, M., Chen, X., Li, X., Ma, B., Vitanyi, P.: The similarity metric. E-print, arxiv.org/cs.CC/0111054 (2002)

23. Li, M., Vitanyi, P.: An introduction to Kolmogorov complexity and its applications. Springer, New York (1997)

24. Zardoya, R., Meyer, A.: Phylogenetic performance of mitochondrial protein-coding genes in resolving relationships among vertebrates. Molecular Biology and Evolution 13, 525–536 (1996)

25. Liò, P.: Phylogenetic and structural analysis of mitochondrial complex I proteins. Gene 345, 55–64 (1999)

26. Larget, B., Simon, D.: Markov chain Monte Carlo algorithms for the Bayesian analysis of phylogenetic trees. Mol. Biol. Evol. 16, 750–759 (1999)

27. Mau, B., Newton, M.A., Larget, B.: Bayesian phylogenetic inference via Markov chain Monte Carlo methods. Biometrics 55, 1–12 (1999)

28. Yang, Z., Rannala, B.: Bayesian phylogenetic inference using DNA sequences: Markov chain Monte Carlo methods. Mol. Biol. Evol. 14, 717–724 (1997)

29. Yang, Z., Nielsen, R., Hasegawa: Models of amino acid substitutions and applications to mitochondrial protein evolution. Mol. Biol. Evol. 15, 1600–1611 (1998)

30. Gascuel, O.: Mathematics of Evolution and Phylogeny. Oxford University Press, USA (2007)

31. Yang, Z.: Computational Molecular Evolution. Oxford Series in Ecology and Evolution. Oxford University Press, USA (2006)

32. Felsenstein, J.: Inferring Phylogenies, 2nd edn. Sinauer Associates (2003)

33. Nielsen, R.: Statistical Methods in Molecular Evolution, 1st edn. Statistics for Biology and Health. Springer, Heidelberg (2005)

34. Liò, P., Goldman, N.: Models of molecular evolution and phylogeny. Genome Res. 8, 1233–1244 (1998)

35. Russo, C.A., Takezaki, N., Nei, M.: Efficiencies of different genes and different tree-building methods in recovering a known vertebrate phylogeny. Mol. Biol. Evol. 13, 933–942 (1996)

36. Cao, Y., Janke, A., Waddell, P.J., Westerman, M., Takenaka, O., Murata, S., Okada, N., Paabo, S., Hasegawa, M.: Conflict among individual mitochondrial proteins in resolving the phylogeny of eutherian orders. J. Mol. Evol. 47, 307–322 (1998)

37. Swofford, D.L., Olsen, G.J., Waddell, P.J., Hillis, D.M.: Phylogenetic inference. In: Hillis, D.M., Moritz, C., Mable, B.K. (eds.) Molecular Systematics, pp. 407–514. Sinauer, Sunderland (1996)

38. Xia, X., Li, W.H.: What amino acid properties affect protein evolution? J. Mol. Evol. 47, 557–564 (1998)

39. Yang, Z.: Maximum likelihood phylogenetic estimation from DNA sequences with variable rates over sites: approximate methods. J. Mol. Evol. 39, 306–314 (1994)

40. Liò, P., Politi, A., Buiatti, M., Ruffo, S.: High statistics block entropy measures of DNA sequences. J. Theor. Biol. 180(2), 151–160 (1996)

41. Kraskov, A., Stögbauer, H., Grassberger, P.: Estimating mutual information. Phys. Rev. E Stat. Nonlin. Soft. Matter. Phys. 69(6 Pt 2), 066138 (2004)

42. Hein, J.: TreeAlign. Methods Mol. Biol. 25, 349–364 (1994)

43. Papetti, C., Liò, P., Ruber, L., Patarnello, T., Zardoya, R.: Antarctic Fish Mitochondrial Genomes Lack ND6. Gene J. Mol. Evol. 65, 519–528 (2007)

44. Sokal, R.R., Rohlf, F.J.: Biometry, 3rd edn. Freeman, New York (1995)

45. Seq-Gen: a program that will simulate the evolution of nucleotide or amino acid sequences along a phylogeny, http://tree.bio.ed.ac.uk/software/seqgen/

Author Index